Radioactive Particles in the Environment

NATO Science for Peace and Security Series

This Series presents the results of scientific meetings supported under the NATO Programme: Science for Peace and Security (SPS).

The NATO SPS Programme supports meetings in the following Key Priority areas: (1) Defence Against Terrorism; (2) Countering other Threats to Security and (3) NATO, Partner and Mediterranean Dialogue Country Priorities. The types of meeting supported are generally "Advanced Study Institutes" and "Advanced Research Workshops". The NATO SPS Series collects together the results of these meetings. The meetings are co-organized by scientists from NATO countries and scientists from NATO's "Partner" or "Mediterranean Dialogue" countries. The observations and recommendations made at the meetings, as well as the contents of the volumes in the Series, reflect those of participants and contributors only; they should not necessarily be regarded as reflecting NATO views or policy.

Advanced Study Institutes (ASI) are high-level tutorial courses intended to convey the latest developments in a subject to an advanced-level audience

Advanced Research Workshops (ARW) are expert meetings where an intense but informal exchange of views at the frontiers of a subject aims at identifying directions for future action

Following a transformation of the programme in 2006 the Series has been re-named and re-organised. Recent volumes on topics not related to security, which result from meetings supported under the programme earlier, may be found in the NATO Science Series.

The Series is published by IOS Press, Amsterdam, and Springer, Dordrecht, in conjunction with the NATO Public Diplomacy Division.

Sub-Series

A.	Chemistry and Biology	Springer
B.	Physics and Biophysics	Springer
C.	Environmental Security	Springer
D.	Information and Communication Security	IOS Press
E.	Human and Societal Dynamics	IOS Press

http://www.nato.int/science
http://www.springer.com
http://www.iospress.nl

Series C: Environmental Security

Radioactive Particles in the Environment

edited by

Deborah H. Oughton
Department of Plant and Environmental Sciences
Norwegian University of Life Sciences
Aas, Norway

and

Valery Kashparov
Ukrainian Institute of Agricultural Radiology (UIAR)
Kiev Region, Ukraine

Springer

Published in cooperation with NATO Public Diplomacy Division

Proceedings of the NATO Advanced Research Workshop on
Hot Particles Released from Different Nuclear Sources
Yalta, Ukraine
7–10 May 2007

Library of Congress Control Number: 2009929399

ISBN 978-90-481-2948-5 (PB)
ISBN 978-90-481-2947-8 (HB)
ISBN 978-90-481-2949-2 (e-book)

Published by Springer,
P.O. Box 17, 3300 AA Dordrecht, The Netherlands.

www.springer.com

Printed on acid-free paper

CONTENTS

PREFACE

Introduction

The relevance of hot particles as a source of radionuclides to the environment has been known since the first atomic weapons tests, when mm particles were observed at the test site. Since that time there has been an increased awareness of the importance of hot particles, and a recognition that discharges associated with particles are the norm rather than exception. Developments in technology have contributed to improvements in characterization, and studies have helped illuminate the relevance for the transfer and impact of particles on the environment and human health. The papers presented at the conference covered all these areas and have been divided into sources and characterization and environmental impacts and effects. The summary below provides an overview of issues discussed in the papers in this book, as well as general conclusions from discussions at the meeting.

Particle Sources and Characterization

The papers presented in this section cover a variety of sites and characterization methods: providing valuable information on hot particles originating from Chernobyl 30 km zone, the Semipalatinsk test site, the Muroroa atoll, DU in Kosovo and Kuwait, the Yennisej river basin, Thule and Palomares accidents as well as sediments in Russian lakes affected by the Chernobyl accident and nuclear weapons fallout.

Techniques to determine the oxidation state, isotopic composition of radionuclides and their abundance in hot particles including matrix information have been developed and reported. Presently results include uranium and plutonium radioisotopes, strontium and caesium but little information is available on more exotic radionuclides such as curium or (T)NORMs. A conclusion from the meeting was the need to explore the availability of samples from waste depositories and related accidental releases into the environment as well as historic samples from Maralinga, the Nevada test site, Lop Nor polygon in China and the Algerian Nuclear test site. Of specific interest are safety test and accidents with nuclear weapons as those might represent the closest-to-reality scenario for dirty bombs. In addition, sampling missions to sites in Russia to different sites were suggested, however it was recognized that these can be difficult to conduct due to high costs involved. It was also proposed to closer investigate particles within the Chernobyl shelter as those might have a considerable impact on inhalation doses to construction workers.

In summary, it would appear that the time is ripe to revisit and group results into different classes of hot particles. This should enable a generalization of information, thus allowing the deduction of generic hypothesis and models related to different release scenarios. SOPs (standard operating procedures) from sampling to isolation and measurement techniques for hot particles should be developed to address the question of representativeness and related uncertainties. Characterization should include the determination of isotopic compositions, matrices, and oxidation status. Such guidelines will be valuable not only for environmental monitoring and emergency preparedness but also for nuclear forensics in general.

Environmental Impact Assessments

The papers demonstrate a number of ways in which hot particles can influence the transfer, mobility and bioavailability of radionuclides. Hot particles impact on the reliability of transfer and dose assessment models and may lead to an over or underestimation in doses. This has been clearly demonstrated for Sr-90 in the Chernobyl area, where models can vary by at least 1–2 orders of magnitude. It would appear that the most sensitive parameters for impact assessments are dissolution rates, distribution coefficients (K_d) and biological concentration factors (BCF). However, there is need to evaluate the impact of particles on the mobility and bioavailability of other radionuclides and other hot particle sources, and particularly to relate particle source characteristics and environment parameters to weathering and dissolution rates.

While there is a general agreement that hot particles may impact on radionuclide bioavailability and exposure, there is a need to gather data through targeted biological experiments. Retention and attenuation are the most important factors influencing the size of the dose to organisms: both human and non-human. Ingestion and inhalation are both important pathways, and radiation physics and modelling of doses can provide valuable data for these exposures, particularly the use of Monte Carlo simulations. However, experimental data is missing on exactly what the biological retention is for various types of particles, in particularly on how this varies with size, and not least on how particles might impact on biological effects. Dissolution techniques can make a valuable contribution to knowledge on solubility and bioavailability assessments, for example, the lung dissolution "model". However it is important that such techniques are standardized, and this includes a need for reference materials for all aspects of the analysis of hot particles.

Finally, major technological developments are expected on size resolution, not least because of the research focus on nanotechnology and nanomaterials. Since one expects that hot particles also exist at the sub-micron level, one would expect that also these particles will also impact on mobility, transfer and bioavailability of radionuclides. This raises questions on the definition of what is a "hot particle", does this include colloids? While at the meeting there was no consensus on the relative importance of this in terms of radiological protection, it is an area where one can expect future developments in scientific knowledge.

ACKNOWLEDGEMENT

The NATO advanced workshop on Hot Particles attracted 32 scientists from 10 countries to Yalta, Ukraine. There was a large proportion from countries of the former Soviet Union, reflecting both the location and the increased research in the area following the Chernobyl accident (more than one third of the papers were concerned with the Chernobyl accident). However, hot particles have an impact on many areas of research, and the multidisciplinary nature of the research was also evident from the participants, which included researchers with a wide variety of backgrounds, ranging from chemists, technology, modellers, biologists and environmental science.

We express our gratitude to the NATO Scientific Committee for their generous support of the conference. We would also like to thank all the participants for making the conference such a success, and to our hosts for ensuring that participants experienced also the history, culture and beauty of the Yalta Region.

Deborah Oughton and Valery Kashparov
Editors

CONTRIBUTORS

ADMON Uri
Israel Atomic Energy Commission, Nuclear Research Centre Negrev (NRCN), Beer Sheva, ISRAEL

AHAMDACH Noureddine
Institute for Radioprotection and Nuclear Safety (IRSN), Fontenay aux Roses, FRANCE

ARDOIS Christophe
Institute for Radioprotection and Nuclear Safety (IRSN), Fontenay aux Roses, FRANCE

BOLSUNOVSKY Alexander
Institute of Biophysics, SB Russian Academy of Sciences, Krashnoyarsk, RUSSIA

BONDAR Yuri
Polessje State Radiation-Ecological Reservation, Hoiniki, Gomel Region, BELARUS

BUGAÏ Dimitry
Institute of Geological Sciences, Kiev, UKRAINE

CHARLES Monty
University of Birmingham, UNITED KINGDOM

CHILD David
Australian Nuclear Science and Technology Organization, Lucas Heights, AUSTRALIA

DAHLGARD Henning
RISOE National Library for Sustainable Energy, Technical University Denmark, Roskilde, DENMARK

DANESI Pier Roberto
Institute for Advanced Studies, AUSTRIA

DUTAU Aliaksandr
Polessje State Radiation-Ecological Reservation,
Hoiniki, Gomel Region,BELARUS

ERIKSSON Mats
IAEA Marine Environment Laboratory, MONACO

FESENKO Sergey
International Atomic Energy Agency (IAEA),
Vienna, AUSTRIA

GLUKHOVSKY Olexiy
Ukrainian Institute of Agricultural Radiology (UIAR),
Chabany, Kiev Region, UKRAINE

GORDEEV Sergey
Institute of Global Climate and Ecology, Moscow,
RUSSIA

HOTCHKIS Michael
Australian Nuclear Science and Technology Organization,
Lucas Heights, AUSTRALIA

IVANOV Yuriy
International Radioecology Laboratory, UKRAINE

JANSSENS KOEN
Chemical Department, University of Antwerpen,
BELGIUM

JERNSTRÖM Jussi
RISOE National Library for Sustainable Energy, Technical
University Denmark, Roskilde, DENMARK

KARHUNEN Tero
Radiation and Nuclear Safety Authority (STUK),
FINLAND

KASHPAROV Valery
Ukrainian Institute of Agricultural Radiology (UIAR),
Chabany, Kiev Region, UKRAINE

KHOMUTININ Yuri
Ukrainian Institute of Agricultural Radiology (UIAR),
Chabany, Kiev Region, UKRAINE

KONOPLEV Alexey
Centre for Environmental Chemistry, SPA "Typhoon"
Obninsk, RUSSIA

KVASINKOVA Elena
Institute of Global Climate and Ecology, Moscow, RUSSIA

LAPTEV Gennady V.
Ukrainian Institute of Hydrometeorology, Kiev, UKRAINE

LAZAREV Mykola
Ukrainian Institute of Agricultural Radiology (UIAR), Chabany,
Kiev Region, UKRAINE

LEVCHUK Svyatolslav
Ukrainian Institute of Agricultural Radiology (UIAR), Chabany,
Kiev Region, UKRAINE

LIND Ole Christian
Department of Plant and Environmental Sciences,
Norwegian University of Life Sciences, Aas, NORWAY

LUNDIN Semen
Ukrainian Institute of Agricultural Radiology (UIAR), Chabany,
Kiev Region, UKRAINE

MALOSHTAN Igor
Ukrainian Institute of Agricultural Radiology (UIAR), Chabany,
Kiev Region, UKRAINE

MATUSEVICH Janna
International Sakharov Environmental University (ISEU),
Minsk, BELARUS

MIRONOV Vladislav
International Sakharov Environmental University (ISEU),
Minsk, BELARUS

NIELSEN Sven P.
RISOE National Library for Sustainable Energy,
Technical University Denmark,
Roskilde, DENMARK

OLSEN Sven K.
RISOE National Library for Sustainable Energy,
Technical University Denmark,
Roskilde, DENMARK

OUGHTON Deborah
Department of Plant and Environmental Sciences
PO Box 5003, Norwegian University of Life Sciences,
1432 Aas, NORWAY

PELIKAN Andreas
Dienstleitungen in der automatischen Datenverarbeitung und
Informationstechnik, AUSTRIA

PÖLLÄNEN Roy
Radiation and Nuclear Safety Authority (STUK),
FINLAND

PRIBYLEV Sergey
International Sakharov Environmental University (ISEU),
Minsk, BELARUS

PROOST Kristof
Chemical Department, University of Antwerpen,
BELGIUM

PROTSAK Valentin
Ukrainian Institute of Agricultural Radiology (UIAR), 08162
Mashinostroiteley street 7, Chabany,
Kiev Region, UKRAINE

ROOS Per
RISOE National Library for Sustainable Energy,
Technical University Denmark,
Roskilde, DENMARK

SALBU Brit
Department of Plant and Environmental Sciences,
Norwegian University of Life Sciences,
Aas, NORWAY

SIISKONEN Teemu
Radiation and Nuclear Safety Authority (STUK),
FINLAND

SMITH Jim
Centre for Ecology and Hydrology,
Winfrith Technology Centre, Dorset,
UNITED KINGDOM

TOIVONEN Harri
Radiation and Nuclear Safety Authority (STUK),
FINLAND

TSCHIERSCH Jochen
Head Radioecology(act.) Institute of Radiation Protection,
Helmholtz Zentrum München,
German Research Center for Environmental Health,
Neuherberg, GERMANY

VAN MEIR Nathalie
Institute for Radioprotection and Nuclear Safety (IRSN),
Laboratorie d'Etude des Transfers dans les Sols et sous-sol,
Paris, FRANCE

VOITSEKHOVITCH Oleg V.
Ukrainian Institute of Hydrometeorology, Kiev, UKRAINE

YOSCHENKO Vasyl
Ukrainian Institute of Agricultural Radiology (UIAR),
Chabany, Kiev Region, UKRAINE

ZABROTSKI Viachaslau
Polessje State Radiation-Ecological Reservation,
BELARUS

ZHURAVKOV Vladislav
International Sakharov Environmental University (ISEU),
Minsk, BELARUS

ZHURBA Marina
Ukrainian Institute of Agricultural Radiology (UIAR), Chabany,
Kiev Region, UKRAINE

PART 1: SOURCES AND CHARACTERISATION

RADIOACTIVE PARTICLES RELEASED FROM DIFFERENT NUCLEAR SOURCES

BRIT SALBU
Department of Plant and Environmental Sciences, Isotope Laboratory, Norwegian University of Life Sciences, P.O. Box 5003, 1432 Aas, Norway

Abstract: Following releases from severe nuclear events such as nuclear weapon tests, use of depleted uranium ammunition and reactor explosions or fire, a major fraction of refractory radionuclides such as uranium (U) and plutonium (Pu) is present as particles, often ranging from submicrons to fragments. Radioactive particles and colloids are also released via effluents from reprocessing facilities and civil reactors, and radioactive particles are identified in sediments in the close vicinity of radioactive waste dumped at sea. Thus, releases of radioactive particles containing refractory radionuclides should also be expected following severe nuclear events in the future.

Radioactive particles in the environment are heterogeneously distributed and can carry substantial amounts of refractory fission products, activation products and transuranics. Samples collected may not be representative and inert particles can be difficult to dissolve. For particle contaminated areas, the estimated inventories can therefore be underestimated. As particle weathering and remobilisation of associated radionuclides will occur over time, these entities represent diffuse sources for future ecosystem transfer. Thus, analytical techniques providing information on particle characteristics such as size distributions, crystallographic structures, oxidation states, being variables influencing weathering, mobility and biological uptake, is essential for impact and risk assessment. The present paper summarizes present documented knowledge on sources that have contributed to radioactive particles in the environment.

Keywords:

1. Introduction

Radioactive particles in the environment are defined as localised aggregates of radioactive atoms that give rise to an inhomogeneous distribution of radionuclides significantly different from that of the matrix background (IAEA CRP, 2001). In water, particles are defined as entities having diameters larger than 0.45 μm, i.e., will settle due to gravity, while particles larges than 1 mm are referred to as fragments. In air, radioactive particles ranging from submicrons in aerosols to fragments are classified according to the aerodynamic diameters, where particles less than 10 μm are considered respiratory. The presence of radioactive species ranging from particles to fragments can easily be identified by autoradiography, reflecting their inhomogeneous distributions in water, soils and sediments (Salbu, 2000).

Following the Chernobyl accident, radioactive particles were identified close to the source (Kuriny et al., 1993) as well as more than 2,000 km from the damaged reactor (Devell et al., 1986). During that time the "hot" particle phenomenon was considered to be a peculiarity of the Chernobyl accident. However, radioactive particles were identified already during the 1950s–1960s in areas where nuclear weapons were tested, for instance within the Nevada Test Site (Crocker et al., 1996). As the technical capabilities to characterize particles were limited to microscopes, only large particles or fragments were described in literature from the 1960s. Thus, the Chernobyl particles re-established the scientific field related to the characterization of radioactive particles and new advanced techniques have been developed. Using sensitive analytical techniques, radioactive particles containing refractory radionuclides have been identified at a series sites affected by radioactive contamination from different nuclear or radioactive sources, and not at all limited to the Chernobyl accident. Therefore, these sites should act as field laboratories, where knowledge on the characteristics and ecosystem behaviour of radioactive particles from previous events can be utilized to improve the preparedness related to accidental releases in the future.

Radioactive particles containing refractory radionuclides such as uranium (U) and plutonium (Pu) were released during severe nuclear events such as nuclear weapons tests (e.g., Maralinga, Mururoa, Nevada Test Site, Marshall Islands), use of depleted uranium ammunition (Kosovo, Kuwait), and nuclear explosions (e.g., Chernobyl) and fires in nuclear reactors (e.g., Windscale) (Crocker et al., 1966; Anspaugh and Church, 1986; Chamberlain, 1987; Cooper et al., 1994; Simon et al., 1995; Kashparov, 1996; Danesi et al., 1998; Salbu et al., 2003, 2004). Radioactive particles have also been released in authorized discharges from nuclear installations to rivers (e.g., Krashnoyarsk, Russia), authorised discharges to sea (Sellafield, UK; La Hague, France), in air emissions during operations (Windscale, early 1950s) and released from radioactive material dumped at sea, for instance at the shores of Novaya Zemlya (Salbu et al., 1993, 1994, 1997; Kjerre, 2006).

The radionuclide composition of released particles will depend on the source, while particle characteristics such as particle size distribution, crystallographic structures and oxidation states will also depend on the release scenarios (Salbu et al., 2001). Particles having U and/or Pu as the carrier matrix are most often formed due to critical or subcritical destruction of nuclear weapon-grade materials (nuclear tests, conventional explosions, impact of depleted uranium ammunition) or nuclear fuel matrices (e.g., reactor explosions, fires, corrosion processes). At high temperature and high pressure release scenarios (reactor explosion, fire), solid materials can liquefy, volatiles will escape, while refractory transuranics, fission and activation products will remain when droplets are solidified. Interactions with construction materials such as moderator and Zircaloy in reactors are also essential for the products formed i.e. the formation of metal-metal crystalline structures such as U-Zr or U-carbide compounds. In the presence of air, U and Pu can also be highly oxidized. At low temperature conditions, corroded U and Pu particles can be released and the characteristics of such particles would be different from those released during high temperature and pressure events (Salbu et al., 2001). Thus, the activity concentration, isotope ratios and burn-up will reflect the emitting source, while release conditions (high temperature and pressure, presence of air) will influence the activity concentrations of volatiles as well as particle characteristics (i.e., size distributions, crystallographic structures, oxidation states) being essential for their ecosystem behaviour.

2. State-of-the art-techniques for characterizing radioactive particles

Radioactive particles deposited in the environment are unevenly distributed (Fig. 1) and samples collected may not be representative (Bunzl, 1997). Furthermore, inert particles can be difficult to dissolve (Oughton et al., 1993). According to IAEA CRP (2001) "The failure to recognize the heterogeneity of radionuclide distribution in the environment leads to analytical inconsistencies, irreproducible results, and erratic conclusions". For particle contaminated areas, the estimated inventories can be underestimated. As particle weathering and remobilisation of associated radionuclides will occur over time, these entities represent diffuse sources for future ecosystem transfer. Thus, analytical techniques providing information on particle characteristics such as size distributions, crystallographic structures, oxidation states, variables influencing weathering, mobility and biological uptake, is essential for impact and risk assessment (Salbu, 2000; IAEA CRP, 2001).

Hot spots are easily recognized by autoradiography. There are several solid state speciation techniques that could be utilized for characterization of individual particles. Today, the most powerful technique for characterizing micrometer sized particles is the combination of scanning electron microscopy (SEM) with x-ray microanalyzer (XRMA) and synchrotron radiation (SR) based X-ray microscopic techniques (Salbu et al., 2001).

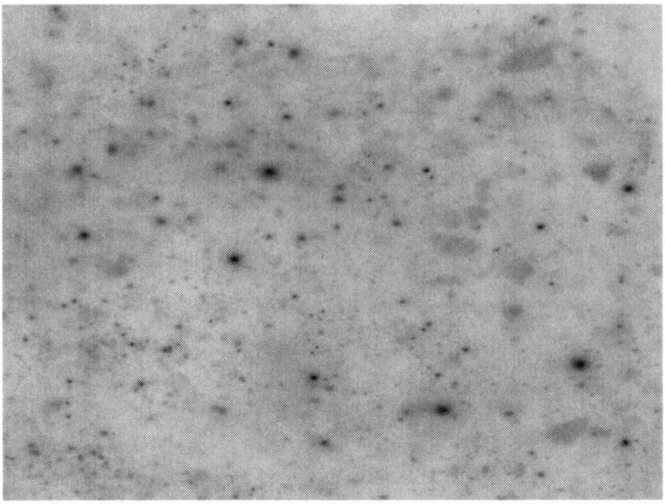

Figure 1. Autoradiography (P imaging) of samples collected at Chernobyl reflects the presence of radioactive particles (Salbu et al., 1993).

Transmission electron microscopy (TEM) with XRMA is also efficient to characterize electron dense colloidal sized structures (nanoparticles). Using SEM, the particle size distribution and the surface structure of individual particles can be characterised, and information on the distribution of high atomic elements on particle surfaces can be obtained using backscattered electrons (BEI mode). From X-ray mapping, the 2 D distribution of elements associated with particle surface can be obtained, while semi-quantitative elemental analysis at specific particle sites can be obtained by X-ray microanalysis (Salbu et al., 1993). SEM is also a preparative mean prior to SR analysis.

Using high flux monochromatic x-ray microbeams, synchrotron radiation (SR) based X-ray microscopic techniques have proved most useful for characterisation of individual particles. Microscopic X-ray fluorescence analysis (μ-XRF) provide information on 2 dimensional or 3 dimensional (μ-tomography) elemental distribution within individual particles, while micro-X-ray diffraction (μ-XRD) gives information on the crystallographic structures of solid particles. X-ray absorption near edge structure spectrometry (μ-XANES) provides information on the oxidation state (Fig. 2). The extended x-ray absorption fine structure analysis (EXAFS) could also potentially provide information on the coordination number and the distance to neighbouring atoms, if detailed knowledge on all atoms involved is available (Oughton et al., 1993; Salbu et al., 2001).

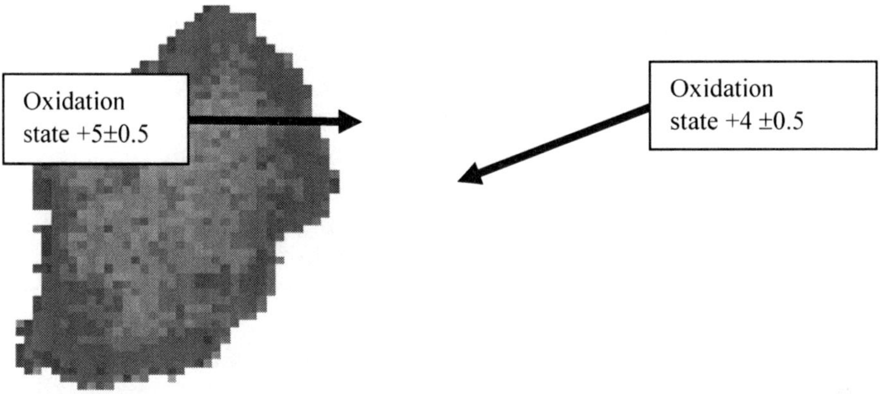

Figure 2. Synchrotron radiation based 2D μ-XANES analysis of uranium in a particle collected to the North of the Chernobyl reactor showing the two dimensional distribution of U oxidation state. The particle has a UO_2 core, while U in the surface is oxidized to U_3O_8 (Salbu et al., 2001).

3. Sources contributing to the release of radioactive particles into the environment

Since 1945, more than 2000 nuclear weapons tests have been performed globally, in the atmosphere, at ground, under ground, and under water UNSCEAR (2000). This is the key source contributing to the release of artificially produced radioactive particles into the environment. In addition, radioactive particles in the environment originate from a series of other nuclear events involving weapon materials or civil nuclear fuel.

4. Particles released during nuclear weapons tests

Hot spots and radioactive particles have been identified at Marshall Island, Nevada Test Site, Maralinga in Australia, Muroroa in French Polynesia, Novaya Zemlya and in Semipalatinsk Test Site in Kazahkstan. Particle releases from Loop Nor in China have been observed in Japan, while particles released at Novaya Zemlya were identified in Norway (unpublished). In general, a large variety of fused or partially fused u and/or Pu-particles, as well as large agglomerates consisting of individual small particles differing in colour, specific activity, density and magnetic properties have been identified. Based on analysis of particles from Marshall Island and Nevada Test Site performed in the 1950s and 1960s, the particle characteristics (size distribution, shape, colour) depended on devices and shot conditions (Crocker et al., 1966), at high altitudes spherical small-sized dense particles with activity distributed throughout the particles were obtained, while at ground surface large irregular shaped particles with lower density and specific activities were observed. From coral-surface bursts Pu particles associated with debris were relatively soluble in water, while beta emitters in sand from air-burst debris were dissolved in 0.1 M HCl. Following safety trials in Maralinga (Cooper et al., 1994), Pu particles from submicrons to several hundred microns and Pu-contaminated soil particles have been reported. Most of the activity was associated with particles within the 250–500 μm fraction, about 5% was present as particles with diameters less than 45 μm, and a respiratory fraction was identified. In addition to ^{239}Pu and U, ^{241}A was also associated with particles. Based on gamma spectrometry and proton-induced X-ray emission spectroscopy, Pu and U were localised on particle surfaces. Leaching experiments using a simulated lung fluid demonstrated the presence of inert particles with low solubility.

In Muroroa, Pu particles with activities up to 1 MBq, ranging from 200 μm to several hundred μm have been identified (Danesi et al., 1998). Using optical microscopy and X-ray micro-fluorescence for analysis of 200–500 μm particles, large differences in the surface structures could be observed; from glassy compact surfaces to conglomerates of small particles with rough appearances. Leaching experiments using human serum demonstrated the presence of inert particles with low solubility. In Semipalatinsk, Pu associated with particles or fragments with activity levels exceeding 50 kBq kg^{-1} have been reported (Yamamoto et al., 1996). Using SEM combined with XRMA a variety of structures, from glassy compact surfaces to conglomerates of small particles have also been observed. Using, and synchrotron radiation μ-XRF tomography, demonstrated that the distributions of U and Pu within particles were overlapping (unpublished). No detailed information on particle releases is available from Novaya Zemlya. However, hot spots in sediments close to Chernaya Bay reflected the presence of particles (Salbu et al., 1997). Radioactive particles have also been identified in air filters from 1961 and 1962 at several sites in Norway (unpublished).

5. Depleted uranium ammunitions

Depleted uranium is frequently used as ammunitions due to the density and hardness of the metal. Large amounts of DU ammunition was applied during the Golf war (1992) and the 1999 Balkan conflict. Following the impact with hard targets, DU is disperse and ignited. Therefore, DU particles ranging from submicrons to several hundred micrometers, mostly in the respiratory fraction, have been observed in sand, soils and in damaged vehicles. Using SEM and synchrotron X-ray microscopic techniques, U in the impacted DU particles was oxidized to UO_2, U_3O_8 or a mixture of these oxidation states. Following a fire in a DU ammunition storage facility, up to mm sized particles with U present in oxidation state +5 and +6, have been identified (Salbu et al., 2003, 2004). Although the source was similar (DU penetrators), the release scenario (detonation, fire) influenced the characteristics of released particles.

6. Nuclear accidents

Radioactive particles have also been released during nuclear accidents such as the fire in the Windscale reactor (1957) and the explosion in the Chernobyl reactor in 1986. Following the fire in Pile No. 1 at Windscale, radioactive particles varying in size within 20–500 μm were released (Kashparov et al., 1996). However, prior to the fire corroded radioactive particles were also released from the fuel via the stack of the air-cooled reactor. Using SEM, the corroded several hundred microns sized U fuel particles had flake-like structures (Salbu et al., 1994) and were inert towards leaching. Following the Chernobyl accident, radioactive particles varying in composition, size, shape, structures and colours have been identified, from compact small-sized crystalline single particles to large amorphous aggregates (Chamberlain, 1987; Kuriny et al., 1993; Kashparov et al., 1999). Fragments and large particles settled close to the site, while small-sized particles were transported more than 2,000 km from the site (Devell et al., 1986). Based on synchrotron-radiation X-ray micro-techniques, inert fuel particles with a core of UO_2 with surface layers of U-C or U-Zr were released during the initial explosion. In contrast, more soluble fuel particles with a UO_2 core and surface layers of oxidised uranium were released during the fire as illustrated in Fig. 1 (Salbu et al., 2001). The particle weathering rate and soil to plant transfer was low for particles released during the explosion and high for particles released during the fire.

7. Conventional detonation of nuclear weapons

Following accidents with nuclear devices and conventional explosions, radioactive particles have been released to the environment. The re-entry of the reactor driven Soviet satellite Cosmos 954 in 1978 resulted in contamination of radioactive particles ranging from submicrons to fragments in large areas in Canada (AMAP, 1997). Following the US B52 aircraft accident at Palomares, Spain in 1966, and the US B52 aircraft accident at Thule, Greenland, 1968, thermonuclear weapons were conventially detonated. The surrounding area was contaminated with submicron to millimeter particles containing ^{239}Pu and ^{235}U (Espinosa et al., 1998; Eriksson, 2002). Based on synchrotron radiation X-ray microtechniques, the Pu-U particles were oxidised (Jernstrøm et al., 2004; Lind et al., 2007).

8. Particles released from nuclear installation

Radioactive particles have been released from European and Russian reprocessing plants during operation. Due to discharges from Sellafield, UK, Irish Sea sediments have been contaminated with uranium containing particles (Kershaw et al., 1999). Due to accidental releases at Dounreay, UK, in the 1960s, large U fuel particles containing a series of fission products are still collected at nearby beaches (unpublished). No detailed information on particles released from La Hague, France, is available, but a major fraction of radionuclides in the effluent from Sellafield and La Hague during normal operation were associated with particles and colloids (Kjerre, 2006). Radioactive particles have also been released into the Techa River from Mayak PA (Kershaw et al., 1999) and into the Yenisey River from Krasnoyarsk Mining and Chemical Combine (Salbu et al., 1994).

9. Particles associated with dumping of waste

During 1959–1991, large amounts of radioactive and nuclear waste, including six submarines with fuelled reactors, were dumped in the Kara Sea and in the fjords of Novaya Zemlya. Using SEM, radioactive particles were identified in sediments collected in the close vicinity of radioactive waste dumped in the Stepovogo bays at Novaya Zemlya (Salbu et al., 1997).

10. Conclusions

Particles containing refractory radionuclides have been identified in connection with all types of serious nuclear events. Therefore, the phenomenon of radioactive particle occurs more frequently than previously anticipated, and if refractory radionuclides are identified in releases from a nuclear source, the presence of particles should be expected. Several years of research demonstrate that the particle composition will depend on the specific source, i.e., reflecting the matrix and the refractory radionuclide composition (e.g., burn-up). The release scenarios (e.g., temperature, pressures, redox conditions) will, however, influence particle characteristics such as particle size distribution, crystallographic structures and oxidation states, being essential for the ecosystem behaviour.

Soil and sediments can act as a sink for deposited particles, and particle contaminated soil and sediments may also act as a potential diffuse source in the future. Therefore, information on particle characteristics and processes influencing particle weathering and remobilisation of associated radionuclides is essential to assess long-term impact from radioactive particle contamination. To produce relevant information on source and release dependent particle characteristics, however, advanced analytical micro-techniques are needed within radioecology.

Acknowledgements

The author will gratefully acknowledge the support provided by: EU 5. and 6. FP (RAFF, ADVANCE), IAEA (CRP, "Characterization of radioactive particles"), The Norwegian Research Council, and The Norwegian Foreign Ministry. Furthermore, the author is inbebdted to O.C. Lind, UMB as well as close collaborating scientists: V. Kashparov, UIAR (Chernobyl), P.R. Danesi, IAEA (Kosovo, Kuwait), M. García León, University of Seville (Palomares), P. Mitchell, Univ. Coll. Dublin/N. Priest, Univ. of Middlesex (Semipalatinsk), P. Dale, SEPA (Dounrey), (Sellafield), S. Lukashenko, Kazakhstan (Semipalatinsk), H. Dahlgaard, Risoe National Laboratory (Thule), Joint Russian-Norwegian Expert Group (Kara Sea, Mayak, Kola) and K. Janssens, Univ. of Antwerp (ESRF, HASYLAB).

References

AMAP-Arctic monitoring assessment programme. (1997) Arctic pollution issues: radioactive contamination. Report from Norwegian Radiation Protection Authority, Oesteraas, Norway.

Anspaugh, L.R. and Church, B.W. (1986) Historical estimates of external γ exposure and collective external γ exposure from testing at the Nevada Test Site. 1. Test series through Hardtack II, 1958. *Health Phys.*, **51**, 35–51.

Bunzl, K. (1997) Probability of detecting hot particles in environmental samples by sample splitting. *Analyst*, **122**, 653–656.

Chamberlain, A.C. (1987) Environmental impact of particles emitted from Windscale Piles, 1954–57. *Sci. Tot. Environ.*, **63**, 139–160.

Cooper, M., Burns, P., Tracy, B., Wilks, M., and Williams, G. (1994) Characterisation of plutonium contamination at the former nuclear weapons testing range at Maralinga in South Australia *J. Radioanal. Nucl. Chem.*, **177**, 161–184.

Crocker, G.R., O'Connor, J.D., and Freiling, E.C. (1966) Physical and radiochemical properties of fallout particles. *Health Phys.*, **12**, 1099–1104.

Danesi, P.R., De Regge, P., La Rosa, J.M., Makarewicz, M.J., Moreno, J.Z., Radecki, Z.E., and Zeiller, E. (1998) Residual plutonium isotopes and americium in the terrestrial environment at the former nuclear test sites of Mururoa and Fangataufa, Proceedings of the 7th International Conference on "Low level measurements of actinides and long-lived radionuclides in biological and Environmental samples", Salt Lake City.

Devell, L., Tovedal, M., Bergstrøm, U., Applegren, A., Chussler, J., and Andersson, L. (1986) Initial observations of fallout from the reactor accident at Chernobyl. *Nature*, **321**, 817–819.

Eriksson, M. (2002) On weapons plutonium in the Arctic environment (Thule, Greenland). *PhD Thesis*, Risø National Laboratory, **3-5-0002**, 1–146.

Espinosa, A., Aragón, A., Hogdson, A., Stradling N., and Birchall, A. (1998) Assessment of doses to members of the public in Palomares from inhalation of plutonium and americium. *Radiat. Protect. Dosimetry*, **79**, 1–4.

IAEA CRP (2001) Co-ordinated research programme on radioactive particle. Report by an International Advisory Committee, IAEA, Vienna.

Jernstrøm, J., Eriksson, M., Osan, J., Tørøk, S., Simon, R., Falkenberg, G., Alsecz, A., and Betti, M. (2004) Non-destructive characterisation of radioactive particles from Irish sea sediment by micro X-ray fluorescence (μ-XRF) and micro X-ray absorption near edge spectroscopy (μ-XANES). *J. Anal. Atom. Spectrometry*, **19**, 1428–1433.

JNREG-Joint Norwegian-Russian Expert Group for Investigation of Radioactive Contamination in the Northern Areas. (1997) Sources contributing to radioactive contamination of the Techa River and areas surrounding the "Mayak" production association, Urals, Russia. Norwegian Radiation Protection Authority, ISBN 82-993979-6-1, 134 p.

Kashparov, V.A., Ivanov, Y.A, Zvarich, S.I., Protsak, V.P., Khomutinin, Y.V., Kurepin, A.D., and Pazukhin, E.M. (1996) Formation of hot particles during the Chernobyl nuclear power plant accident. *Nucl. Tech.*, **114**, 246–253.

Kashparov, V.A., Oughton, D.H., Protsak, V.P., Zvarisch, S.I., Protsak, V.P., and Levchuk, S.E. (1999) Kinetics of fuel particle weathering and ^{90}Sr mobility in the Chernobyl 30 km exclusion zone. *Health Phys.*, **76**, 251–259.

Kershaw, P.J., Denoon, D.C. and Woodhead, D.S. (1999) Observations on the redistribution of plutonium and americium in the Irish Sea sediments. 1978 to 1996: concentrations and inventories. *J. Environ. Radioact.*, **44** (3), 191–221.

Kjerre, L. (2006) Characterization of radioactive particles from Krasnoyarsk-26, Russia. *Thesis*, Norwegian University of Life Sciences, Aas, Norway.

Kuriny, V.D., Ivanov, Y.A., Kashparov, V.A., Loschilov, N.A., Protsak, V.P., Yudin, E.B., Zhurba, M.A., and Parshakov, A.E. (1993) Particle associated Chernobyl fall-out in the local and intermediate zones. *Ann. Nucl. Energy*, **20**, 415–420.

Lind, O.C., Salbu, B., Janssens, K., Proost, K., Garcia-Leon, M., and Garcia-Tenorio, R. (2007) Characterization of U/Pu particles originating from the nuclear weapon accidents at Palomares, Spain, 1966 and Thule, Greenland, 1968 *Sci. Total Environ.*, **376**(1–3), 294–305.

Oughton, D.H., Salbu, B., Brand, T.L., Day, J.P., and Aarkrog, A. (1993) Under-determination of strontium-90 in soils containing particles of irradiated uranium oxide fuel. *Analyst*, **118**, 1101–1105.

Salbu, B. (2000) Speciation of radionuclides. *Encyclopaedia Analytical Chemistry*, John Wiley & Sons Ltd., Chichester, pp. 12993–13016.

Salbu, B., Bjørnstad, H.E., Sværen, I., Prosser, S.L., Bulman, R.A., Harvey, B.R., and Lovett, M.B. (1993) Size distribution of radionuclides in nuclear fuel reprocessing liquids after mixing with seawater. *Sci. Tot. Environ.*, **130/131**, 51–63.

Salbu, B., Janssens, K., Lind, O.C., Proost, K., and Danesi, P.R. (2003) Oxidation states of uranium in DU particles from Kosovo. *J. Env. Radioact.*, **64**, 163–167.

Salbu, B., Janssens, K., Lind, O.C., Proost, K., Gijsels, L. and Danesi, P.R. (2004) Oxidation states of uranium in depleted uranium particles from Kuwait. *J. Envir. Radioact.*, **78**, 125–135.

Salbu, B., Krekling, T., Lind, O.C., Oughton, D.H., Drakopoulos, M., Simionovici, A., Snigireva, I., Snigirev, A., Weitkamp, T., Adams, F., Janssens, K., and Kashparov, V.A. (2001) High energy X-ray microscopy for characterisation of fuel particles. *Nucl. Instr. and Meth. A*, 467, **21**, 1249–1252.

Salbu, B., Krekling, T., Oughton, D.H., Østby, G., Kashparov, V.A., Brand, T.L., and Day, J.P. (1994) Hot particles in accidental releases from Chernobyl and Windscale Nuclear installations. *Analyst*, **119**, 125–130.

Salbu, B., Nikitin, A.I., Strand, P., Christensen, G.C., Chumichev, V.B., Lind, B., Fjelldal, H., Bergan, T.D.S., Rudjord, A.L., Sickel, M., Valetova, N.K., and Føyn, L. (1997) Radioactive contamination from dumped nuclear waste in the Kara Sea – results from the Joint Russian-Norwegian expeditions in 1992–1994. *Sci. Tot. Environ.*, **202**, 185–198.

Simon, S., Jenner, T., Graham, J., and Borcher, A. (1995) A comparison of macro- and microscopic measurements of plutonium in contaminated soil from the Republic of the Marshall Islands. *J. Radioanal. Nucl. Chem.*, **194**, 197–205.

UNSCEAR (2000) Sources and effects of ionizing radiation. The United Nations Scientific Committee on the Effects of Atomic Radiation. United Nations, New York.

Yamamoto, M., Tsumara, A., Katayama, Y., and Tsukatani, T. (1996) Plutonium isotopic composition in soil from the former Semipalatinsk nuclear test site. *Radiochimica Acta*, **72**, 209–215.

SINGLE PARTICLES HANDLING AND ANALYSES

URI ADMON
Israel Atomic Energy Commission, Nuclear Research Centre, Negev, P.O. Box 9001, Beer-Sheva 84190, Israel
phone: +972-8-6278835, e-mail: uadmon@netvision.net.il
In collaboration with:
E. Chinea-Cano, D. Wegrzynek, H. Aigner, A. Markowicz, D. Donohue – IAEA, Seibersdorf, G. Tamborini – ITU, Karlsruhe, F. Ruedenauer, W. Costin – ARC, Seibersdorf

Abstract: A single environmental hot particle carries a wealth of information about its radiological, chemical and metallurgical history, including clues about its release-scenario. The physical and chemical characteristics of the particle bear upon its mobilization and long-term behavior in the ecosystem.

The ability to locate, re-locate, handle and analyze single, isolated particles is essential in the arenas of nuclear forensics, safeguards investigations and environmental research. Single particles should be analyzed rather than agglomerates, because the latter invariably display averaged values, such as speciation or radionuclide dispersion in environmental samples or enrichment levels in safeguards samples.

A rainbow of advanced microanalytical techniques is presently available, often in remote laboratories. Full characterization of an individual particle is a multiple-instrument task that requires the ability to repeatedly re-locate and micromanipulate particles in the sub-micrometer size-range and up. Within the framework of the Hot Particles Coordinated Research Project (CRP), launched by the IAEA (NAAL, Seibersdorf) in 2001, methods have been developed, implemented, and tested on real samples for single particle re-location, ex-situ micromanipulation and relocation (transfer) between source- and target-instruments. The two, three, and six reference marks re-location algorithms have been written. Two types of ex-situ particle micromanipulation systems based on light microscopes (binocular- and stereo-microscope types) have been built. They were used in a variety of cases, such as loading single particles on TIMS filaments, or attaching them to needles for synchrotron measurements. FIB micro-surgery of single particles, exposing their inner structure, has been demonstrated.

These methods, along with other methods for single particles handling and analyses will be discussed in this article.

D.H. Oughton and V. Kashparov (eds.), *Radioactive Particles in the Environment*,

Keywords: particle analysis, re-location, relocation, micromanipulation, nuclear forensics, hot particles, environmental particles, micro-XRF tomography, micro-XANES, micro-XRD.

1. Introduction

Every nuclear activity, of any type and scale, declared or clandestine, creates a trail of "hot" particles. The physical, chemical and microstructural characteristics of these particles are source related and release-process dependent, and dictate their weathering, mobilization, transfer and long-term behaviour in the ecosystem. A single "environmental" hot particle, tiny as it may be, carries a wealth of information about its radiological, chemical and metallurgical history, including clues about its release-scenario. Detailed analyses of individual particles utilizing a variety of methods and instruments, often available in different laboratories, are of prime importance in environmental research on the one hand, and in nuclear forensics on the other.

The ability to locate, re-locate, handle and analyze single, isolated particles-of-interest (POIs) is essential in these two arenas. Single particles should be analyzed rather than agglomerates, because the latter invariably display averaged values, such as speciation or radionuclide dispersion in environmental samples, and isotopic composition or enrichment level in nuclear forensic and safeguards samples.

Air borne particles, in the micron- and sub-micron size range, tend to agglomerate or to adhere to atmospheric dust particles or to soil grains. Particles in marine environment mix and interact with sediments and under-water organisms. These limitations accentuate the need for tools for handling single particles enabling their characterization by multiple, complementary microstructural and microanalytical high-level methods.

The Hot Particles CRP (IAEA, 2001) was launched by the IAEA in 2001. Within the framework of this project methods and equipment have been developed and tested on real samples for the re-location, ex-situ micromanipulation and relocation (transfer) of single particles from a source instrument to a target instrument (Admon et al., 2002a,b, 2005) At the same time, in-situ SEM micro-manipulation technique has been developed at JAERI (Sakurai, 2005, 2006; Esaka, 2004, 2007).

The developed methods have been successfully applied in a variety of real cases, and a range of particle sizes (down to sub-micrometer). POIs' initial identification and location have been done by SEM, μ-XRF, fluorescence light microscopy or fission track analysis (FTA), and the final analyses by SEM, SIMS, TIMS, ICP-MS, Synchrotron μ-XRF-tomography. The latter technique allows virtual slicing of a particle. Real slicing through single needle-mounted particles, exposing their inner structure, was also carried out using the dual beam SEM/FIB technique (Admon et al., 2006).

The particle re-location method (Admon et al., 2002a,b, 2005) has been implemented in several laboratories including the IAEA, the ITU, the Chinese Institute of Atomic Energy (CIAE), and others.

This article will be focused on the handling methods of single particles that enable their multiple-instrument analyses.

2. What is a single particle?

The term "single particle" cannot be uniquely and unequivocally defined. A particle may be a homogeneous entity by one property yet heterogeneous by another; it can be of any size and shape; it can be isolated, adhere to other particles, or even form agglomerates; in other words, the definition whether a particle is "single" or not is arbitrary and depends on the connotation.

Our interest is focused on hot particles released from different sources in a variety of mostly-violent scenarios. Hence, from practical reasons, a broad definition of the term "a single particle" should be applied. The definition has to be based on a-priori discernible characteristics and not on a-posteriori information, namely, properties that become known only after the completion of the analyses. Our definition is rather simplistic: a particle is called "single" if it is mechanically sturdy enough as not to disintegrate while being repeatedly micromanipulated, transferred between source- and target-instruments, and analyzed.

Figure 1*a* shows a grain of sand (SiO_2) decorated with numerous DU-oxide particles adhering to its surface. The sample was collected in Kuwait after 12 years of weathering in desert environment. According to our definition, this grain is regarded as a single particle, and so is any isolated DU-oxide particle that can be detached from it. This can readily be done by means of needle micromanipulation.

Figure 1*b* displays a reactor-core particle collected 8 km SW from ChNPP (Kashparov, 2001), evidently carried by the west-plume during the initial stages of the Chernobyl accident. Despite its pronounced heterogeneity, this is a single particle.

Figure 1*c* is a picture of the MOX fuel powder from the 1994 Munich Airport seizure (Tamborini and Betti, 2000; Wallenius et al., 2007). Seen are agglomerates of Pu-oxide platelets and rods, and a U_3O_8 faceted particle. In-situ SEM micromanipulation can partly separate these agglomerates to isolated single particles.

From the point of view of nuclear forensics, particularly when isotopic ratios are used to determine the age and radiological history of a hot particle, it is essential to avoid any mixing, hence interference, between different particles. If this happens, the interference between particles leads inevitably to averaging of their isotopic ratios, thus falsifying the results. It has been shown by Sakurai et al. and Esaka et al. (Sakurai, 2005; Sakurai et al., 2006; Esaka et al., 2004, 2007), using the in-situ SEM micromanipulation technique developed in JAERI, that the "purity" and dispersion of the results can be greatly improved by separating μm-size particles clinging to each other, that would otherwise be picked-up together, and be wrongly analyzed as one.

In-situ SEM micromanipulation, because of its high-resolution on-line imaging power and nanometric-precision needle movement capability, can indeed help solving this problem, enabling a true single particle analysis.

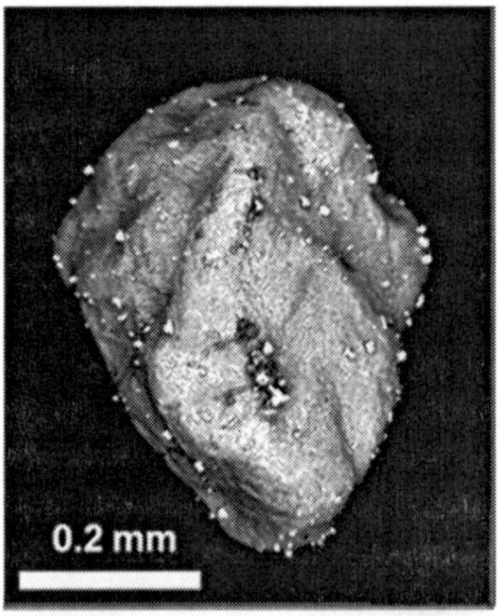

Figure 1a. A grain of sand (SiO_2) brought from Kuwait (2002) decorated with numerous DU-oxide particles adhering to its surface. According to our definition, this grain is regarded as a *single particle*, and so is any isolated DU-oxide particle that can be detached from it.

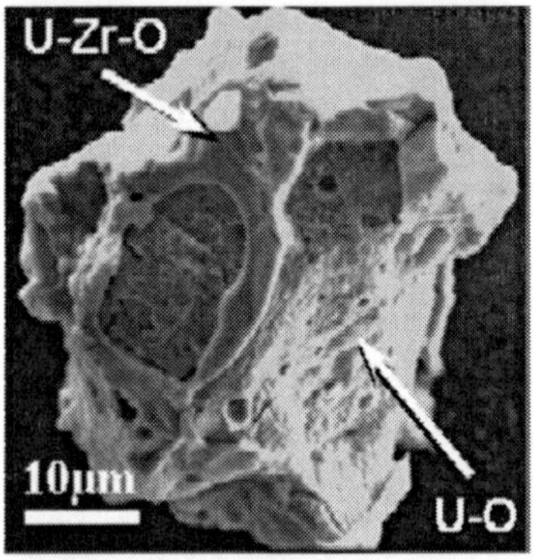

Figure 1b. A ChNPP particle found 8 km SW of Unit 4. Despite its pronounced heterogeneity, this is a single particle.
(Courtesy of Valery Kashparov, picture from the UIAR-ChNPP Hot Particles Data-Base).

Figure 1c. Agglomerates of Pu-oxide platelets (A) and rods (B), and U_3O_8 faceted particle (C), from the Munich Airport seizure (1994) of MOX fuel. In-situ SEM micromanipulation can partly separate them to isolated single particles.
(Courtesy of Ian Ray and Klaus Mayer [ITU]).

3. Location, re-location and relocation of single particles

Environmental, forensic and safeguards samples are usually collected on filters or on cotton swipes, or brought to the laboratory in buckets of contaminated soil, marine sediments, estuary deposits or water from reservoirs, rivers, lakes and other sources. The particles-of-interest (POIs) are almost invariably mixed within very large populations of non-POIs, such as atmospheric dust, biota debris, various fibers, etc. The particles are recovered (e.g., suspended by ultrasonication), then spread over special substrates (e.g., graphite or silicon planchettes). The available techniques require a lot of skill, and are beyond the scope of this article. To avoid cross-contamination, the sample preparation, handling and following analyses must be carried out in clean laboratory environment.

Three major difficulties are encountered before a particle can be actually analyzed:

- ***Locating*** and identifying POIs within large populations of non-POIs in the source-instrument.
- ***Re-locating*** selected POIs in target-instruments, to where they have been moved from the source-instrument, and verifying their presence and integrity.
- ***Relocating*** (lifting, transferring, depositing) and handling single POIs while keeping their tracks, avoiding accidental loss.

These three issues, represented in a circular diagram in Fig. 2, are discussed below.

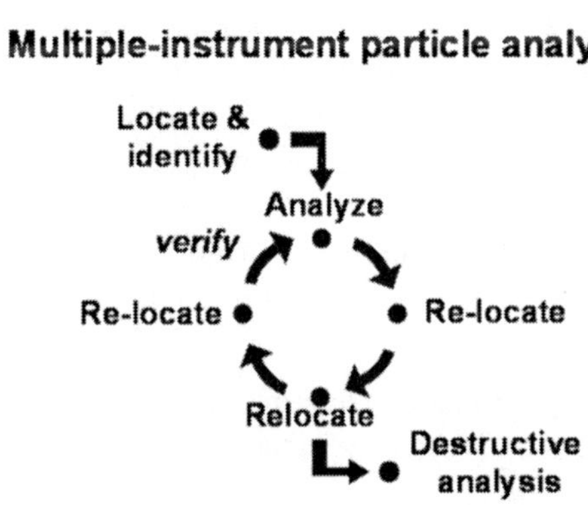

Figure 2. The multiple-instrument particle analysis circle.

Initially, POIs are identified, *located* and listed by their coordinates in the *source instrument*.

For further analyses, selected POIs need to be *re-located*, then micromanipulated and *relocated* to the *target instrument*, where they are *re-located* once again and analyzed.

The circle terminates when the particle is annihilated in a destructive analysis such as mass-spectrometry or FIB micro-surgical slicing.

3.1. PARTICLE LOCATION

Locating and identifying POIs, the first step in particle analysis, can be done by a variety of methods (Vajda, 2001). There are passive methods, such as alpha-track radiography (Boulyga and Becker, 2001), autoradiography (Kerkápoly et al., 2005a,b), or binary partitioning and gamma-counting (Burns et al., 1995), or active methods aimed at stimulating the emission of characteristic signals that make POIs be distinguishable from all others. Methods of the latter type include fluorescing by violet and ultraviolet light, μ-XRF scanning (Eriksson et al., 2005), SEM-EDS automated search (a method that evolved from the Gun-Shot-Residues routines widely used by Police Forensic laboratories, when nuclear terrorism, illicit trafficking and illegal possession started becoming a global threat), and Fission-Track-Analysis, which is usually the first method to be used for the detection of particles containing fissile isotopes, particularly ^{235}U.

The particle search is carried out in the source instrument. The search yields a list of POIs and their corresponding coordinates in the coordinate-system of the specimen stage of the source instrument. The accuracy and reproducibility of the stage mechanism are crucial factors, because any error will propagate (as discussed below) into the predicted location of the particle in the target instrument, where it is sought afterwards for further analysis.

Having the sample scanned and the particles-of-interest identified as such and located, we are ready to proceed to the next steps (Fig. 2). If the source-instrument is the SEM, then the POIs selected for further analyses are invariably first re-visited at their listed coordinates for a thorough SEM characterization by imaging and microanalysis. Then, as well as in all other cases, the particles are repeatedly re-located in a target instrument by applying one of the re-location algorithms, micromanipulated and relocated (transferred) to another target-instrument, and so on. The final step is killing the particle in a destructive analysis (any kind of mass-spectrometry); alternatively, it can by subjected to micro-surgery (slicing in the FIB).

The focal point of this article is the re-location and relocation of single particles, and they are discussed in the following sections.

3.2. PARTICLE RE-LOCATION – THE GENERAL CASE: THE 3-POINTS ALGORITHM (THE TRIANGULATION METHOD)

The 3-points algorithm is used in the general case for re-locating POIs on a planar substrate in the target instrument, according to their stage-coordinates measured in the source instrument. It is implicitly assumed that no non-linear distortion of the sample takes place during the transfer between the two instruments. This coordinate-transformation is named the triangulation method.

A detailed discussion of the 3-points algorithms for particle re-location, including error-evaluation and a few real-cases testing is given by Admon et al. (2005). Here a brief summary will be presented. The method has been implemented and is currently used by several laboratories.

Let **A**, **B**, and **C** be three non-colinear reference point marks on the sample, Fig. 3.

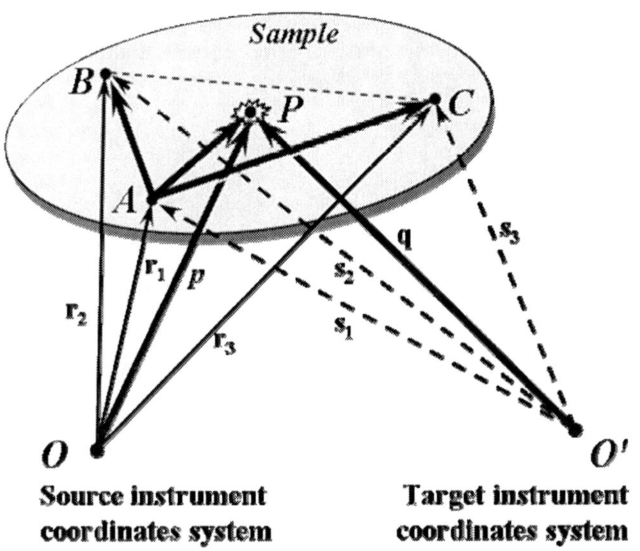

Figure 3. 3-D vector diagram representing the 3-points re-location method. It is aimed at predicting the stage-coordinates of a POI, ***P***, in the *target instrument, **O′*** (e.g., SIMS) based on its measured coordinates in the *source instrument, **O*** (e.g., SEM). ***A,B,C*** are three reference marks on the sample surface, identifiable and measurable in both instruments. For clarity, the sample is shown as being transparent. (See Admon et al., 2005).

Let ***O***, ***O′*** be the stage (Cartesian) coordinates systems of the *source instrument* (e.g., SEM) and the target instrument (e.g., SIMS), respectively.

It can be readily shown that any point P co-planar with A, B, and C may be expressed as the vector-sum

$$\vec{p} = \sum_{i=1}^{3} m_i \vec{r}_i \qquad \text{(in coordinate system } \boldsymbol{O}\text{)}$$

or

$$\vec{q} = \sum_{i=1}^{3} m_i \vec{s}_i \qquad \text{(in coordinate system } \boldsymbol{O'}\text{)}$$

where the components $\{m_i\}$ are independent of the coordinates system used, $m_1+m_2+m_3=1$.

By analogy, ***P*** can be regarded as the centre-of-mass of 3 point-masses $\{m_i\}$, at ***A***, ***B***, and ***C***. $\{m_i\}$ may be positive as well as negative. The location of the centre-of-mass of a set of masses is a property of the set, independent of the choice of coordinates system. In other words, ***P*** is uniquely defined by $\{m_i\}$ with respect to ***A, B, C*** in any coordinates system.

Let $\{x_i,y_i\}$ and $\{u_i,v_i\}$ be the measured stage coordinates of the reference marks ***A,B,C*** in the source and in the target instruments, respectively. Remembering that ***A,B,C,P*** are co-planar we can write

$$\begin{pmatrix} x_P \\ y_P \\ 1 \end{pmatrix} = \begin{pmatrix} x_1 & x_2 & x_3 \\ y_1 & y_2 & y_3 \\ 1 & 1 & 1 \end{pmatrix} \begin{pmatrix} m_1 \\ m_2 \\ m_3 \end{pmatrix} = R \begin{pmatrix} m_1 \\ m_2 \\ m_3 \end{pmatrix} \tag{1}$$

The reference-marks-matrix, ***R,*** is defined in this equation for later use.

The target stage coordinates of any given POI, (u_P,v_P), can be calculated from its known source coordinates, (x_P,y_P), by the relation

$$\begin{pmatrix} u_P \\ v_P \end{pmatrix} = \begin{pmatrix} u_1 & u_2 & u_3 \\ v_1 & v_2 & v_3 \end{pmatrix} \begin{pmatrix} m_1 \\ m_2 \\ m_3 \end{pmatrix} = \begin{pmatrix} u_1 & u_2 & u_3 \\ v_1 & v_2 & v_3 \end{pmatrix} \begin{pmatrix} x_1 & x_2 & x_3 \\ y_1 & y_2 & y_3 \\ 1 & 1 & 1 \end{pmatrix}^{-1} \begin{pmatrix} x_P \\ y_P \\ 1 \end{pmatrix} \tag{2}$$

This equation is the basis of the triangulation method.

It is worth mentioning, that the (3 × 3) matrix is only regular if the three points **A**, **B** and **C** are not colinear. Also, the system can be expanded to a statistical GLM (General Linear Model), if more than three reference points are used.

Figure 4 shows how the 3-point re-location method is used in practice. Three copper ‘finder’ grids, widely used for TEM biological samples, are glued to the planchette forming a triangle. As will be explained later (Section 3.5), the highest accuracy is obtained with the maximal-area triangle contained within the usable borders of the planchette, namely, an equilateral triangle. Certain features on the ‘finder’ grids, such as the inner corners of the ~5 μm triangle inside the “W” or “A” may serve as adequate, convenient reference marks. They are readily discernible in all possible source- and target instruments thus allowing the implementation of the 3-point algorithm (Admon et al., 2002a,b, 2005).

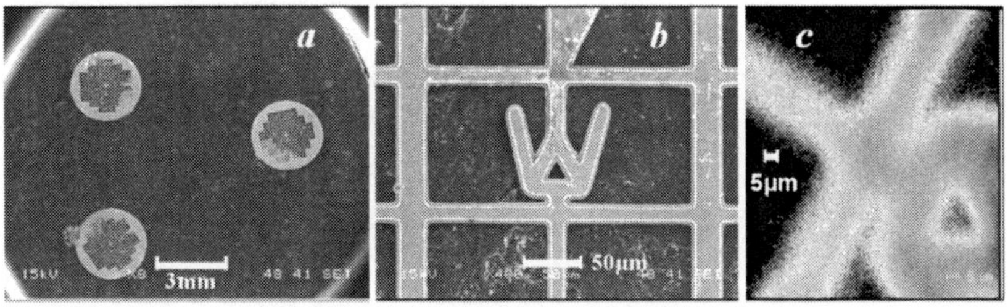

Figure 4. Practicing the triangulation method (Admon et al., 2002a,b, 2005): *(a)* The three reference markers (TEM 'finder' grids) glued to the polished graphite planchette. *(b)* Close-up SEM picture of one of the letters in the finder grid, W, the triangular hole in which serves as a convenient reference point for both the SEM and the SIMS. *(c)* SIMS copper-image of the same W. The triangular hole is easily discernible, allowing measuring of its coordinates. (The pictures were taken by Liu, 2005).

3.3. PARTICLE RE-LOCATION: THE 2-POINT ALGORITHM

In certain cases (e.g., in SEM-to-LM experiments) the sample plane is perpendicular to the viewing direction (z) in both the source and the target instruments. In such cases the dimensions measured on the sample do not require tilt-distortion correction, an inherent feature of the 3-point algorithm. Hence, two reference marks, ***A*** and ***B***, are sufficient for re-locating any particle, ***P***, in the sample plane (Fig. 5).

If (x,y) are coordinates measured in the source instrument and (u,v) in the target instrument, then

$$\begin{pmatrix} u_P \\ v_P \end{pmatrix} = \begin{pmatrix} x_0 \\ y_0 \end{pmatrix} + \begin{pmatrix} W & Z \\ -Z & W \end{pmatrix} \begin{pmatrix} x_P \\ y_P \end{pmatrix} \tag{3}$$

where

$$\delta x = x_B - x_A, \delta y = y_B - y_A, \delta u = u_B - u_A, \delta v = v_B - v_A$$

$$W = \frac{\delta x.\delta u + \delta y.\delta v}{\sqrt{(\delta x^2 + \delta y^2)(\delta u^2 + \delta v^2)}}$$

$$Z = k\sqrt{1 - W^2}$$

$$x_0 = u_A - Wx_A - Zy_A$$

$$y_0 = v_A + Zx_A - Wy_A$$

and where $k = +1$ if the expression $(\delta y \cdot \delta v - \delta x \cdot \delta u)$ is positive, and $k = -1$ otherwise.

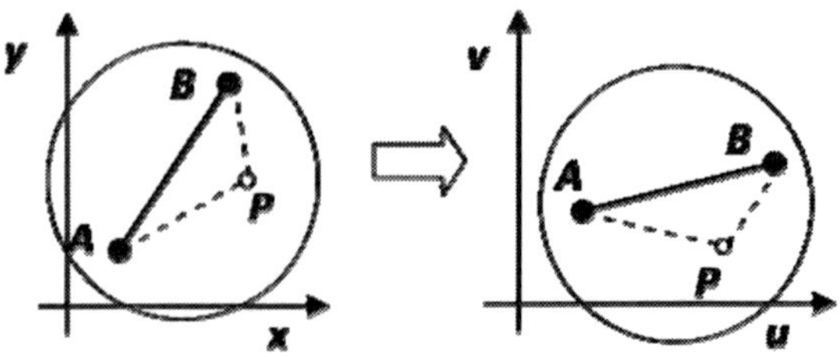

Figure 5. In the 2-point case the sample plane is perpendicular to the viewing direction (z); hence, the transformation from the *source*- to the *target-instrument* reduces to 2-dimensional *shift and rotation.*

3.4. RE-LOCATION AND μ-MANIPULATION IN FISSION TRACK ANALYSIS: THE 6-POINT ALGORITHM

Fission Track Analysis (FTA) is a commonly used technique for the search of fissile particles, namely, particles containing a fissile isotope, such as ^{235}U, ^{239}Pu or others. Their presence is revealed by the effects associated with the energetic fission fragments ejected from the particle as a result of fission events, usually induced by neutron irradiation. Alpha emission and spontaneous fission will also trigger these effects. It is a sensitive, reliable and relatively fast method (for details, see the caption to Fig. 6). However, it has three drawbacks: first, it requires access to neutron-irradiation of the samples in a nuclear reactor. Second, it is insensitive to non-fissile isotopes and cannot detect them. Third, the particles are invariably embedded in an organic fixating "carrier" film (e.g., collodion or lexan), that is prohibitive for SEM examination unless the organic material is removed.

In the FTA method the fissile POIs, usually hiding within large populations of non-POIs embedded together in the "carrier" film, are located according to the "star"-like track patterns created in the "detector" as a result of neutron irradiation. The "stars" are actually the local latent damage zones induced in the "detector" substrate when hit by the energetic fission fragments emitted from the particles in each fission event (Fig. 6b). They are exposed by rinsing the "detector" in a strong etchant (NaOH, 6N). Figure 6c shows that all fission-fragment tracks actually emerge from a 2 μm particle at the centre of the "star", resting slightly above the "detector" surface. The "stars" are typically a few tens μm across, easily discernible by a light microscope; the particles themselves are frequently invisible or barely visible.

After the irradiation, and before etching, the "carrier" and the "detector" are separated. The "star" pattern exposed in the "detector" is, in fact, a map showing the locations of all fissile POIs in the "carrier". Using this map, there are two ways to locate the particles in the "carrier". In the first, used in some laboratories, as JAERI, AFTAC and STUK, the "detector" and "carrier" are re-joined and aligned so that the assumed location of the particle can be directly and readily seen, albeit with some alignment uncertainty; then, using a scalpel, a tungsten-carbide needle or a laser-beam, a small square (typically 100 × 100 μm^2) around this location is cut out from the "carrier" and lifted away for further analysis.

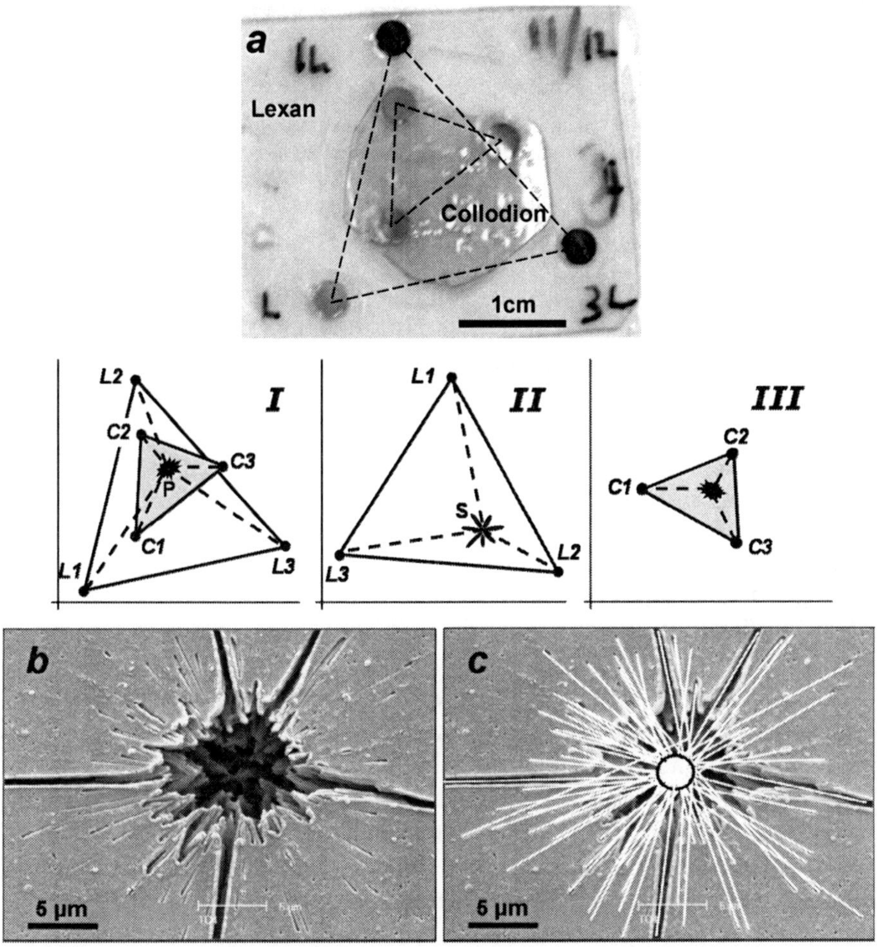

Figure 6. The 6-point algorithm, used for the location of fissile particles by the FTA method: (*a*) The particles are embedded in a "carrier" fixating film (e.g., collodion or lexan) lain on top of the "detector", usually a poly-carbonate (e.g., lexan) substrate. Two triplets of reference marks (TEM "finder" grids) are glued: one to the "carrier" and one to the "detector", and their coordinates measured jointly (*case I*), as described in the text. After neutron irradiation the "carrier" film is separated from the "detector", the latter rinsed in NaOH for etching and exposing the "star"-like track patterns of local damage induced by of energetic fission fragments, (*b*). In this case all tracks seem to have emerged from a 2 μm particle at the centre of the "star", (*c*). The "stars" coordinates are measured in the "detector" reference system (*case II*); applying the 6-point algorithm allows calculating the predicted locations of the corresponding particles in the "carrier" (*case III*). (*Continues on the next page*)

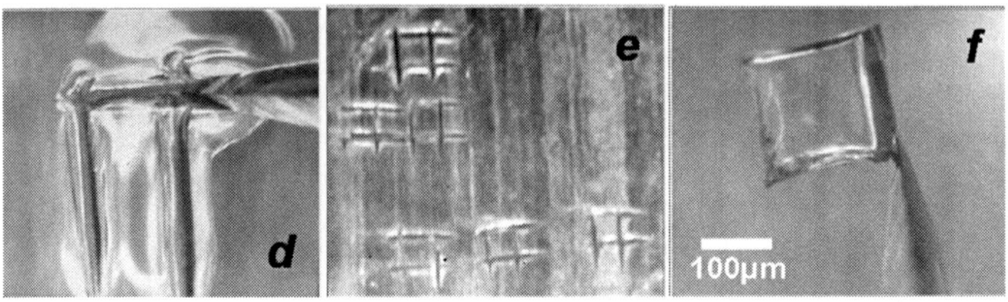

Figure 6 (continued). (d)–(f) Finally, utilizing the micromanipulation system, 100 × 100 μm^2 squares are cut in the "carrier" around the calculated location of each particle. Then, they are lifted and taken for further analyses to the TIMS or ICP-MS. The size of these squares is several times bigger than the uncertainty in the predicted particle locations. (Note: in JAERI and AFTAC a different methodology is used).

The second method, named the 6-points algorithm (Fig. 6), is based on a double-use of the 3-points algorithm described above. Three triplets of "finder" grids are attached: one triplet to the "detector", the other to the "carrier". Let $\mathbf{L}_\mathbf{I}$, $\mathbf{L}_\mathbf{II}$ be the reference marks matrices (defined in Eq. 1) of the "detector" marks ($\mathbf{L}_1$, $\mathbf{L}_2$, $\mathbf{L}_3$) in the I and II coordinate systems, respectively. Similarly, let $\mathbf{C}_\mathbf{I}$, $\mathbf{C}_\mathbf{III}$ be the matching matrices of the "carrier" marks ($\mathbf{C}_1$, $\mathbf{C}_2$, $\mathbf{C}_3$) in systems I and III. Bearing in mind that the locations of the particle and the star-of-tracks coincide in system I, it follows that $\mathbf{p}_\mathbf{III}$, the coordinates of the particle in system III, can be expressed as the following transformation of the "star" coordinates $\mathbf{s}_\mathbf{II}$ measured in system II:

$$\vec{p}_{III} = \left(C_{III} C_I^{-1} L_I L_{II}^{-1}\right) \vec{s}_{II} \tag{4}$$

Here, $\mathbf{L}^{-1}$ is the inverse matrix of $\mathbf{L}$, etc. Thus, measuring the coordinates $\mathbf{s}_\mathbf{II}$ = (x_s,y_s) of the "stars" in system II allows predicting the POIs' locations $\mathbf{p}_\mathbf{III} = (u_p,v_p)$ in system III. Cutting and lifting, as in the first method, complete the task.

It is noteworthy that the CEA (Baude, 2001, 2005) perfected a third method, by which the "detector" and "carrier" are laid side-by-side on a glass plate on the precision specimen stage of an inverted light-microscope. Before separation, the "detector" and "carrier" are simultaneously perforated with a sharp needle to produce the required reference marks pattern. Then, the coordinates of the tracks "stars" are measured on the "detector", transformed to the stage co-ordinates where the particles should be found in the "carrier", and the process is completed by moving the stage to these locations, cutting around the particles, and lifting them. The advantage of this method is that the free-working distance is not limited, but on the other hand, transmitted light illumination (advantageous for some samples) is not integral, and hence cannot be used effectively.

3.5. PARTICLE RE-LOCATION: ACCURACY AND UNCERTAINTY

The inherent uncertainty involved in the particle re-location method, namely, the possible deviation of the real location of a particle in the target instrument (e.g., SIMS) from its predicted location, which is calculated on the basis of its measured coordinates in the source instrument (e.g., SEM), was thoroughly analyzed and evaluated by Admon et al. (2005).

The origin of the uncertainty is instrumental (optical, mechanical), human-, and sample-related, depending on the accuracy of the instruments involved as well as the skill of the operator. Applying the principles of the error propagation method (GUM, 1995), the re-location expanded uncertainty, 2σ, of any given POI can be calculated as a function of its geometric position (xP,yP) with respect to the reference marks. It was thus possible to draw "uncertainty-field maps" displaying the spatial re-location uncertainty for POIs distributed over the entire sample area. Maps in the SEM-to-SIMS case are shown (Fig. 7) for three configurations of the reference marks on the planchette. The uncertainty-field distribution is determined by the geometry and size of the reference marks configuration. Evidently, the best selection (lowest uncertainties) is attained when the reference marks form the maximal-size equilateral triangle contained within the useable planchette borders. In fact, the smaller the area of the triangle formed by the reference marks, the larger the re-location uncertainty.

Results of SEM-to-SEM and SEM-to-SIMS re-location experiments, using the 3-points algorithm and involving instruments in three laboratories (IAEA-Seibrsdorf, ITU-Karlsruhe, and the Technical University-Vienna) and, are presented in the form of bull's-eye diagrams (Fig. 8). The results are in good agreement with the calculation. For example, in the SEM-to-SEM experiments 66% of the 77 measured particles deviated by less than 6 μm, 94% by less than 12 μm, from their predicted locations. These values are in full agreement with the estimation.

Figure 9 displays a few of the 25 particles identified and located in the SEM at IAEA, and then re-located in the SIMS at ITU (Admon et al., 2002a,b, 2005). The bull's-eye diagram for this experiment (Fig. 8c) shows that the deviations between the locations where the particles were actually found and their predicted ones are under 20 μm, as estimated by the model.

It is interesting to mention that the 3-point method has been implemented by Liu Guorong, Li Jinghuai and Li Jing (Liu, 2005) in the China Institute for Atomic Energy (CIAE). They measured over 160 particles in a SEM-to-SIMS experiment, and the results are in good agreement with the predicted uncertainty. The obtained re-location precision was better than 50 μm.

The 2-point particle re-location method has been used by Sangely et al. (2006) at the CEA in SEM-to-SIMS experiments. The sharp corners of two triangular aluminum foil pieces served as reference marks, and the obtained re-location precision was better than 50 μm. It should be noted, however, that the obtainable precision can be improved to less than 20 μm by applying the 3-point algorithm, as was demonstrated by Admon et al. (2002a,b, 2005) in a practically identical SEM and SIMS configuration (Fig. 7).

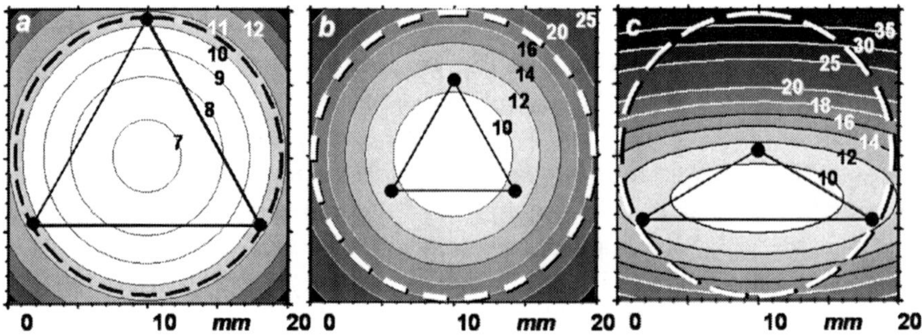

Figure 7. Maps of the expanded uncertainty, (2σ, in μm-units), of the particle predicted location in the SIMS, based on the propagation of measurement errors in the SEM (±1 μm for the marks, ±1.5 μm for the POIs), and in the SIMS (±7.5 μm for the marks). Dashed circles, 20 mm in diameter, indicate the borders of the usable area for particle dispersion on the planchette. Maps show the effects of size (*b*), and geometry (*c*), of the three reference marks configuration. Best precision is achieved when the marks form the triangle of maximal area (*a*), confined within the sample borders. If a particle is within the circle, or ellipse, confined by the three reference marks, then the uncertainty in its predicted location will be less than 11 μm (Admon et al., 2002a,b, 2005). (In collaboration with G. Tamborini [ITU] and H. Aigner [IAEA]).

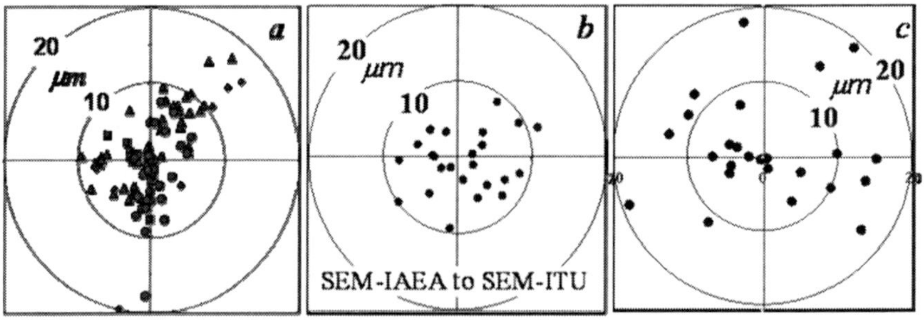

Figure 8. Deviation (in μm units) of the actual from the predicted POIs' locations in SEM-to-SEM and SEM-to-SIMS experiments, based on the triangulation method. (*a*) 77 particles in four SEM-to-SEM experiments: 66% deviated by less than 6 μm, 94% by less than 12 μm. In (*b*) and (*c*) the source instrument was the SEM-IAEA, and the target instruments – the SEM-ITU and the SIMS-ITU, respectively; the same 25 particles were identified and re-located in both laboratories. (Admon et al., 2002a,b, 2005). (In collaboration with G. Tamborini [ITU] and H. Aigner [IAEA]).

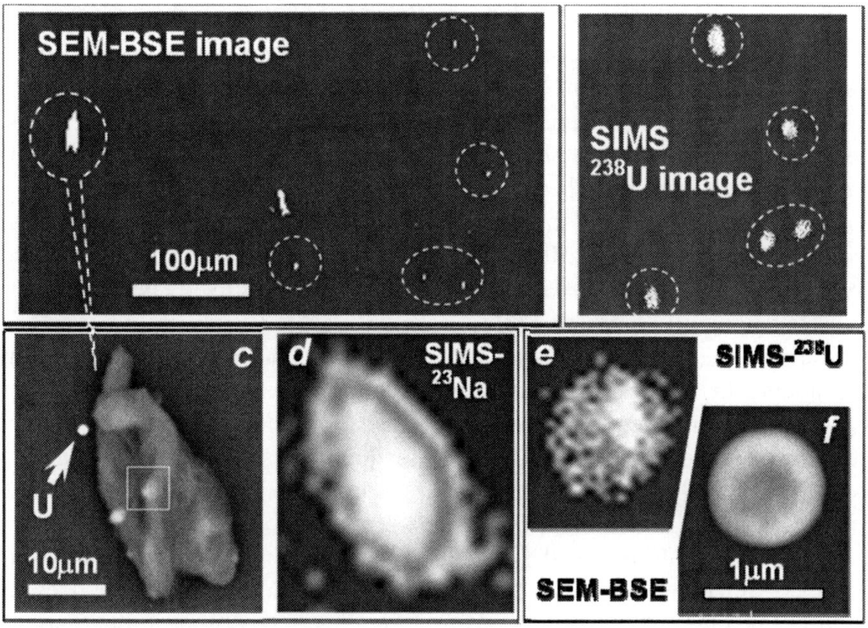

Figure 9. From the SEM in IAEA to the SIMS in ITU (Admon et al., 2002a,b, 2005): The same particles (U_3O_8 and NaCl) were located, identified and analyzed in the two laboratories, demonstrating the effectiveness of the particle re-location method. (*a*) SEM-BSE image showing five 1 μm U_3O_8 particles (encircled) and a large NaCl particle. (*b*) The same U_3O_8 particles in a SIMS-^{238}U image. The NaCl particle is seen in the SEM-image (*c*) and in the SIMS-^{23}Na image (*d*). The U_3O_8 particle seen in (*c*) is shown in the close-up pictures (*e*, *S*IMS) and (*f*, *S*EM). The NaCl particle covers two additional U_3O_8 particles (bright spots). (In collaboration with G. Tamborini [ITU]).

3.6. PARTICLE RELOCATION

The art of single particle micromanipulation and relocation (lifting, transferring, depositing and handling) will be discussed in detail in Chapter 5.

4. Using laser-ablation ICP-MS for isolated particle analysis

LA-ICP-MS has the potential of becoming a key-tool in particle analysis. It has a laser-beam (e.g., Nd:YAG @ 213 nm, but others are in use too) that can be usefully focused down to a 10 μm spot. This actually represents the obtainable spatial resolution. In other words, two 1 μm particles can be 10 μm apart and still be resolved by the LA-ICP-MS. It is noteworthy that the crater-like damage zone visible on the substrate after a particle is ablated is similar in size to the beam.

Once the coordinates of all identified POIs are measured, either by SEM, μ-XRF, FTA or another method, the sample can be transferred to the LA-ICP-MS. The 3-points algorithm is applied, predicting the new coordinates where the POIs can be found. Then, the laser-beam is aimed and shot at the particles one-by-one. The beam-size should be set somewhat larger than the re-location extended uncertainty, which can be estimated roughly as 20 μm following the propagation-of-error calculation (Admon et al., 2005). This limits the spatial resolution; however, it hardly creates a real limitation when environmental samples are concerned, because of the scarcity of POIs dispersed over the substrate among all other non-POIs.

Chiappini et al. (2006) and Hubert et al. (2007) reported preliminary experiments in LA-ICP-MS analyses of micrometer-size uranium particles, using the 1,030 nm femtosecond laser. They concluded that the results appear very promising because the LA-ICP-MS technique is sensitive, rapid, direct and easy to use, and shows advantages with respect to the TIMS, thus lends itself as an efficient alternative method for particle analysis.

Becker et al. (2008) used LA-ICP-MS for geochronology of Precambrian zircon ($ZrSiO_4$) particles from the Baltic Sea, determining their age according to the radiogenic lead (daughter) to uranium (parent) isotope ratio. The same principles are applied in nuclear forensic investigations.

Elish et al. (2007) demonstrated another micro-scale application of LA-ICP-MS, close to particle analysis, and of high relevance to nuclear forensics. They mapped the longitudinal profile of uranium concentration and isotopic composition along a single hair strand of a person that experienced a short-term exposure to uranium. Since the hair, growing at an average rate of 17 μm/h (12 mm/month) is a very efficient bio-concentrator for uranium in the human body, the longitudinal concentration profile can reveal the near-history of "occupational" exposure-to-uranium events, with a time-resolution of hours, depending on the bio-kinetics of the person involved.

The main advantage of LA-ICP-MS over other MS methods is that it is applied directly to the sample, saving preparatory steps as μ-manipulation. In FTA, for example, the particle can be ablated together with its encapsulating "carrier". A careful usage of the re-location algorithm and a precise alignment of the optical system is all that is required. This combination of FTA search and LA-ICP-MS has the potential of becoming the ultimate method for the analysis of U- and Pu-containing particles.

5. Micromanipulation under a light microscope: methodology

Particles in the size range from slightly less than 1 μm and above can be micromanipulated in systems based on a light microscope (LM). Depending on the optical performance requirements, either a "uniaxial" or a "stereo" LM can be used. The former is the conventional binocular light microscope (BLM). It offers a better image resolution, higher useful magnification and contrast than the stereo-microscope (SLM), but this is at the cost of ease of operation and shorter free working distance (FWD) and depth of field. The SLM enables easier operation and a higher free working distance, but a lower resolution and, to some extent disturbing, image parallax.

The human aspects of the micromanipulation (MM) system operation, utilizing either a BLM or an SLM, cannot be underestimated. A high degree of eye-hand co-ordination and "soft finger" touch are required. Altogether, the operator has to control 6 degrees of freedom in moving the particle and needle, while bringing them to a mutual touch in the focal plane of the LM. A proper set-up of the system, as will be described, helps simplifying the complexity of this task, thus reducing the efforts involved.

Let us discuss now a few basic features of the BLM and the SLM, that bear upon the capabilities and limitations of the MM system, and on their employment modes for particle micromanipulation. It should be noted that in a few cases, the detailed explanations included in the legends to the figures are not repeated in the text.

The two MM systems described in the following chapters are manually-operated and can be self-assembled at a relatively low cost. Computer-controlled systems offering excellent performence are also commercially available, but at a much higher cost.

5.1. THE BINOCULAR LIGHT MICROSCOPE (BLM)

In the BLM the objective lens (OL) forms a single image, which is split by prisms and simultaneously displayed by the left and right oculars and by the CCD camera. All three images are identical; the line of sight is parallel to the optical axis of the microscope and hence perpendicular to the sample plane. Consequently, no image parallax is formed. This is true in bright-field (BF) and dark-field (DF) operation modes, as well as in transmitted- and reflected-light illumination conditions. A high Numerical Aperture (NA) is allowed, hence a rather small depth of field and a short free working distance. The FWD, namely, the spacing between the sample and the lower end of the objective lens, restricts the accessibility and maneuverability of needles, blades or other tools in the vicinity of the sample.

The objective lenses of a BLM-based MM system must be of the long-focal-distance (LFD) type. It is recommended that a set of dual-use BF/DF para-focal objectives: ×5, ×10, ×20, ×50 will be installed. The FWD of these objectives is typically about 9–10 mm, allowing a convenient inclination angle (10–20°) for the needle. The outer diameter of the objectives must be small enough to avoid collisions (see below); some companies produce BF/DF objectives that are prohibitively large.

Changing the magnification of the BLM is done primarily by changing the objective lens by rotating the objectives-turret. An additional magnification by an adjustable factor of ×1-×2.5 is usually achievable by auxiliary lenses inserted in the ray path above the deflection mirror of the reflected-light illuminator. Having the objectives moving in-and-out in such a close proximity to the needle implies that the x- and y-axes of the precision-stage and of the needle holder should be co-aligned at 45° to the normal setting, as shown in Fig. 10. This positioning reduces the risk of collision between the needle and a moving objective, yet allows a convenient reach and operation of all control knobs of the stage, needle holder and microscope.

It is noteworthy that the CCD set-up shown in Fig. 10 co-aligns the x and y movements of the precision specimen stage, as well as those of the needle, with the horizontal and vertical directions on the CCD screen. This has proven itself to be extremely beneficial for even a trained operator, improving his performance and reducing his fatigue after a prolonged particle-hunt. Apparently, it is much easier for the operator's brain and hand to control the movement of the particle to an arbitrary point on the screen, where the needle-tip is waiting, by x and y displacements (driven by a single micrometer at a time) that are parallel to the horizontal and vertical axes of the displayed image, respectively.

The same principle applies to MM systems based on BLM and SLM.

In many cases the operator prefers looking through the eyepieces rather than watching the CCD monitor. It is advantageous, therefore, to equip the BLM with an "erecting" ocular unit. This unit actually re-inverts the image, so that the "up", "down", "right" and "left" directions in the sample coincide with those in the image. In conventional ocular units these directions are inverted, forcing the operator to move opposite to his natural tendency.

These problems do not exist in computerized MM systems, where the movements of the needle and sample are controlled by mouse-clicking on the screen.

5.2. THE STEREO LIGHT MICROSCOPE (SLM)

The modern SLM is of the Common Main Objective (CMO) type. It comprises a single large diameter objective lens that forms an image at infinity, and two independent optical tubes, one for each eye, that look at the sample through the objective at somewhat different angles. The axes of these tubes are inclined with respect to the main optical axis, and by 10–12° to one another; hence, each eye receives a slightly different image, thus giving the observer a stereoscopic perception. However, this inclination is also responsible for image parallax: the up-down focusing movement of the microscope creates a lateral shift of the image (Fig. 11).

A CCD camera receives its image either from a third, separate optical tube or by a beam splitter through one of the eye-tubes. The separate tube may be co-axial with the CMO, so the image is parallax-free, but in most cases it is inclined, so a parallax is formed; however, it is not in the same direction as the eye-parallax.

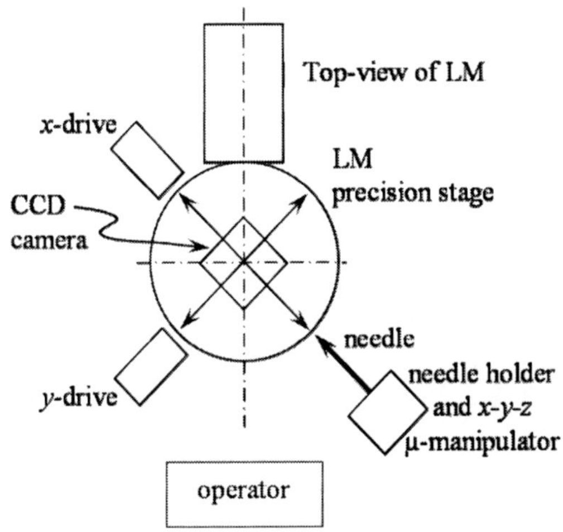

Figure 10. The optimal set-up of a micromanipulation system comprising a LM, precision *x-y-z (z=F)* stage, a needle attached to an *x-y-z* micromanipulator and a CCD camera. This set-up allows an easy operation and reduces the risk of collision between the needle and a moving objective of the BLM or SLM. The *x* and *y* movements of the stage coincide, respectively, with the *horizontal* and *vertical* directions on the CCD screen.

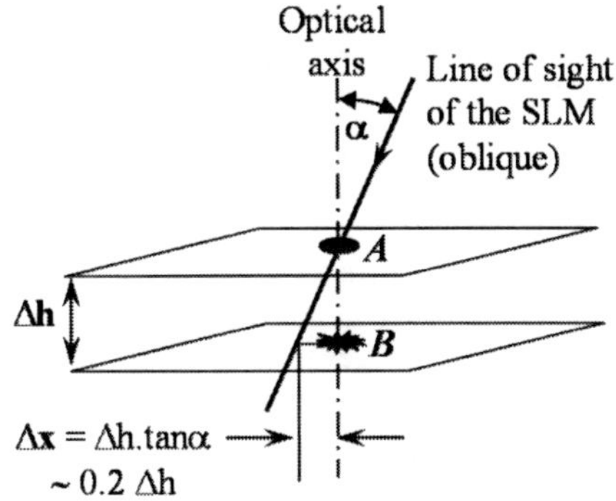

Figure 11. The origin of parallax in the SLM: the two on-axis vertically-displaced features ***A*** (the needle-tip) and ***B*** (the particle) look in the SLM as if they are horizontally-displaced by $\Delta x = \Delta h.\tan\alpha \sim 0.2\Delta h$, as a result of the oblique line-of-sight of each tube of the SLM. The parallax is directional, and must be considered during the "closing range" operation. In the BLM the line-of-sight is vertical and paraxial (parallel to the optical axis), and there is no parallax.

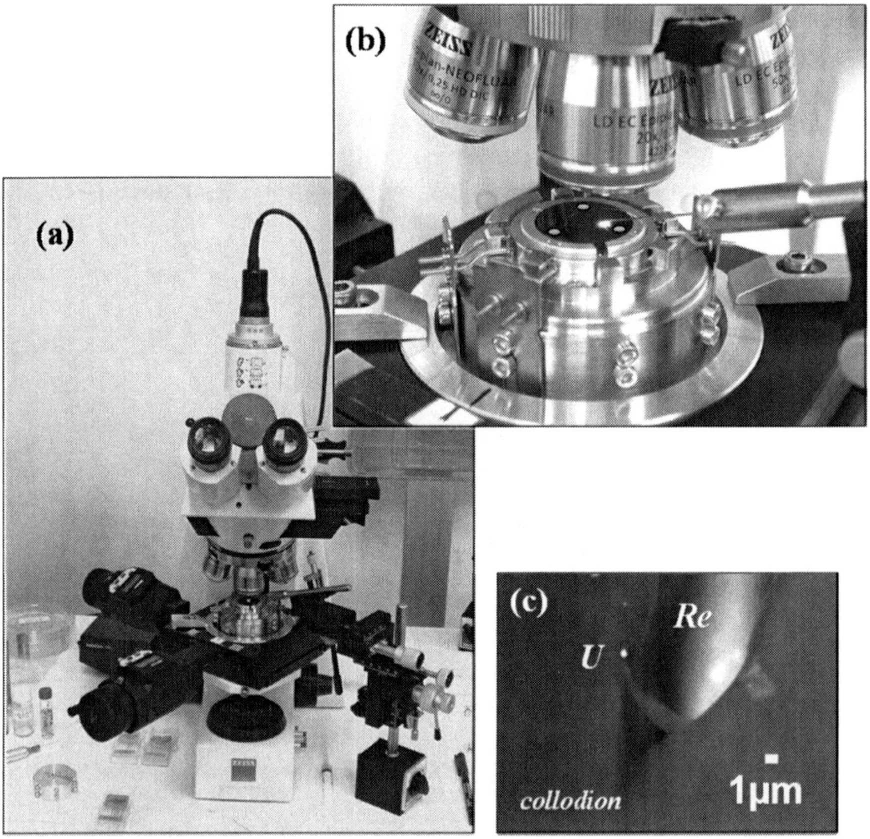

Figure 12. BLM-based MM system (NAAL-IAEA, Seibersdorf).

(a) General view, showing the Zeiss microscope, precision *x-y* Maerzhauser MT50 × 50 stage mounted on the focusing mechanism (µm-precision *z*-control) of the microscope; the needle holder and *x-y-z* Narishige micromanipulator; the CCD camera. The system is set-up to pick-and-load single particles onto TIMS (Rhenium) filaments.

(b) Up to six filaments (two are shown) can be attached to the turret base in waiting positions around the planchette. The three reference marks (TEM finder-grids) are glued to the planchette near its rim.

(c) After being lifted, a particle is transferred from the 1 µm needle-tip to the filament by stabbing into a tiny semi-dry flexible-collodion droplet on the filament. As a result of the stabbing, a single uranium particle has been deposited adhering to the sticky collodion, that micro-cleaved exposing the underlying rhenium filament.

5.3. THE PROBLEM OF PARALLAX IN THE SLM

The operator must be aware of the parallax, when present (Fig. 11), and particularly of its direction. This is true when the "closing range" process is monitored through the CCD, and equally true when the oculars are used, or moreover, combination of both. Most people do have one dominant eye, an enhancing factor of the effect of the parallax. Before touchdown, namely, when the needle and the particle are still vertically separated, their perceived horizontal relative positions are falsified by the parallax. Unless taken into account, unwanted accidental collisions may occur.

5.4. MICROMANIPULATION SYSTEM BASED ON A BLM

Figure 12 shows a micromanipulation system based on a BLM, built by the author at the IAEA laboratories, Seiberdorf. It comprises an upright binocular light microscope (Zeiss, Axioplan), a precision x-y specimen stage (Maerzhauzer, MT50 × 50) driven by two 1 μm precision, 50 mm range micrometers (Mitutoyo, 164). The needle holder is attached to an x-y-z micromanipulator (Maerzhauzer, MMJ); the micromanipulator seen in the picture was recently replaced by a hydraulic joy-stick controlled unit (Newport, NSA12-NSC200) with 100 nm precision.

The *z*-movement of the specimen stage is provided by the coarse and fine focusing mechanism of the BLM, with precision better than 1 μm. This precision, necessary for the needle-to-particle "closing-range" operation (as will be described later) is actually a requirement set by the optics of the BLM, namely, by its limited depth-of-field.

This MM system is designed for re-locating and lifting a selected POI from a polished graphite planchette, and immediately depositing it on a standing-by rhenium filament of the TIMS analyzer (Fig. 12*b*). The deposition is done by stabbing the loaded needle into a tiny semi-dry flexible collodion droplet on the filament. As a result of the stabbing (Fig. 12*c*) the particle is detached from the needle and adheres to the sticky collodion. It is noteworthy that the presence of carbon atoms (originating from the collodion) in the vicinity of the particle atoms during the thermal vaporization process in the TIMS enhances their ionization efficiency.

Figure 13 displays the "scar" (displaced particles zone) left on the planchette after a normal particle-lifting operation. It is about 20 μm across; it should be assumed that every particle within this zone has been co-lifted with the selected POI, and may be included and averaged in the measurement. However, in view of the scarcity of POIs in environmental samples, a "scar" of this size does not seem to pose a significant problem. The "scar" is created by occasional vibrations of the needle caused by unstable workbench and by the duel between the needle and the particle until the latter surrenders and can be conveniently picked-up. The size of the "scar" can be considerably reduced by taking measures to damp down the vibrations by using an adequate workbench, and stiffening the needle by inserting and gluing most of its free length into the metal tube of a surgical needle.

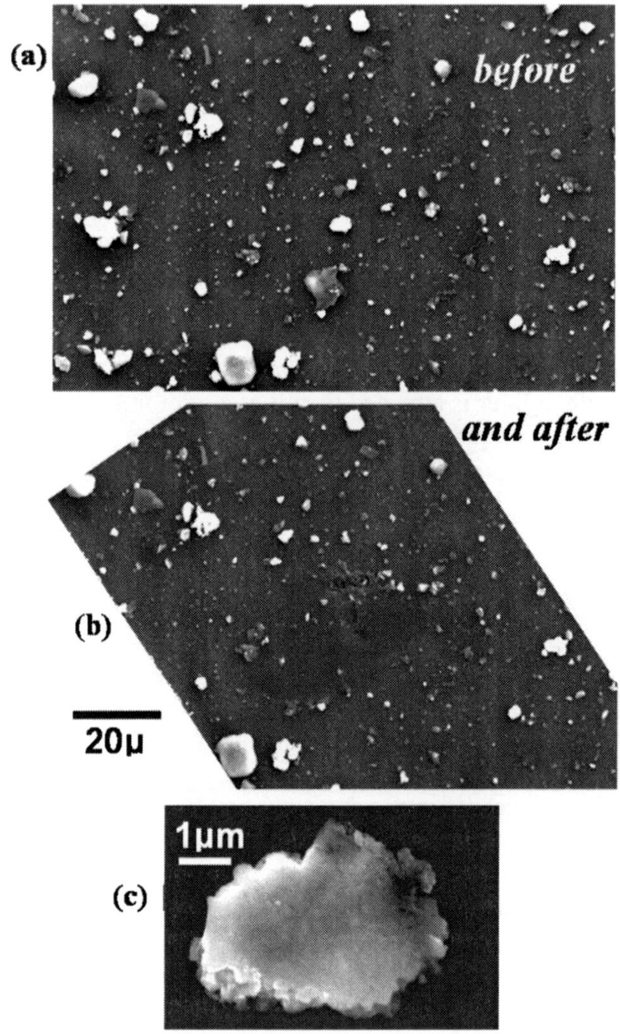

Figure 13. A thorium-rich particle-of-interest *(a)* has been located by an automatic SEM search on a graphite planchette, then re-located and photographed in the SEM. *(c)* Microanalysis showed it is a thorium oxide particle. Later, the planchette was moved to a BLM-based MM system, the particle re-located once again, lifted by a needle and transferred to the ICPMS for (destructive) isotope analysis. *(b)* The "scar" (zone of displaced particles) left on the planchette after the particle-lifting operation. It is created by occasional vibrations of the needle and by the struggle with the particle. Being ~20 µm (or mostly, less) across, it does not create a significant problem when environmental samples are studied.

5.5. MICROMANIPULATION SYSTEM BASED ON A SLM

The SLM based MM systems are less expensive than the BLM-based ones, and are useful in cases where the optical requirements can be compromised, namely, for particle size in the range 5–10 μm and above.

Figure 14 shows a micromanipulation system based on a SLM, built by the author at the IAEA laboratories, Seiberdorf. It comprises a SLM (Leica, FL-III) equipped with an additional focusing unit (F2 in Fig. 14), which supports a precision x-y specimen stage (Maerzhauzer, MT50 × 50) driven by two 1 μm precision, 50 mm range micrometers (Mitutoyo, 164). The needle holder is attached to a x-y-z micromanipulator (Narishige, MMN). The additional focusing mechanism (F2) is identical to the one supporting the SLM (F1), and is mounted on the same column. Consequently, and regardless of the parallax, the up-down movements of the SLM and the sample are co-axial. This is an important feature for the "closing range" procedure. The needle, once set at the centre of the field-of-view, need not be re-aligned.

In the close-up picture (Fig. 14*c*), one can see the particles spread over a mylar tambourine, pre-coated with a thin semi-transparent gold, platinum, or other conductive layer, to overcome electrostatic charge effect. These effects can be severe and disrupt the whole procedure; the needle, specimen stage and sample need all to be well-hooked to ground potential.

Before bringing the sample to the MM system it has been scanned to identify and locate DU particles by the μ-XRF system, developed in the Instrumentation Unit in IAEA-Seibersdorf (Wegrzynek et al., 2005). Applying the 3-point algorithm to the coordinates measured in the μ-XRF system it was possible to determine the predicted coordinates and re-locate the particles under the SLM.

A DU-oxide particle on top of a manually-sharpened 0.5 mm diameter graphite needle is displayed in Fig. 14*d*. The needle was partially inserted into the metal tube of a surgical needle No. 21 (same inner diameter) for reinforcement, and brought to the synchrotron for μ-XRF tomography. In the synchrotron, the needle is firmly held by a gripper, and rotated on its axis. The graphite needle is used because of its low atomic number, and the metallic tube – to strengthen it.

5.6. NEEDLE-TO-PARTICLE "CLOSING RANGE" AND PARTICLE LIFT-OFF PROCEDURES UNDER THE BLM AND SLM

Once visually re-located on the substrate, the particle can be picked-up by the needle. The "closing range" operation between the needle and the particle is a delicate procedure, aimed at avoiding accidental collisions, damaging the needle, or losing the particle.

The BLM and SLM differ in their focusing principle. In the BLM the sample is brought to the stationary focal point of the objective lens by moving it up-and-down. In the SLM it is the whole optical column that is raised and lowered for focusing on the sample, while the sample remains at a fixed position. Consequently, in each system a different "closing range" procedure is used.

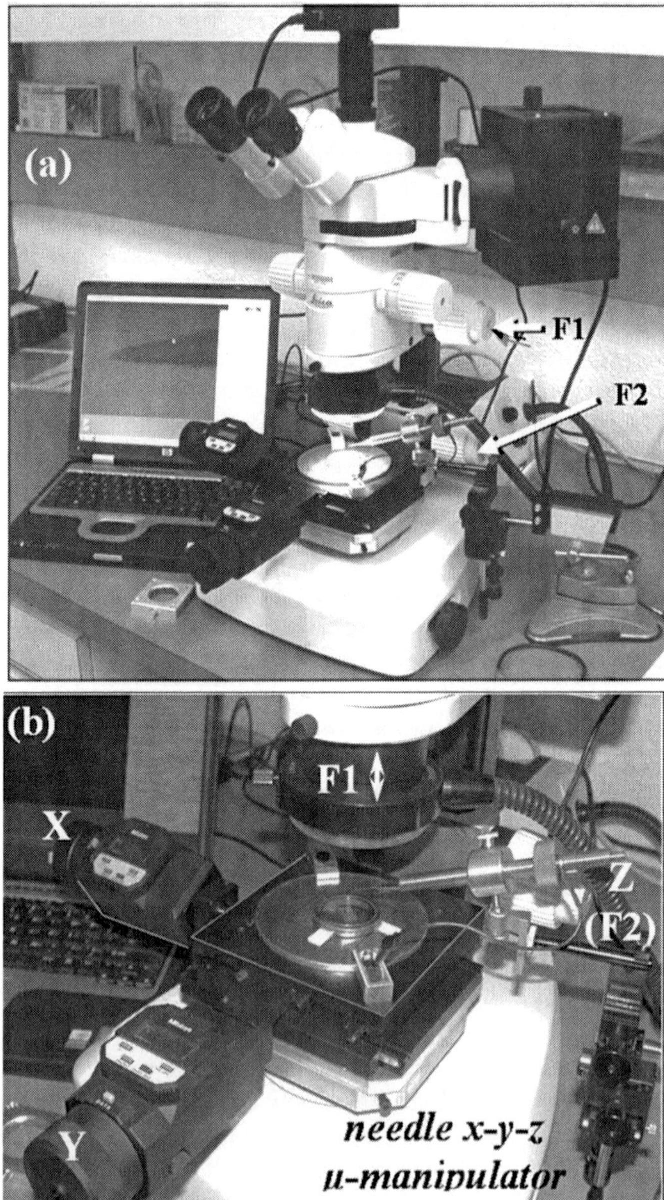

Figure 14 (continues on the next page). SLM-based MM system (NAAL-IAEA).

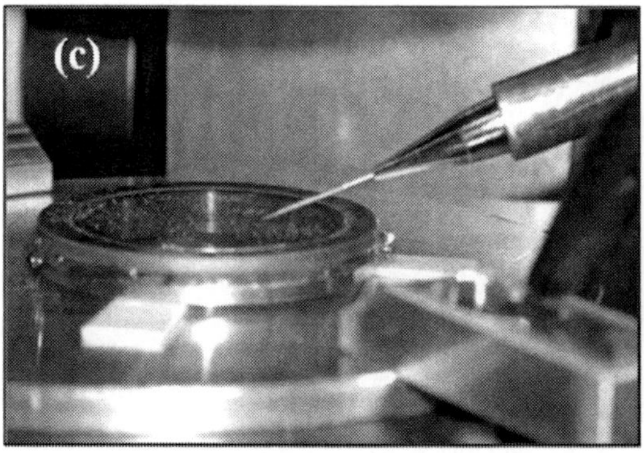

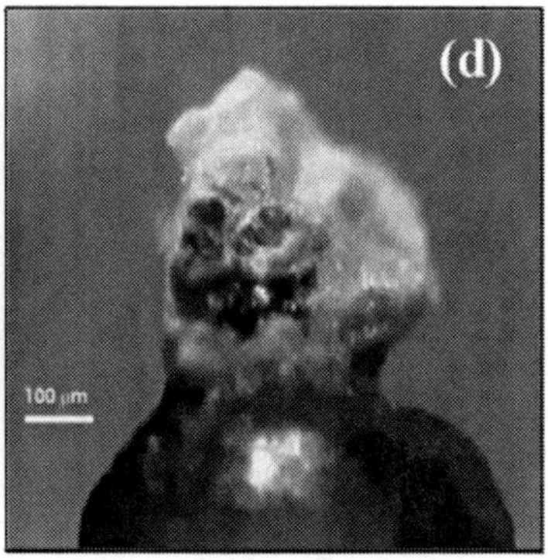

Figure 14 (continued). SLM-based MM system (NAAL-IAEA, Seibersdorf).

(*a*) General view. A second focusing unit (lowest white knob, F2) has been added to the Leica steromicroscope, to enable *z*-movement with μm-precision of the *x*-*y* Maerzhauser MT50 × 50 stage. The needle has an independent *x*-*y*-*z* movement, but actually needs to be aligned just once in the SLM-based system, to be – when in focus – at the centre of the field-of-view. The particle is brought to the stationary needle (and not vice-versa) by manipulating F2 and the *x* and *y* micrometers of the stage.

(*b*)–(*c*) Close-up on the heart of the system, comprising the needle above the mylar tambourine over which the particles are dispersed, together with the (glued) three reference markers; the *x*-*y*-*z* micromanipulator, holding the needle; the *x* and *y* micrometer drivers of the precision stage; the *z* control (=F2) added to the stage; the objective lens and ring illuminator of the SLM, which is raised and lowered by its fine-focusing mechanism (F1).

(*d*) DU particle on a graphite needle, showing a mixture of oxide phases (yellow and black).

The "closing range" and particle lifting step-by-step procedures for the BLM- and SLM-based MM systems are shown and explained in detail in Figs. 15 and 16, and demonstrated in Fig. 17. The explanations, given in the legends, are not repeated here.

The principle of operation of both systems is to "bring the particle to the needle", and not vice-versa, as is invariably being done in procedures like artificial egg-fertilization. The advantage of this approach is that the needle and the particle are always kept at the centre of the field of view even in the high magnifications used. The fine movements, particularly at the final "closing range" steps, are applied through the x,y and z (=F) precision drives of the stage, rather than through those of the needle micromanipulator; the former are finer and allow higher sensitivity and stability. In the BLM the z-driver of the stage is the built-in focusing-knob, while in the SLM it is an additional focusing mechanism (F2), added to the system, that supports the x-y stage.

It is worthy to re-emphasize that in the SLM, the parallactic effect may create a misleading perception of the relative positions of the needle and the particle before they actually touch one another, and the operator has to bring this into account.

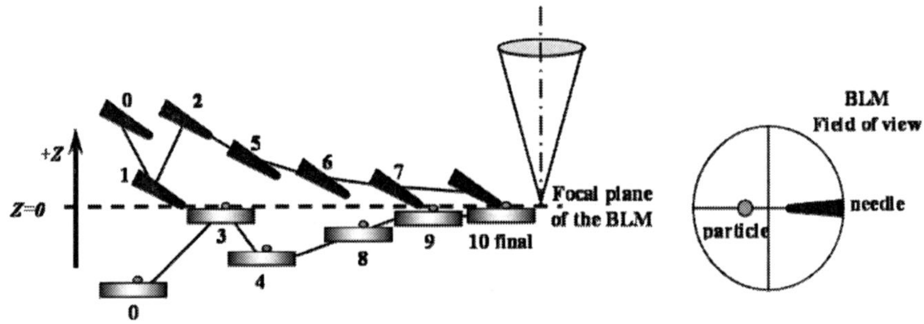

Figure 15. *"Closing range" between the needle and the particle under the BLM: (0) initial positions of the particle and needle; (1) needle is lowered to focus at Z = 0 and positioned close to the* centre of the field of view, as shown; (2) needle is raised to $+Z_2$ and becomes blurred or invisible; (3) stage is raised bringing the particle to focus at $Z = 0$ near the centre of the field of view, as shown; (4) stage is lowered to $-Z_4$, the particle becomes blurred but visible; (5)–(7) the needle is gradually lowered to in-focus position; (8)–(10) stage is carefully manipulated, gradually raising the particle to focus until contact is reached and the particle picked by the needle. Then, the stage is lowered and moved away, and the "receiver" (e.g., TIMS filament) is brought to be under the needle. Finally, using a similar procedure, the needle and receiver are brought to mutual touch and the particle detached and deposited.

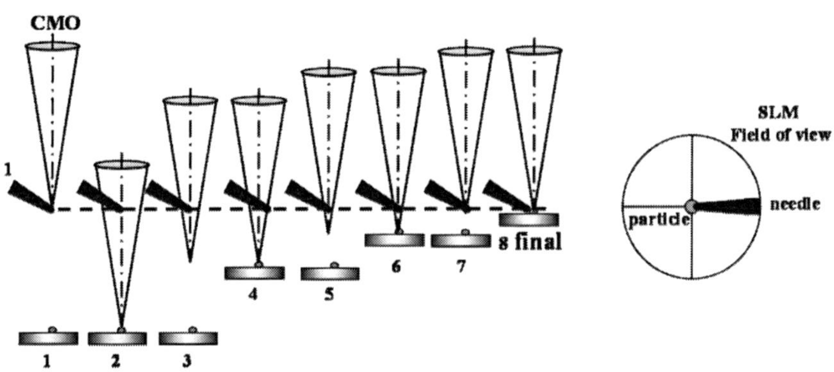

Figure 16. "Closing range" between the needle and the particle under the SLM: (1) Initial positions: needle-tip is in the focal point at centre of the FOV of the SLM, while particle is out-of-focus, way below the needle; (2) SLM is lowered by F1 and focused on the particle which is also moved to the centre of the FOV. This aligns both particle and needle along the optical axis of the SLM; (3)–(8) using F1 and F2 the stage and the SLM are sequentially and gradually raised, with possible slight x-y corrections, until the particle and the needle level with one another (both come to focus) and contact is made. Then, the stage is gently manipulated until the particle is picked-up. Alternatively, before contact, the coordinates of the stage are noted, and the stage is lowered and driven away to the location where the needle-tip can be dipped into a standing-by freshly-prepared droplet of glue. Finally, the stage is driven back without delay to the previously marked coordinates, and the lift-off operation is completed. Glue is necessary for safe lifting particles larger than ~5 μm, and is recommended whenever the loaded needle itself (and not the particle alone) is transferred to the target instrument (e.g., the SR).

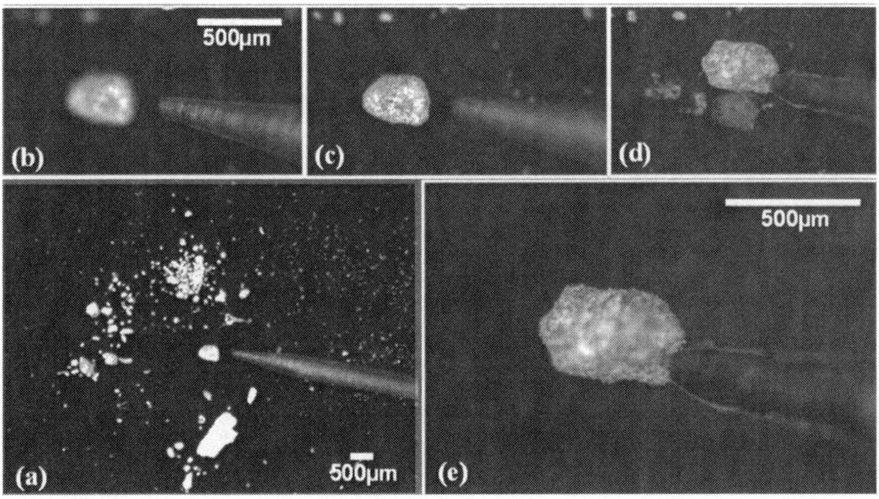

Figure 17. "Closing range" between the needle and the particle in the SLM:

(*a*) The battlefield, in dark-field illumination, viewed through the SLM: seen are the needle (a sharpened 0.5 mm graphite rod) and a dispersion of particles over a mylar tambourine;

(*b*) Focus is on the needle tip, the particle is way below the needle and hence out-of-focus;

(*c*) As in *b*, but focus is on the particle. Now the needle is out-of-focus;

(*d*) Touchdown: the particle and the needle are finally at the same level, ready for lift-off. The *x* and *y* coordinates of the stage are noted. To improve adherence, the particle has been rotated by 180°; then, because of the large size of the particle, glue had to be applied to the needle-tip. To do this, the stage was slightly lowered, and safely driven all the way to bring a droplet of fresh glue, standing-by close enough to the sample, to touch-and-wet the tip of the needle. Then, the stage was driven back to the previously marked particle coordinates, contact resumed, and after letting the glue to cure for a few minutes, the lifting process could be completed.

(*e*) The particle attached to the needle, while the sample holder is lowered away. The transparent glue is visible at the needle tip.

To enable safe lifting of particles larger than ~5 μm, and to reduce the risk of particle loss when the loaded needle carrying the particle is transferred and inserted as-is for measurements into the target instrument (e.g., in the synchrotron), it is recommended to apply glue to the needle-tip before contact is made. This is done in the following way. With the candidate-for-lifting particle at the centre of the field-of-view (FOV) the coordinates of the stage are noted, and the stage is lowered and driven away, bringing a standing-by freshly-prepared glue droplet to the FOV, where the needle-tip is dipped in. Finally, the stage is driven back without delay to the previously marked coordinates, contact is resumed, and after letting the glue to cure for a few minutes, the lift-off operation is completed.

5.7. VIEWING A PARTICLE FROM ALL SIDES: THE ROTARY HORIZONTAL NEEDLE HOLDER

In synchrotron μ-XRF-tomography and μ-XRD examinations it is required that the particle, at the tip of the needle, can be axially rotated through 360° around the needle axis, perpendicular to the beam.

To facilitate a thorough check-up of the particle and needle, after being prepared in the MM system but before being brought to the SR, a simple, inexpensive, yet efficient rotary horizontal needle holder has been designed and built by the author, Fig. 18 (Admon et al., 2006). The device fits both BLM- and SLM-based systems, and is aimed to enable gripping up to 60 mm long needles, allowing their full axial rotation perpendicular to the vertical optical axis of the microscope. This check-up is needed per se, and necessary in view of the high cost of the SR beam time.

The device is attached on top of the precision x-y-z stage in the MM system. The accuracy and reproducibility of the angular positioning of the particle is within 2°.

Two types of needle grippers were built. The front end of the first is identical to that of a syringe, and hence can accept standard surgical needles, into the bore of which graphite or metal needles are inserted. For example, the inner diameter of surgical needle No. 21 is 0.5 mm, fitting commercially available 0.5 mm needles. The second type ends with a small 4-chuck gripper that can hold firmly 0.5 to 1 mm needles, namely, either naked needles or inserted for reinforcement into the metal tube of a surgical needle whose plastic base has been removed. Needles of the latter type also fit the holder at the μ-XRF-tomography setup used at the ANKA synchrotron.

Figure 18 displays four pictures taken from four viewing directions of a 100 μm DU particle, glued to a graphite needle. Each picture is by itself a reconstruction (using the *ImageJ* software) of 4–6 through-focus serial pictures, a technique used for overcoming the limited depth-of-field of the LM. The four pictures shown are taken from a series of 64, covering a full 360° rotation, that can be displayed sequentially as a movie.

6. Micromanipulation *in-situ* the SEM

Micromanipulation in-situ the SEM can indeed solve a lot of problems, associated with the transferring of the selected to the target instruments for further analyses.

The automated particle search in the SEM, controlled by any of the commercially available dedicated software, yields a list of stage-coordinates of potentially interesting particles, candidates for higher level analyses. Obviously, first in the line are re-visiting and thoroughly analyzing each POI, before the sample is removed from the SEM. In this case, re-location is straightforward and no coordinate transformation is needed.

If the SEM is equipped with an internal x-y-z, or preferably, x-y-z-θ micromanipulator (Fig. 19), the next step – namely, preparing the selected POIs for higher level analyses – can be readily performed *in-situ*. The method (Fig. 20) has been developed and perfected in JAERI by Sakurai et al. (2005, 2006) and Esaka et al. (2004, 2007). The target instruments in this case are the SIMS and the ICP-MS. The selected POIs are lifted one-by-one from the source planchette, transferred and deposited on an adjacent target planchette. The deposition takes place in an array of well-separated, pre-defined coordinates on the target planchette; it becomes now an easy task to re-locate and analyze these particles in the SIMS, or dissolve them individually for the ICP-MS.

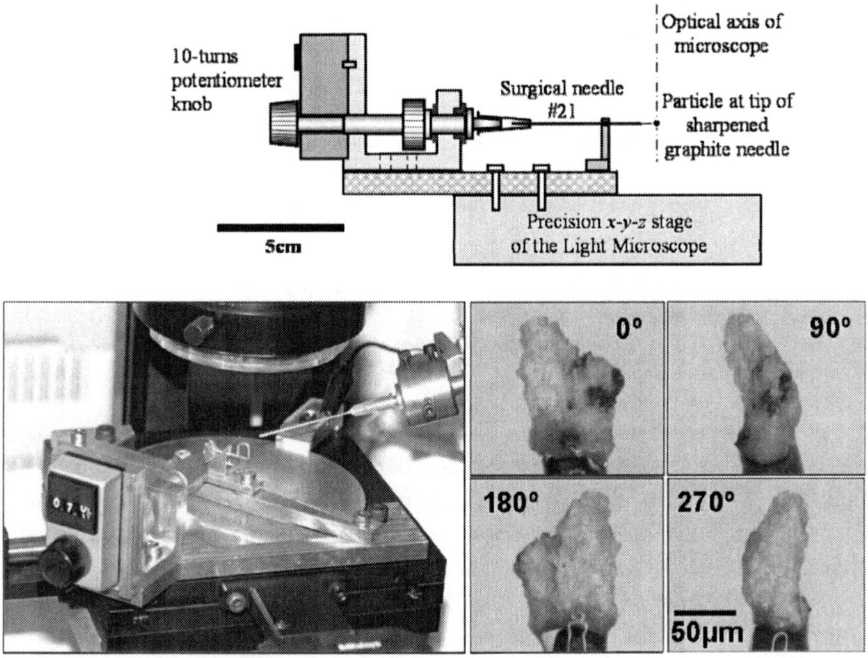

Figure 18. The *rotary horizontal needle holder* designed and built by the author to enable all-sides-viewing by axial rotation of a particle fixed to the tip of needle. The pictures (see text for details) are of a mostly yellow DU-oxide particle. The holder can grip needles with diameter up to 1 mm. The same viewing direction is used in synchrotron microtomography setup.

The device is attached on top of the precision *x-y-z stage* in the MM system, and allows ~2° reproducibility and accuracy of the angular positioning of the particle. The second needle on the right (with an inserted graphite needle) is ready for the next particle pick-up.

The pictures on the right show a mostly yellow DU-oxide particle, viewed from four different directions. Through-focus series of 4–6 pictures were used to reconstruct each of the displayed pictures, to overcome the limited depth-of-field of the microscope. A series of 64 such pictures were obtained, covering a 360° rotation, that can be displayed as a movie.

Figure 19. In-situ particle μ-manipulation in the SEM: the Kleindiek *x-y-z* micromanipulator installed inside the specimen chamber of an SEM, and three steps in particle pick-up (top). An earthed tungsten needle is used. The particle is attracted to the needle by electrostatic forces. The free working distance is 10 mm, as in regular SEM operation.
(Courtesy of S. Kleindiek [Nanotechnik] and G. Tamborini [ITU]).

It has been shown (Esaka et al., 2007) that in-situ micromanipulation allows separation between adjacent micron-size particles, that would otherwise be lifted as one. This helps improving the "purity" of the results, a critical factor in nuclear forensics, by avoiding cross-effects such as mixing, averaging, and interference between different particles.

Micromanipulation of air-borne sub-micron pollutant particles have been performed in-situ the ESEM (Kaegi and Holzer, 2003), for TEM sample preparation, by lifting and depositing individual particles onto TEM grids supporting a thin carbon film. The film was not ruptured, indicating the delicacy of the task.

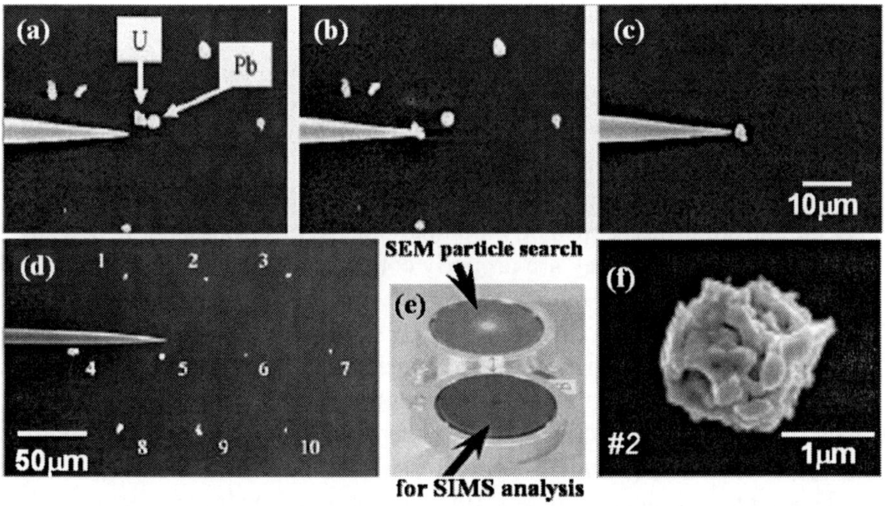

Figure 20. Identifying and isolating Particles-Of-Interest *in-situ* the SEM for further analysis by SIMS or ICP-MS. This technique, developed and perfected in JAERI by Sakurai et al. (2005, 2006) and Esaka et al. (2004, 2007), helps improving the "purity" of the results by avoiding cross-effects such as mixing, averaging, and interference between particles. Selected POIs, located in a SEM search on the *source* planchette, are lifted, transferred, and deposited on the *target* planchette.

(a)–(c) In-situ separation between two adjacent micron-size particles, lifting the U particle alone.

(d)–(e) Depositing an array of well-separated POIs, at pre-defined coordinates on a planchette, ready to be re-located and analyzed by SIMS, or individually dissolved for ICP-MS analysis.

(f) Close-up SEM images and microanalyses of isolated single particles, as the one shown (#2 in the array), is carried out before the destructive MS analysis.

(Courtesy of S. Sakurai, F. Esaka et al. [JAERI]).

It is noteworthy that Kaplan-Ashiri et al. (Kaplan-Ashiri et al., 2004, 2006) pioneered in performing tensile measurements of a single WS_2-nano-tube, stretching it between two micromanipulated needles in-situ the SEM. The WS_2-NT has been picked up and "glued" to the needles (silicon cantilever) at both ends in-situ the SEM. The "glue" was actually electron-beam induced build-up of amorphous carbon contamination under the SEM-chamber vacuum conditions. The source of the carbon is the oil vapor from the pumps, and its deposition rate can be controlled by regulating the beam accelerating voltage. Although this method is seemingly impractical for picking-up several particles, each one on a separate needle as would be required for particle analysis, it is cited here as an example to the strength of in-situ micromanipulation.

7. The needles

As a rule, the needle for particle micromanipulation should be conductive and well-grounded, as should be the sample, to avoid undesirable electrostatic repulsion effects. This is true for *ex-situ* SLM or BLM micromanipulation (in normal laboratory environment, where sporadic charging may occur) as well as for *in-situ* SEM micromanipulation. In the SEM, the needles, the particles and the substrate are constantly bombarded by electrons and will charge-up instantaneously unless they are properly connected to ground potential.

Most particles spread over a (more or less conductive) substrate are partially or fully oxidized and do charge to some extent under the electron beam in the SEM. The negative charges result in electric field lines to ground, which create local positively charged areas surrounding the particles, visible as dark halos in secondary electrons (SE) images. This seems to be a stable arrangement which is why particles stay put under the beam.

This situation changes drastically when a non-conductive or un-grounded needle enters the scene and comes to close proximity with the particles.

Figure 21 displays the electrostatic charging effect that occurred in-situ the HV (high vacuum) SEM when an un-grounded, electrically floating needle was used. Particles are repelled and swept away like by a tsunami wave, created by the electrostatic field in front of the sharp tip of the needle charged-up by the 20 kV electron beam. The particles were evenly dispersed over the graphite planchette before the tungsten needle moved in; grounding the needle alleviated this problem.

In the LV (Low Vacuum) operation mode of a VP (Variable Pressure) SEM, the surface negative charges are neutralized by the positive-ion skirt surrounding the electron beam. Consequently, there are no electrostatic charging effects as the one described above, and non-conductive (e.g., uncoated glass) needles can be used.

Various types of needles may serve in the MM system: metallic, prepared by controlled dissolution of metal wires while being pulled-out from an electrolytic or chemical bath (the most commonly used is tungsten, but many other metals are commercially available), or non-metallic, such as glass, usually prepared on-site from glass capillaries by a micropipette puller, or others like tungsten-carbide which is relatively more sturdy and durable than most needles.

Sakurai et al. (2005, 2006) and Esaka et al. (2007) report the use of gold-coated fine glass needles for *in-situ* HV-SEM particle separation and sample preparation for SIMS analysis. Kaegi and Holzer (2003) used uncoated glass needles for TEM sample preparation *in-situ* VP-SEM (ESEM) micromanipulation of sub-micrometric air-pollutant particles, collected on filters near busy highways. Kaegi advocates the use of glass needles because of their high elasticity, durability and good adhesion properties: particles grabbed from the filter adhere to the needle until they are placed on a suitable holder for subsequent analysis.

When metallic needles are used, care should be taken to thoroughly rinse and clean them to remove production-process residues, such as sodium-tungstate from tungsten needles, because they may contaminate the particles and generate isobaric interference effects.

It is interesting to note that in one of the first attempts of lifting single hot particles from a substrate, carried out in ITU in the early '90s, an eye-lash was used as a needle, Fig. 22 (Ian Ray, ITU, private communication).

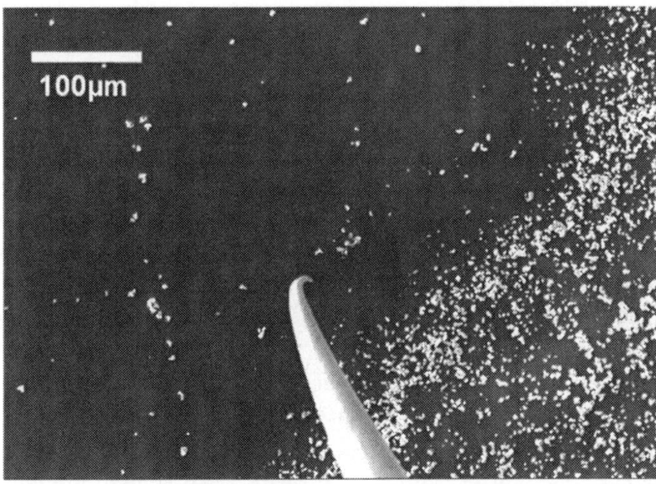

Figure 21. Electrostatic charging effect in-situ the HV (high vacuum) SEM: particles are repelled and swept away like by a tsunami wave, created by the electrostatic field in front of the tip of an un-grounded, floating needle charged-up by the 20 kV electron beam. The particles were evenly dispersed over the graphite planchette before the tungsten needle moved in; grounding the needle alleviated this problem. In the LV (low vacuum) operation mode of a VP (variable pressure) SEM, the surface negative charges are neutralized by the positive-ion skirt around the electron beam.

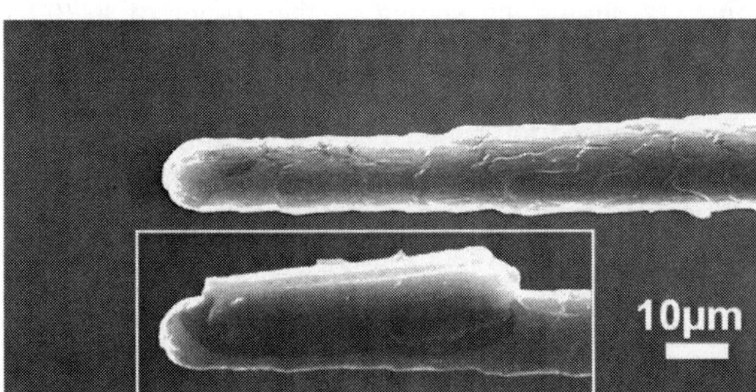

Figure 22. Eye-lash was used as a needle in the early '90 in one of the first attempts of lifting single hot particles from a substrate. The picture and the insert speak for themselves. (Ian Ray, ITU, private communication).

8. Exposing the inner microstructure: virtual and real micro-surgery of single particles

The mechanisms governing the "environmental" behavior of a hot particle, for instance its weathering kinetics in the atmosphere, leaching in the soil, corroding in marine water or dissolving in body fluids, are determined to a great extent by its internal microstructure. Open- or close-channel porosity and the presence of inclusions are the dominant factors. Any type of modeling calls for exposition of this hidden information.

The inner microstructure of a particle can be revealed either by virtual slicing, namely, X-ray fluorescence or absorption μ-tomography methods using the intense beam of the synchrotron, or by real slicing, by means of an energetic focused-ion-beam (invariably, Ga^+ ions) in one of the rapidly developing FIB dual-beam instruments.

In the synchrotron, the needle is axially rotated holding the particle in the intense X-ray beam line. Sharpened graphite rods, partially inserted for reinforcement into the metal tubes of stripped-base surgical needles, should be used because of their low atomic number (relative to, e.g., tungsten needles).

A beautiful 3-D reconstruction and μ-XRF-tomography virtual slicing of ChNPP U-fuel particles have been demonstrated by Salbu et al. (2001) and Lind (2006), Fig. 23, combining SR μ-tomography with SEM imaging. Similar technique has been used by Wegrzynek et al. (2005) and by Errikson et al. (2005) to study particles recovered from the Mururoa Atoll French nuclear test site and from the Thule accident site.

In the dual-beam FIB the particle (fixed to the tip of a μ-manipulator-driven needle) can be imaged and microanalyzed using electron- and ion-beams, as well as a variety of contrast mechanisms. Of a special merit is the ion-beam induced secondary-electrons (SE) imaging mode. Presently, the FIB has become the ultimate technique for ultra-thin slicing, mainly used for TEM sample preparation. Figure 24 shows the slicing of the corner of a 200 μm DU-oxide particle, corrosion product of a DU munition fragment collected in Kuwait. While being dynamically viewed under the electron beam, the particle is sliced by the sub-micron focused ion beam. The ion-milling (invariably, Ga^+ ions) can be done in any pattern, with nanometric precision, and only an insignificant fraction of the particle-mass is being sputtered away and lost. Before the slicing is started, a "welding" bridge (e.g., platinum) is deposited between the particle and a second μ-manipulator driven needle, which is brought to close proximity with the particle. The deposition is actually a localized, electron-beam-induced reduction of an organo-metallic gas injected near the welding zone. After completion of the slicing, both surfaces can be examined in electron- and ion-beam modes, and the fragments (each on its own needle) taken for further analyses.

Examination of the cut surface of this mostly-oxidized particle reveals, Fig. 24*e*, high inter-granular porosity, that explains its apparent fragility. The sliced surface of another, partially-oxidized particle (Fig. 24*f*) exposes a titanium-rich inclusion in the DU-0.8 wt% Ti alloy core. The composition of the particle is a "certificate of attribution", and the presence of the inclusion tells us about the metallurgical history of its parent alloy.

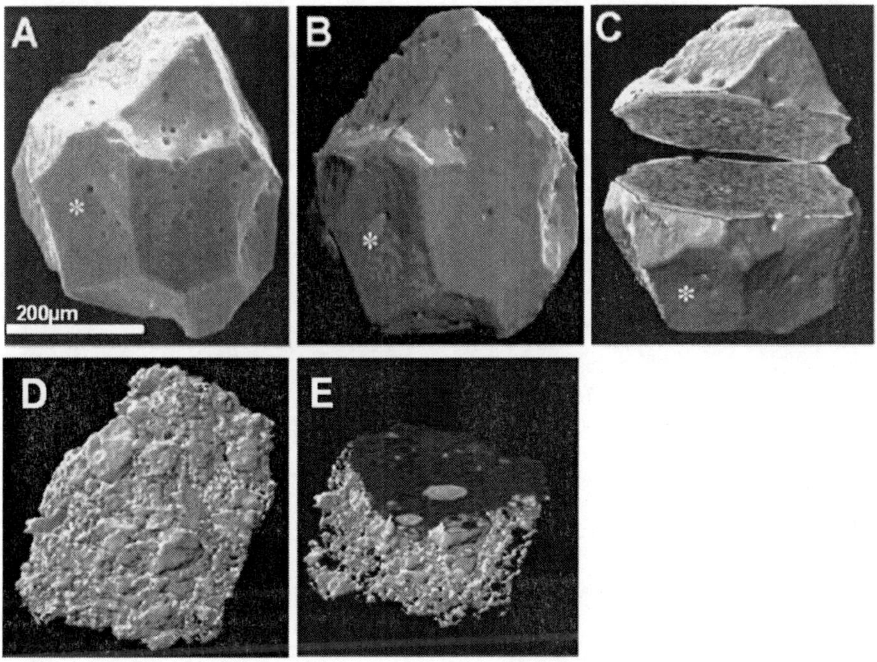

Figure 23. Virtual slicing by µ-XRF-tomography, exposing the inner structure of a single U-fuel particles from the Chernobyl accident:

(A) SEM image of the fuel particle containing non-oxidized uranium, released during the initial explosion and carried by the westward plume.

(B) 3-D reconstruction created by 500 tomographic slices spanning 180° axial rotation of the particle in the 17 keV X-ray beam of the synchrotron.

(C) Virtual slicing of the particle, showing its inner structure and "open channels" porosity.

(D) X-ray Absorption µ-tomography (µ-XAS) of an oxidized fuel particle released during the fire that followed the explosion and carried by the northbound plume.

(E) Virtual slicing through the particle exposing its heterogeneous inner structure.

These pictures, that look much nicer in their original colour presentation, are shown here by courtesy of B. Salbu and O.C. Lind (Salbu et al., 2001; Lind, 2006).

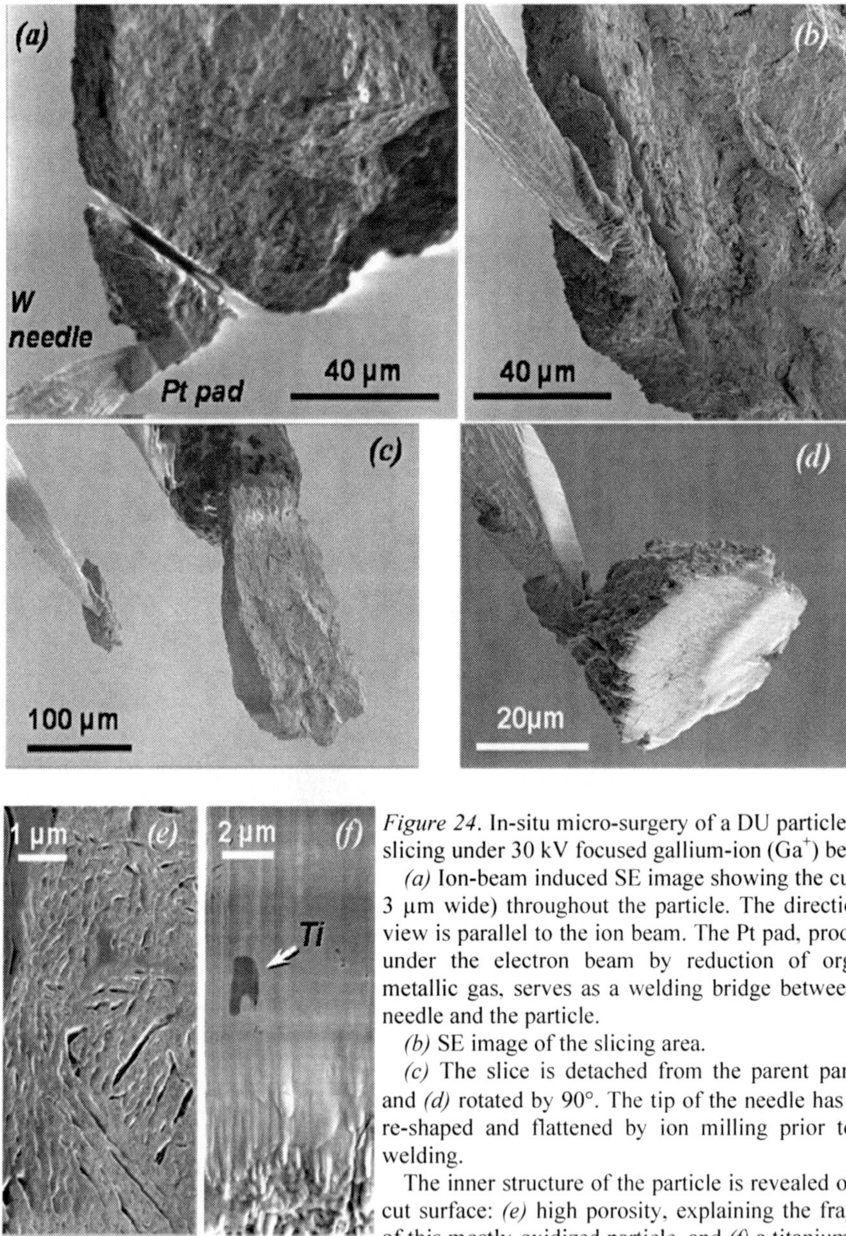

Figure 24. In-situ micro-surgery of a DU particle: FIB slicing under 30 kV focused gallium-ion (Ga^+) beam.

(a) Ion-beam induced SE image showing the cut (2–3 μm wide) throughout the particle. The direction of view is parallel to the ion beam. The Pt pad, produced under the electron beam by reduction of organo-metallic gas, serves as a welding bridge between the needle and the particle.

(b) SE image of the slicing area.

(c) The slice is detached from the parent particle, and *(d)* rotated by 90°. The tip of the needle has been re-shaped and flattened by ion milling prior to the welding.

The inner structure of the particle is revealed on the cut surface: *(e)* high porosity, explaining the fragility of this mostly-oxidized particle, and *(f)* a titanium-rich inclusion in the DU-0.8 wt% Ti core-alloy of another, partially oxidized particle.

(a) and *(e)* are ion-beam induced SE images, whereas the others are ordinary SE(ET) images.

It should be stressed that the FIB technique is straightforward, fast and direct, albeit invasive; it may be considered non-destructive in the sense that only a small fraction of the total particle mass is lost in the ion-milling process, and the remaining fragments can be still subjected to other analyses, such as mass-spectrometry. The SR virtual slicing technique, on the other hand, is time consuming, indirect, requires a lot of computation, offers inferior spatial resolution, and is much more expensive and technically complicated. The advantage of the SR over the FIB, however, is that in the same facility additional unique SR techniques, namely, μ-XANES and μ-XRD, can be applied to the same particle.

And last but not least: FIB slicing has proven itself as a key-tool in TEM sample preparation in several areas of material sciences. Through-surgery of a particle exposing its inner core can be very beneficial, when combined with surface analysis techniques, in particle analysis.

9. Combined FTA – SEM analysis

The initial search for locating fissile POIs (namely particles containing uranium or plutonium) within large population on non-POIs is usually done by fission track analysis (FTA), as it is faster, more efficient and more reliable than other methods, e.g., SEM search. The drawback of the FTA technique, however, is that as part of the procedure the particles are all embedded in a thin polycarbonate layer (10–50 μm collodion or lexan). The often invisible particles are located by their coordinates; small (~0.1 × 0.1 mm^2) squares are blindly-cut from the polymer around their calculated locations. Then, the squares are lifted and brought, for instance, to TIMS analysis.

One step is clearly missing from this procedure. It would be desirable to first characterize the particle thoroughly in the SEM, before it is annihilated during the (destructive) MS analysis. To be able to do so, the particle must be first stripped from its polymer shroud. This can be done by either plasma ashing or by solvent-vapor re-fluxing. Preliminary experiments by the author, that will be reported elsewhere, have shown that both approaches are promising. In particular, 9 out of 10 particles, embedded in ~30 μm lexan layer and lain on a silicon wafer, were found in their original locations, the lexan completely stripped off, after being exposed to 6 h of plasma ashing in oxygen atmosphere.

10. Conclusions and practical considerations: how far should we go?

We are now able to locate, re-locate, and relocate single hot particles from the sub-micrometric size range and up. Air-borne particles in the size-range 0.5–10 μm have been the most studied ones because of their global-scale environmental impact on the one hand, and significance in nuclear forensic and safeguards cases on the other. Individual particles out of this size range are either too small to yield significant measurement results, or too heavy to be air-borne to large distances from their release site. A rainbow of microanalytical techniques is now available for particle characterization. Combining the handling methods with the chain of analytical capabilities allows subjecting the very same particle to a variety of multiple-analyses.

The problem, of course, is that information has a price. The beam-time cost of the synchrotron, a multi-megabuck installation, is several k$ per particle. It is prohibitively high for using μ-XRF-tomography or other SR-techniques for any statistically significant number of particles, as would be required for meaningful medical and biological studies.

SEM or even the more expensive SEM/FIB characterization, per particle, may cost only a few percent of SR analyses. It is widely accepted that at the single particle level, particularly in nuclear forensic cases where only a few particles may be available in the first place, SEM/EDS (or WDS) characterization is essential and usually also sufficient before killing the particle in the final, destructive MS analysis.

The micromanipulation techniques enable us to pre-select, characterize and evaluate a single particle *before* it is brought to the synchrotron or to other instrument. It allows "purifying" isotopic-ratio measurements by alleviating the inevitable averaging associated with the tendency of small particles to agglomerate. Multiple-analyses of the very same particle, in different instruments and laboratories, are made possible.

A problem that calls for a better understanding is the origin and nature of forces acting between the particle and the needle or the substrate, that make the particle adhere to either one of them. Generally speaking, they can be electrostatic, Van-der-Waals, or even capillary forces. Practically, every surface exposed to human environment is covered with a few molecular layers of water or condensable organic molecules that waft around in the air; the small particles are even more likely to have this crud on them, which seems to act like glue, sticking particles to surfaces. A better understanding of the participating forces can ease the lift-and-deposit operations, and reduce the risk of particle loss.

The dual-beam FIB instrument is rightfully conquering its place as a key-tool in material sciences. It is being used primarily for precision micro-slicing as in TEM sample preparation; the same capability can be used in the arena of particle analysis, providing electron- and ion-beam induced imaging, and exposing the inner structure for SEM, TEM, μ-XANES or μ-RAMAN examination.

The scanning helium ion microscope (e.g., ALIS/Zeiss), now emerging in the market, may become an important tool in particle analysis because of its potentially much higher resolution and unique signal modes compared to the best SEM instruments.

Acknowledgement

The author kindly acknowledges the many illuminating discussions, advises and contributions of several people, and particularly thanks Eli Boblil, Zvi Berant, Alex Wolf, Anton Nirschl, Mladen Bogovac, Stephan Vogt, David Joy, Dale Newbury, David Simons, Cynthia Zeissler, Stephane Baude and Maria Betti. Special thanks are given to Brit Salbu, Ole-Christian Lind, Ian Ray, Klaus Mayer, Satoshi Sakurai, Fumitaka Esaka, Valery Kashparov, Stephan Kleindiek and Liu Guorong for allowing showing their results.

References

Admon, U., Ruedenauer, F., Donohue, D.L., Chinea-Cano, E., and Aigner, H. (2002a): Precise relocation of radioactive particles in SEM/SIMS analysis of environmental samples, *Proceedings of the 43th INMM Annual Meeting*, June 23–27, 2002, Orlando, Florida.

Admon, U., Donohue, D.L., Aigner, H., Tamborini, G., Bildstein, O., and Betti, M. (2002b): From the SEM (IAEA) to the SIMS (ITU): Re-location experiment of μm-size particles – first results, *Proceedings of the International Conference. Advances in Destructive and Non-Destructive Analysis for Environmental Monitoring and Nuclear Forensics*, Oct. 21–23, 2002, Karlsruhe, Germany.

Admon, U., Donohue, D.L., Aigner, H., Tamborini, G., Bildstein, O., and Betti M. (2005): Multiple-instrument analyses of single micron-size particles, *J. Microsc. Microanal.*, **11**, 354–362.

Admon, U., Chinea-Cano, E., Wegrrzynek, D., Costin, W., Ruedenauer, F., Bogovac, M., and Markowicz, A. (2006): The art of single particle μ-manipulation, or: the 'Hide and Seek' Game, *Technical Meeting on Analytical Methods for Characterization of Hot Particles and their Impact on Environment*, March 6–10, 2006, ICTP, Trieste.

Baude, S. (2001): Isotopic measurements on microscopic particles: The French experience to detect fissile materials, *Proceedings of the IAEA Symposium on International Safeguards*, Vienna, Austria, Oct. 29–Nov. 2, 2001, IAEA-SM-367/10/05, pp. 1–5.

Baude, S. (2005): Overview of CEA operations and developments, *Presentation at 2005 Particle Analysis CGM*, IAEA Vienna, (Nov. 2005).

Becker, J. Sabina., Sela, H., Dobrowolska, J., Zory, M., and Becker, J. Susanne. (2008): Recent applications on isotope ratio measurement by ICP-MS and LA-ICP-MS on biological samples and single particles, *Int. J. Mass Spectrometry* **270**, 1–7.

Boulyga, S.F. and Becker, J.S. (2001): Determination of uranium isotope composition and ^{236}U content of soil samples and hot particles using inductively coupled plasma mass spectrometry, Fresenius, *J. Anal. Chem.*, **370**, 612–617.

Burns, P.A., Cooper, M.B., Lokan, K.H., Wilks, M.J., and Williams, G.A. (1995): Characteristics of Pu and Am contamination at the former U.K. atomic weapons test ranges at maralinga & emu, *J. Appl. Radiat. Isot.*, **46**(11), 1099–1107.

Chiappini, R., Pointurier F., Hubert, A., and Marie, O. (2006): New developments and methodology for actinide measurements at ultra trace levels, *Proceedings of the IAEA Symposium on International Safeguards: Addressing Verification Challenges*, Vienna, Austria, 16–20 Oct. 2006, pp. 821–829.

Elish, E.J., Karpas, Z., and Lorber, A. (2007): Determination of uranium concentration in a single hair strand by LA-ICP-MS applying continuous and single pulse ablation, *J. Anal. Atom. Spectrom.*, **22**, 540–546.

Eriksson, M., Osa'n, J., Jernstrom, J., Wegrzynek, D., Simon, R., Chinea-Cano, E., Markowicz, A., Bamford, S., Tamborini, G., Torok, S., Falkenberg, G., Alsecz, A., Dahlgaard, H., Wobrauschek, P., Streli, C., Zoeger, N., and Betti, M. (2005): Source term identification of environmental radioactive Pu/U particles by their characterization with non-destructive spectrochemical analytical techniques, *Spectrochimica Acta, Part B*, **60**, 455–469.

Esaka, F., Esaka, K.T., Lee, C.G., Magara, M., Sakurai, S., Usuda S., and Watanabe, K. (2007): Particle isolation for analysis of uranium minor isotopes in individual particles by secondary ion mass spectrometry, *Talanta*, **71**, 1011-1-15.

Esaka, F., Watanabe, K., Fukuyama, H., Onodera, T., Esaka, K., Magara, M., Sakurai, S., and Usuda, U. (2004): Efficient isotope ratio analysis of uranium particles in swipe samples by total-reflection x-ray fluorescence spectrometry and secondary ion mass spectrometry, *J. Nucl. Sci. Tech.*, **41**, 1027–1032.

GUM (1995): *Guide to the Expression of Uncertainty in Measurement* – BIPM, IEC, IFCC, ISO, IUPAC, IUPAP and OIML, Geneva, Switzerland: International Organization for Standardisation, ISBN 92-67-10188-9.

Hubert, A., Pecheyran, C., Pointurier, F., Claverie, F., and Marie, O. (2007): Uranium isotope analysis of particles by laser ablation and inductively coupled plasma mass spectrometry, *Esarda 29th Annual Meeting, Symposium On Safeguards and Nuclear Material Management*, Aix-en-Provence, May 22–24, 2007.

IAEA, (2001): Coordinated research project (CRP) on radiochemical, chemical and physical characterization of radioactive particles in the environment, first *Research Coordination Meeting (RCM)*, Vienna, 7–11 May 2001.

Kaegi, R. and Holzer, L. (2003): Transfer of a single particle for combined ESEM and TEM analyses, *J. Atmos. Environ.*, **37**, 4353–4359.

Kaplan-Ashiri, I., Cohen, S.R., Gartsman, K., Ivanovskaya, V., Heine, T., Seifert, G., Wiesel, I., Wagner, H.D., and Tenne, R. (2006): On the mechanical behavior of WS2 nanotubes under axial tension and compression, *Proc. Natl Acad. Sci. USA.*, **103**(3), 523–528.

Kaplan-Ashiri, I., Cohen, S.R., Gartsman, K., Rosentsveig, R., Seifert, G., and Tenne, R. (2004): Mechanical behavior of individual WS2 nanotubes, *J. Mater. Res.*, **19**(2), 454–459.

Kashparov, V. (2001): Contamination of the ChNPP 30 km zone hot particles database, Ukrain Institute of Agricultural Radiology (UIAR).

Kerkápoly, A., Vajda, N., and Pintér, T., (2005a): Film autoradiography used for hot particle identification, *J. Radioanal. Nucl. Chem.*, **265**(3), 423–429.

Kerkápoly, A., Vajda, N., Pintér, T., and Pintér-Csordás, A., (2005b): Analysis of hot particles originating from failed and damaged fuels, *Central Eur. J. Phys.*, **3**(1), 1–12.

Lind, O.C. (2006): Characterisation of radioactive particles in the environment using advanced techniques, PhD Dissertation, Norwegian University of Life Sciences, (UMB).

Liu, G., (2005): Study on relocating microsize particles from SEM to SIMS, M.Sc. Thesis, China Institute for Atomic Energy (CIAE), Beijing.

Sakurai, S. et al. (2005): R&D on safeguards environmental sample analysis at JAERI, *Proceedings of the GLOBAL-2005 Conference*, Tsukuba, Japan, Oct. 9–13, Paper No. 459.

Sakurai, S., Magara, M., Esaka, F., Hirayama, F., Lee, C.G., Yasuda, K., Inagawa, J., Suzuki, D., Iguchi, K., Kokubu, Y.S., Miyamoto, Y., Shinohara, N., and Usuda S. (2006): Development of safeguards environmental sample analysis techniques at JAERI as a network laboratory of IAEA, *Proceedings of the IAEA Symposium on International Safeguards: Addressing Verification Challenges*, Vienna, Austria, 16–20 Oct. 2006, pp. 791–799.

Salbu, B., Krekling, T., Lind, O.C., Oughton, D.H., Drakopoulos, M., Simionovi, A., Snigireva, I., Snigirev, A., Weitkamp, T., Adams, F., Janssens, K., and Kashparov, V.A. (2001): High energy x-ray microscopy for characterisation of fuel particles, *Nucl. Instrum. Methods Phys. Res.*, **A 467-468**, 1249–1252.

Sangely, L., Marie, O., Diallo, S., Pointurier, F., and Baude, S. (2006): Particle analysis for uranium isotopics on samples using new generation cameca IMS-7f-SIMS supported by SEM automated uranium detection, *Proceeding of the IAEA Symposium on International Safeguards: Addressing Verification Challenges*, Vienna, Austria, 16–20 Oct. 2006, pp. 813–819.

Tamborini, G., and Betti, M. (2000): Characterization of radioactive particles by SIMS, *Mikrochimica Acta*, **132**, 411–417.

Vajda, N. (2001): Radioactive particles in the environment; occurrence, characterization, appropriate analytical techniques (Review), *1st Research Coordination Meeting of the Coordinated Research Project on Radiochemical, Chemical and Physical Characterization of Radioactive Particles in the Environment*, IAEA, Vienna, 7–11 May, 2001.

Wegrzynek, D., Markowicz, A., Bamford, S., Chinea-Cano, E., and Bogovac, M. (2005): Micro-beam x-ray fluorescence and absorption imaging techniques at the IAEA laboratories, *Nucl. Instrum. Methods Phys.*, **B 231**, 176–182.

Wallenius, M., Luetzenkirchen K., Mayer, K., Ray, I., Aldaves-de-las-Heras L., Betti, M., Cromboom, O., Hild, M., Lynch, B., Nicholl, A., Ottmar, H., Rasmussen, G., Schubert, A., Tamborini, G., Thiele, H., Wagner, W., Walker, C., and Zuleger, E. (2007): Nuclear forensic investigations with a focus on plutonium, *J. Alloys Compounds*, **444,445**, 57–62.

CHARACTERISATION OF DU PARTICLES FROM KOSOVO AND KUWAIT

OLE CHRISTIAN LIND[1], BRIT SALBU[1], KOEN JANSSENS[2], KRISTOF PROOST[2], PIER ROBERTO DANESI[3]
[1]*Norwegian University of Life Sciences, Norway*
[2]*University of Antwerpen, Belgium*
[3]*Institute of Advanced Studies, Italy*

Abstract: Large amounts of DU material has been released to the environment in Kosovo and Kuwait due to the use of depleted uranium (DU) ammunition during the 1991 Gulf war and the 1999 Balkan conflict. Following the impact of metallic DU penetrators on solid targets, DU particles are dispersed and ignited, and uranium in the particles is oxidized. Following inhalation or ingestion of DU particles, the amount of DU that is absorbed into the blood and distributed to tissues and organs will largely depend on the particle size distribution and associated weathering rates (solubility) both in the respiratory system as well as the GI tract. Particle weathering rates are related to particle characteristics such as specific surface area, size distribution, elemental composition, U oxidation states and crystallographic structure.

In order to improve the basis for health impact assessments related to contamination of DU in Ceja Mountains, Kosovo and Al Doha, Um Al Kwaty as well as Manageesh Oil fields in Kuwait, DU particles originating from these sites have been isolated and characterised in detail using advanced analytical techniques. Information on particle size, surface structure and surface elemental distribution were obtained by means of electron microscopy, whereas elemental distribution, oxidation state and crystallographic structure of U were determined using synchrotron radiation (SR) based x-ray microscopic techniques. Inductively Coupled Plasma Mass Spectrometry (ICP-MS) and Accelerator Mass Spectrometry (AMS) were employed to obtain U isotope ratios for source identification purposes.

Characteristics of DU particles in soil/sand from Kosovo and Kuwait contaminated during the Balkan conflict and Gulf war varied significantly depending on the release scenario. Atom ratios ($^{236}U/^{235}U$) obtained from ICP-MS and AMS varied between individual particles, but demonstrated that U in the DU-particles originated from reprocessed fuel. The bright yellow, highly oxidised DU particles associated with a fire in a DU ammunition storage facility showed a wide particle size distribution; from submicron to several

D.H. Oughton and V. Kashparov (eds.), *Radioactive Particles in the Environment*,

hundred micrometers. Furthermore, the micro-XRD signals closely resembled those of schoepite, dehydrated schoepite and metaschoepite. DU in sand samples contaminated with these particles is expected to dissolve quite rapidly in biological fluids. In contrast, the potential mobility of smaller and less oxidised DU particles, originating from the corrosion of DU penetrators or collected in tanks hit by DU ammunition should be significantly lower. The presence of respiratory DU particles where U is more oxidised than UO_2, and exhibiting high weathering rates, reflects that DU particles are potentially more bioavailable than natural uranium. Thus, environmental or health impact assessments for areas affected by DU ammunitions should take into account oxidation states, their corresponding weathering rates and the subsequent mobilisation of U from oxidized DU particles.

Keywords: depleted uranium particles, oxidation states, XANES, SEM-XRMA, Kosovo, Kuwait.

1. Introduction

Depleted uranium (DU), a waste product of U enrichment, has several civilian uses (e.g., counterweights in commercial aircraft, fork lifts and keels of sailing yachts) and military applications (e.g., armour-piercing ammunition). Military use of DU is the most important source of DU particles dispersed in the environment. However, in the aircraft crash and subsequent fire that occurred in Amsterdam on 4th October 1992, DU used as counterweights was released into the environment (De Haag et al., 2000). During the Golf war and Balkan conflicts large amounts of DU ammunition were expended.

Several reports on DU contamination in Kosovo (Danesi et al., 2003b; Salbu et al., 2003), Serbia-Montenegro (McLaughlin et al., 2003), Bosnia and Herzegovina (UNEP, 2003), Kuwait (Salbu et al., 2004a) and Iraq (Gerdes et al., 2004) have been published. DU particles were identified at all sites except in Iraq where no information on particles appears to be available at the present time. Large amounts of DU ammunitions have also contaminated military proving grounds and their surroundings (Sowder et al., 1999).

When hitting hard targets such as tanks, penetrators made from metallic DU alloyed with small amounts of Ti or Al generate a cloud of DU dust (aerosol) that ignites spontaneously and create a fire. Unspent DU penetrators deposited in the field will corrode over time and fragments and particles are formed. Accidental dispersal of DU particles originating from ammunition may also occur as demonstrated in Al Doha, Kuwait in 1991 when a fire in a US Army ammunitions storage caused local dispersion of DU (Salbu et al., 2004a).

Depleted U contamination is generally localised close to the source, but can be transported in the environment due to resuspension and agricultural practices. Furthermore, due to weathering of fragments and particles and the subsequent remobilisation of U, ecosystem transfer of U from soil/sand to the food chain may take place.

Assessment of the risk of impact from most radionuclides is based on the total radiological dose rate to the organism of concern. However, for natural U and certainly for DU where ^{235}U is depleted, the risk from the chemical toxicity of metal uranium itself is considered greater than the radiological toxicity (Sheppard et al., 2005). Inhalation of DU particles is considered the major pathway for DU exposure, but direct ingestion of DU should also be taken into account (Bleise et al., 2003). About 95% of inhaled particles with larger than 10 μm aerodynamic equivalent diameter (AED) deposit in the upper respiratory tract. Most of these clear to the pharynx, and to the GI tract. Particles <10 μm can reach deeper pulmonary regions (bronchioles and alveoli) and stay there for considerable time (Bleise et al., 2003). The amount of DU that is absorbed into the blood and distributed to tissues and organs will largely depend on particle characteristics such as specific surface area (Chazel et al., 1998), size distribution, elemental composition, U oxidation states and crystallographic structure which ultimately influence weathering rates (solubility) both in the respiratory system as well as the GI tract.

In the present work, we summarize, compare and discuss results from previous investigations performed on individual DU-particles using SEM-XRMA, SR based micro x-ray techniques as well as ICP-MS and AMS (Salbu et al., 2003, 2004a), to improve the scientific basis for environmental impact assessments in areas contaminated with DU particles. It is hypothesised that characteristics of particles originating from the same type of source (i.e., DU penetrators) should differ according to different release scenarios. Therefore, focus was put on the identification of differences in particle characteristics that could be related to various release scenarios (i.e., impact, fire and corrosion of DU penetrators), although environmental conditions also can influence the solid state speciation.

2. Materials and methods

In 1999, the International Atomic Energy Agency (IAEA) joined a UN mission to Kosovo and collected soil samples contaminated with DU. In 2002, the IAEA organised a field mission to collect samples at selected sites in Kuwait, where DU ammunition had been used. Soil and sand samples collected at a target site at Ceja Mountain (a former Serbian army and anti-aircraft position) in south west Kosovo, and at Al Doha, Manageesh and Um Al Kwaty in Kuwait were air dried. The Kosovo samples were sieved into two fractions, a "coarse" fraction (>3 mm) and a "fine" fraction (<3 mm), which were made available to

UMB. Kuwaiti dried soil and sand samples as well as swipe samples (tissues) from inside impacted tanks were also made available to UMB for the characterisation of particles without further treatment. An overview of the particles isolated and characterised in the present work is provided in Table 1.

TABLE 1. Investigated DU samples

	Site	Site description	Characteristics
Kosovo	Ceja Mountain	Soil from underneath jackets	Small-sized DU particles
	Al Doha	Fire in a DU ammunition storage	Small to very large particles of DU
Kuwait	Manageesh	Below a DU penetrator	Small-sized DU particles
	Um Al Kwaty	Swipe from inside damaged tank	Corroded material containing DU. Small-sized DU particles

2.1. SEM-XRMA

Soil or sand samples containing depleted uranium particles were mounted on carbon tapes and subjected to scanning electron microscopy (SEM) using a JEOL JSM 840 instrument interfaced with X-ray micro-analyser (ISIS 300, Oxford Instruments). Swipe sample particles were transferred to carbon tape by touching the swipe cloths. In Secondary Electron Imaging (SEI) mode information on the surface structures of particles is obtained. Using Backscattered Electron Imaging (BEI) mode, bright areas reflect the distribution of high atomic number elements. Using X-ray mapping, the distribution of individual elements such as U within specific particles is attained, while XRMA provides semi-quantitative information on individual elements at the electron beam spot, such as the concentration of U and Ti at specific particle sites. By superimposing X-ray-mapping with SEI or BEI images, individual particles containing U are identified prior to μ-XANES analysis.

2.2. SR-BASED MICRO-XANES

Particles characterized by SEM with XRMA were subjected to μ-XANES analysis using the X-ray microscopic facility at beamline L, HASYLAB, Hamburg (Salbu et al., 2003, 2004a). μ-XRF was utilized to determine the concentration of elements within individual particles using the μ-XANES set-up (HPGe detector). Using μ-XANES, information on the oxidation state of U in DU particles was obtained by tuning the monochromatic, focused (20 × 20 μm beam by using a polycapillary lens) X-ray beam over the U L_{III} absorption edge (17.163 keV), while keeping the beam position on individual DU particles. The flux at the beam spot on the sample was about 10^9 photons per second at 17.1 keV. The incident and transmitted beam intensity (I_0, I) were measured by ionisation chambers, while the U L_{III} fluorescence intensity was recorded by means of a HPGe detector having an area of 30 mm^2 mounted at 45° to the incident beam and 30 mm from the sample. The HPGe detector was well collimated (2 mm diameter pinhole). The μ-XANES spectra were collected at 1 eV increments over a 300 eV energy range (extending from about 80 eV below and 220 eV above the U L_{III}). Based on well-defined U oxidation state standards (UO_2, Institute of Energy Technology, Kjeller; U_3O_8, Institute of Energy Technology, Kjeller; $UO_2Ac_2 \times 2\ H_2O$ p.a., Riedel-De Haën AG, Seelze-Hannover; $UO_2(NO_3)_2 \times 6\ H_2O$ p.a., Merck, Darmstadt) the μ-XANES spectra were recorded and the inflection point energies for the oxidation states of U established. The μ-XANES profile shapes and energy shifts (inflection point energies) of U in DU particles were compared with those of the standards. For non-stoichiometric U oxides, linear regression was employed to determine the oxidation state (Proost, 2005).

2.3. SR-BASED MICRO-XRD

The XRD data was collected at the ESRF ID18F end-station (Salbu et al., 2004a). During the experiment, the monochromator was tuned to 28 keV and a compound refractive lens focused the monochromatic beam. The samples were mounted in slide frames and moved through the beam by means of a motorized XYZ sample stage. Perpendicular to the direction of the primary beam, an energy-dispersive X-ray detector allowed the collection of fluorescent signals from the sample. Furthermore, a Photonics Science X-ray diffraction camera (CCD camera) was used to record the diffraction signals. Measurements included point measurements (5 × 10 micrometer, single XRF spectrum and XRD image collected during typically 10–100 s) and line scans (horizontal or vertical lateral movement of the particle through the primary beam with increment of typically 2–5 micrometer and the collection of a XRF-spectrum and/or XRD-image at all positions along the line).

2.4. ICP-MS AND AMS MEASUREMENTS OF ISOTOPE RATIOS

Following the characterization of particles by non-destructive techniques, individual particles were sacrificed in order to identify the source of the particles. Individual particles, certified reference U ore materials (EC nuclear Reference Material 113 and 114, respectively) and a small (0.12 g) soil sample collected at Ceja Mountain, Kosovo were leached in aqua regia (3:1 mix of concentrated HCl and HNO_3). Uranium concentrations and $^{235}U/^{238}U$ isotope ratios were determined by ICP-MS (Perkin-Elmer ELAN 6000) after radiochemical separations based on selective sorption on anion exchange resins (Dowex AG 1 × 8) to separate U and Am from Pu and TruSpec to separate U from Am (Clacher, 1995). The $^{235}U/^{238}U$ and $^{236}U/^{235}U$ isotope ratios were also determined by accelerator mass spectrometry (AMS) using the 14UD tandem accelerator at the Australian National University, Canberra, according to the procedure described by Marsden et al. (2001).

3. Results and discussion

3.1. IDENTIFICATION AND CHARACTERISATION

Despite the low radioactivity of DU, particles containing DU were easily identified. Using SEM in BEI-mode, U particles can be directly identified as bright spots on the monitor (Fig. 1). The observed particle size distributions ranged from sub-micron to 30 μm in Kosovo samples and sub-micron to several hundred μm in Kuwaiti samples. DU particles collected close to impacted DU penetrators or corroded 'unspent' DU penetrators or inside attacked tanks (swipes) tend to be quite small, whereas DU particles released during the fire in a DU ammunition facility at Al Doha ranged from submicron to several hundred μm.

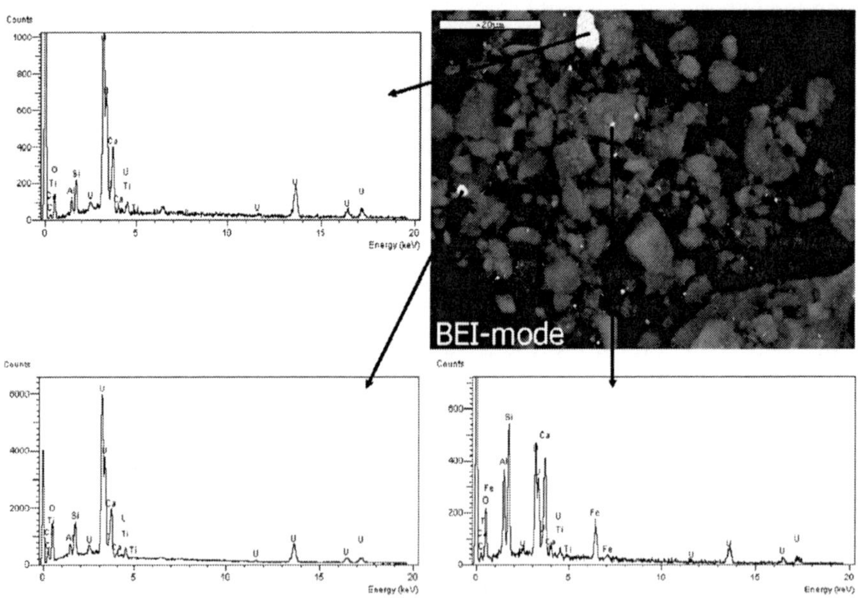

Figure 1. Scanning electron microscopy of DU-particle contaminated soil collected at Ceja Mountain, Kosovo. Backscattered Electron Imaging (BEI) with bright areas reflecting high atomic number elements are clearly visible. XRMA spectra from spot measurements on bright areas (yellow arrows) confirm the presence of U and Ti. Bar = 20 μm.

Aliquots of the sample from Ceja Mountain were also investigated by Danesi et al. (2003a,b). Their results indicated the presence of hot spot areas where hundreds of thousands of particles could be found in a few milligrams of DU contaminated soil. The particle size distribution showed that most of the DU particles were <5 μm in diameter and more than 50% of the particles had a diameter <1.5 μm. Thus, a large fraction of the DU at sites investigated in the present work was associated with respiratory particles. Recent studies at other sites confirm these findings (Chazel et al., 2003; Mitchel and Sunder, 2004).

The oxidation states of U within investigated DU particles, as revealed by comparison with a well-defined standard using μ-XANES, showed that all the DU particles were in an oxidised form (Salbu et al., 2003, 2004a). About 50% of the DU particles from Kosovo were characterised as UO_2 (U(IV)), while the remainder were present as two groups of uranium oxides with higher average oxidation states (+4.6 ± 0.5 and +5.3 ± 0.5). No higher oxidation state for U was observed for DU particles from Kosovo. Kuwaiti DU particles (n = 2) collected inside attacked tanks (swipes) were characterized as hyper-stoichiometric uranium oxides with average oxidation states +4.4 ± 0.5 and 4.7 ± 0.5 (Proost, 2005). Considering the Kuwaiti DU particles collected close to 'unspent' DU penetrators, the XANES profile of one of them coincided with those of the uranyl (+6) standards. The other particle that was subjected to XANES analysis exhibited an average oxidation state (+4.6 ± 0.5) similar to that observed in DU affected areas in Kosovo. Tetravalent U is sparingly soluble with solubility products (k_{sp}) of $10^{-53.44}$ for UO_2 (am) (Rai et al., 1997) and as low as $10^{-60.2}$ (Rai et al., 2003) for UO_2 (c). Uranium chemistry predicts that particle weathering rate (year^{-1}) increases with the oxidation state for U and should be higher for U-particles with average oxidation state +4.6 and +5.3 than for UO_2 (Kashparov et al., 1999).

The large DU particles released during the fire in the DU ammunition storage appeared with a strong yellow colour (Fig. 2), typical of uranyl compounds, and a crystalline structure (Fig. 3) which was quite different from all DU particles observed in Kosovo and other sites in Kuwait, as well as from that of U particles released during the Chernobyl accident (Salbu et al., 1998, 2001, 2004b). XANES spectra of these particles were similar to solid uranyl standards. Recent μ-XRD results (Lind et al., 2009), demonstrate that U is present as schoepite ($UO_3 \cdot 2.25\ H_2O$), dehydrated schoepite ($UO_3 \cdot 0.75\ H_2O$) and metaschoepite ($UO_3 \cdot 2.0\ H_2O$), i.e., crystallographic forms of U with relatively high solubility product constants ranging from $10^{4.70}$ to $10^{6.33}$ (Giammar and Hering, 2004). In fact, the solubility of schoepite at near neutral pH (~10 mg/l) is about five orders of magnitude higher than for UO_2 (~0.1 μg/l) (UNEP, 2001). Thus, the solid state speciation and consequently the potential mobility and bioavailability of DU particles released during the storage fire and subsequently subjected to weathering in the Kuwaiti desert for ~10 years are significantly different to DU particles released during impact ~5 years prior to sampling.

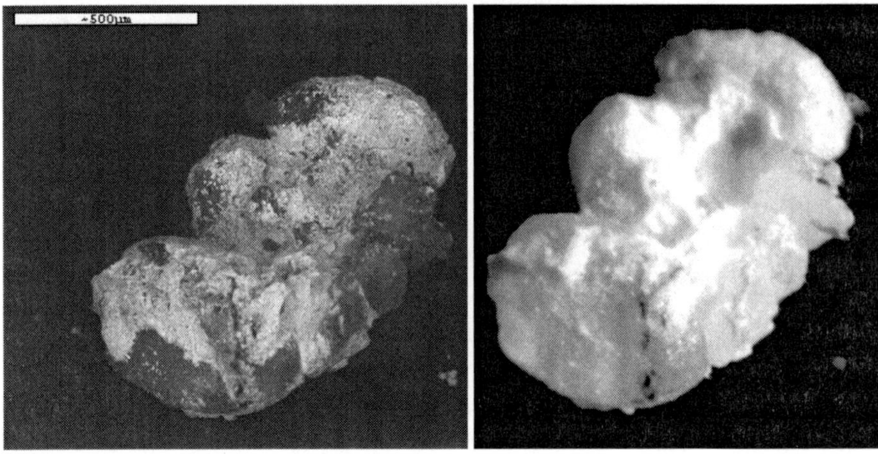

Figure 2. Grain of sand partly covered with depleted U originating from the ammunition storage fire in Al Doha, Kuwait. (Left) Scanning electron micrograph recorded in BEI mode. (Right) Light microscopy image of the same particle.

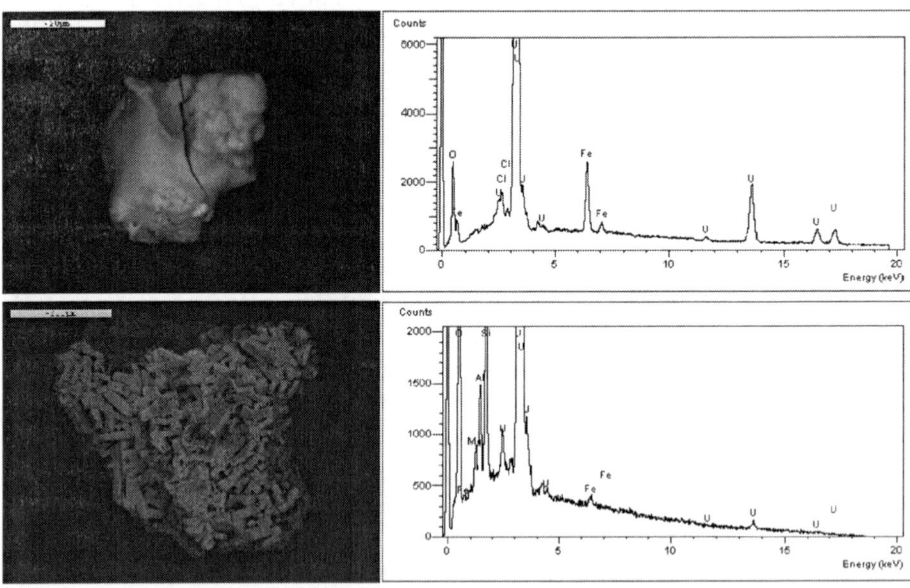

Figure 3. Scanning electron microscopy (BEI-mode) of DU-particles collected on swipe at Um Al Kwaty (above) and from soil at Al Doha (below), Kuwait (Salbu et al., 2004a). The corresponding XRMA spectra from spot measurements on particles (right) confirm the presence of U and Fe, but no Ti. Bar = 20 μm.

3.2. SOURCE IDENTIFICATION

Based on ICP-MS and AMS, the $^{235}U/^{238}U$ atom ratio in individual DU particles from Kosovo and Kuwait was rather constant, at approx. 0.002, which was in good agreement with previously reported values (Salbu et al., 2003, 2004a). In contrast, the $^{236}U/^{235}U$ atom ratio measured in DU particles from Kuwait varied with sampling site, with values of $\sim 10^{-2}$ measured in particles associated with the DU ammunition storage facility fire, and values of $\sim 10^{-3}$ measured in particles collected inside disabled tanks. Using ICP-MS and AMS, Danesi et al. (2003a) confirmed that the $^{236}U/^{235}U$ atom ratios in bulk samples of the highly contaminated Ceja Mountain soils were similar to those found for individual particles ($\sim 10^{-2}$). The presence of ^{236}U can only originate from the use of U recycled from spent fuel or to handling of natural U with equipment previously contaminated in the course of recycling operations in enrichment plants.

4. Conclusions

The present results show the importance of the source term and the release scenario associated with a nuclear event for the solid state speciation of U and for the potential uptake in man and biota. As the particle weathering rate is expected to be higher for U(VI) and UO_{2+x} ($x > 0$) than for UO_2, the presence of respiratory uranyl, their corresponding weathering rates and the subsequent remobilisation of U should be included in environmental impact assessments of areas contaminated with DU and in health implication assessments arising from the use of this type of ammunition. Atom ratios ($^{236}U/^{235}U$) obtained from ICP-MS and AMS demonstrated that U in individual DU-particles originated from reprocessed fuel.

Acknowledgements

The authors are indebted to Prof. D. H. Oughton and associate professor L. Skipperud from the Isotope Laboratory, Norwegian University of Life Sciences and professor L. K. Fifield, National University Canberra for measurements of U isotope ratios. The work has been funded by EU (contract no. FIGE-CT-2000-00108 ADVANCE), and the Norwegian Research Council (contract no. 141479/720).

References

Bleise, A., Danesi, P.R., and Burkart, W. (2003) Properties, use and health effects of depleted uranium (DU): a general overview. *Journal of Environmental Radioactivity* **64**, 93–112.

Chazel, V., Gerasimo, P., Dabouis, V., Laroche, P. et al. (2003) Characterisation and dissolution of depleted uranium aerosols produced during impacts of kinetic energy penetrators against a tank. *Radiation Protection Dosimetry* **105**, 163–166.

Chazel, V., Houpert, P., and Ansoborlo, E., (1998) Effect of U_3O_8 specific surface area on in vitro dissolution, biokinetics, and dose coefficients. *Radiation Protection Dosimetry* **79**, 39–42.

Danesi, P.R., Bleise, A., Burkart, W., Cabianca, T. et al. (2003a) Isotopic composition and origin of uranium and plutonium in selected soil samples collected in Kosovo. *Journal of Environmental Radioactivity* **64**, 121–131.

Danesi, P.R., Markowicz, A., Chinea-Cano, E., Burkart, W. et al. (2003b) Depleted uranium particles in selected Kosovo samples. *Journal of Environmental Radioactivity* **64**, 143–154.

De Haag, P.A.M.U., Smetsers, R.C.G.M., Witlox, H.W.M., Krus, H.W. et al. (2000) Evaluating the risk from depleted uranium after the Boeing 747-258F crash in Amsterdam, 1992. *Journal of Hazardous Materials* **A76**, 39–58.

Gerdes, A., Weyer, S., Brey, G., Durakovic, A. et al. (2004) Monitoring depleted uranium contamination in the biosphere of Iraq using MC-ICP-MS. *Geochimica et Cosmochimica Acta* **68**, A506.

Giammar, D.E. and Hering, J.G. (2004) Influence of dissolved sodium and cesium on uranyl oxide hydrate solubility. *Environmental Science and Technology* **38**, 171–179.

IAEA (2003) Radiological Conditions in Areas of Kuwait with Residues of Depleted Uranium, IAEAL 03-00324, International Atomic Energy Agency, pp. 1–73.

Kashparov, V.A., Oughton, D.H., Protsak, V.P., Zvarisch, S.I. et al. (1999) Kinetics of fuel particle weathering and ^{90}Sr mobility in the Chernobyl 30 km exclusion zone. *Health Physics* **76**, 251–259.

Lind, O.C., Salbu, B., Skipperud, L., Janssens, K. et al. (2009) Solid state speciation and potential bioavailability of depleted uranium particles from Kosovo and Kuwait. *Journal of Environmental Radioactivity* **100**, 301–307.

McLaughlin, J.P., Vintro, L.L., Smith, K.J., Mitchell, P.I. et al. (2003) Actinide analysis of a depleted uranium penetrator from a 1999 target site in southern Serbia. *Journal of Environmental Radioactivity* **64**, 155–165.

Mitchel, R.E.J. and Sunder, S. (2004) Depleted uranium dust from fired munitions: Physical, chemical and biological properties. *Health Physics* **87**, 57–67.

Proost, K. (2005) Development and application of microscopic polycapillary-based X-ray absorption near-edge spectroscopy (μ–XANES). Thesis/Dissertation, University of Antwerpen, pp. 1–247.

Rai, D., Felmy, A.R., Sterner, S.M., Moore, D.A. et al. (1997) The solubility of Th(IV) and U(IV) hydrous oxides in concentrated NaCl and $MgCl_2$ solutions. *Radiochimica Acta* **79**, 239–247.

Rai, D., Yui, M., and Moore, D.A. (2003) Solubility and solubility product at 22 degrees C of UO_2(c) precipitated from aqueous U(IV) solutions. *Journal of Solution Chemistry* **32**, 1–17.

Salbu, B., Janssens, K., Lind, O.C., Proost, K. et al. (2003) Oxidation states of uranium in DU particles from Kosovo. *Journal of Environmental Radioactivity* **64**, 167–173.

Salbu, B., Janssens, K., Lind, O.C., Proost, K. et al. (2004a) Oxidation states of uranium in depleted uranium particles from Kuwait. *Journal of Environmental Radioactivity* **78**, 12–135.

Salbu, B., Krekling, T., Lind, O.C., Oughton, D.H. et al. (2001) High energy X-ray microscopy for characterisation of fuel particles. *Nuclear Instruments and Methods in Physics Research Section A* – Accelerators Spectrometers Detectors and Associated Equipment **467**, 1249–1252.

Salbu, B., Krekling, T., and Oughton, D.H. (1998) Characterisation of radioactive particles in the environment. *Analyst* **123**, 843–849.

Salbu, B., Lind, O.C., and Skipperud, L. (2004b) Radionuclide speciation and its relevance in environmental impact assessments. *Journal of Environmental Radioactivity* **74**, 233–242.

Sheppard, S.C., Sheppard, M.I., Gallerand, M.O., and Sanipelli, B. (2005) Derivation of ecotoxicity thresholds for uranium. *Journal of Environmental Radioactivity* **79**, 55–83.

Sowder, A.G., Clark, S.B., and Field, R.A. (1999) The transformation of uranyl oxide hydrates: The effect of dehydration on synthetic metaschoepite and its alteration to becquerelite. *Environmental Science and Technology* **33**, 3552–3557.

UNEP (2001) *Depleted Uranium in Kosovo Post-Conflict Environmental Assessment*, United Nations Environment Programme, pp. 1–188.

UNEP (2003) *Depleted Uranium in Bosnia and Herzegovina: Post-Conflict Environmental Assessment*, ISBN No: 92 1 158619 4, United Nations Environment Programme, pp. 1–303.

FORMATION OF RADIOACTIVE AEROSOL PARTICLES DURING THE WILDLAND FIRES IN CHERNOBYL ZONE AND THEIR RADIOECOLOGICAL IMPACT

VASYL YOSCHENKO[1], VALERY KASHPAROV[1], SVYATOSLAV LEVCHUK[1], SEMEN LUNDIN[1], VALENTIN PROTSAK[1], YURI KHOMUTININ[1], OLEXIY GLUKHOVSKY[1], IGOR MALOSHTAN[1], JOCHEN TSCHIERSCH[2]

[1]Ukrainian Institute of Agricultural Radiology, Mashinobudivnykiv str.7, Chabany, Kyiv-Svjatoshin Distr., Kyiv reg. 08162, Ukraine

[2]GSF-National Research Center for Environment and Health, Institute of Radiation Protection, 85764 Neuherberg, Germany

Abstract: Controlled grassland and forest fires have been carried out at the experimental plots in the Chernobyl zone in order to determine the parameters of the radioactive aerosols resuspension, transportation and deposition under well-described conditions. Radionuclide airborne concentrations, resuspension factors, deposition intensity and velocities and dispersal composition of the radioactive aerosols were measured. Doses to firemen from inhalation of radionuclides were calculated. A mathematical model for the initial plume rise was developed, while the further transportation of radioactivity in the atmosphere was described by the Gauss model. A Monte–Carlo algorithm was applied for modeling the radioactive aerosols resuspension/transportation in order to take into account the stochastic nature of the plume rise input parameters. In the model exercises the total radioactivity releases from the burning plots were estimated.

Keywords: radioactive aerosol, wildland fire, resuspension, deposition, dispersal composition, inhalation dose, initial plume rise, modeling.

1. Introduction

Inside the Chernobyl exclusion zone a huge amount of radionuclides were deposited. Detailed contamination maps have been published (Kashparov et al.,

2001). Some of these nuclides were deposited during the Chernobyl accident directly to trees and bushes; others were transferred by root uptake from the contaminated soil to the vegetation. The exclusion zone is covered mostly (almost 2/3 of the total area) by forests and grasslands. It is estimated that several percent of the initially deposited radionuclides are contained in the biomass of the grasslands and even more in the biomass of the forests (see the characterization of the experimental plots in this report). These radionuclides currently fixed in the biomass potentially may be mobilized by the wildland fires; the increases of the radionuclides airborne concentration in the Chernobyl zone during the fires are periodically reported.

The area is no longer cultivated and is mostly abandoned. Frequently the wildland fires which occur are detected and combated late. The number of the detected fires varies in the range from 42 to 116 per year. The fires have already destroyed thousands of hectares forest. The problem is expected to increase in future, because the absence of any measures for maintenance of the forests, meadows and peatlands will lead to an uncontrolled growing of the vegetation.

The aims of the present study were to estimate the radioecological impact of the fires in the ecocenoses of the Chernobyl zone (i.e., the resuspension parameters and atmospheric transportation scale of the formed radioactive aerosols) and to evaluate the potential risk to the firemen/population related to the inhalation of the radioactive aerosols.

2. Methods

2.1. CONTROLLED FIRES DESIGN

The controlled fires in the grassland and forest sites were carried out in 2001–2003 (Yoschenko et al., 2006a,b) in order to determine the radioactive aerosols parameters in the surface air layer and the radiological impact of the fires in the Chernobyl zone. For these researches, three sites (two in the grasslands and one in the forest) had been selected in the Exclusion zone and radioecologically described before the active experiments. Methodology of the active experiments involved the measurements of the controlled parameters at the spatial grid and detail record of the meteorological parameters during the experiments. The measurements along the radioactive plume axis included measurements of the radionuclides airborne concentration in the surface air layer by means of isokinetic aerosol samplers, measurements of the fallout densities at the horizontal plates and determination of the radioactive aerosol dispersal composition by multi-cascade impactors (see example of the equipment items installation in Fig. 1).

The radionuclides activities in the collected samples of soil, vegetation and at the filters were measured in the laboratory conditions by the standard γ-spectrometric (^{137}Cs) and radiochemical (^{90}Sr and plutonium isotopes) methods. Obtained experimental results were used for evaluation of the working conditions of the firefighters in the Chernobyl zone. Also, these results were applied for development of the mathematical models of the radioactivity release and spreading during the fires.

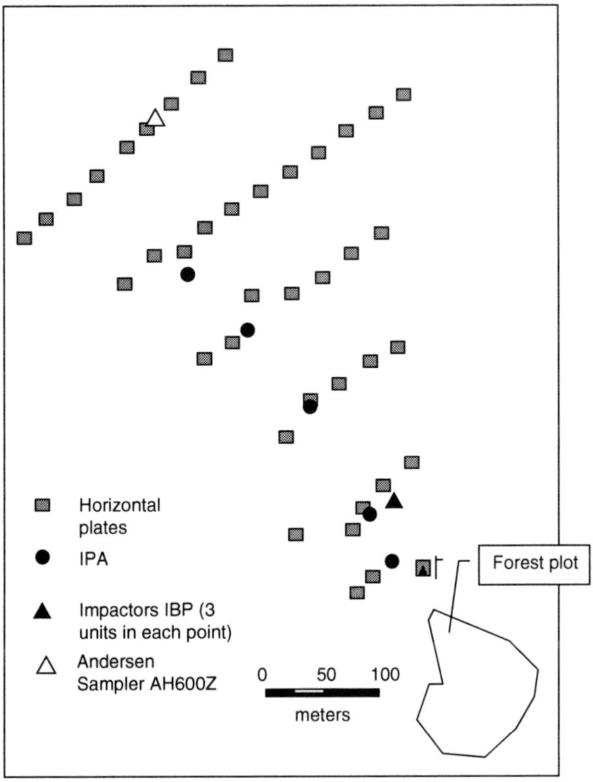

Figure 1. Scheme of deployment of the sampling equipment during the forest fire experiment.

2.2. DESCRIPTION OF THE EXPERIMENTAL SITES

Both grassland fires were carried out within the same experimental area near the former village Chistogalovka, at a distance of approximately 3 km west of the Chernobyl reactor site. The forest site was located 5 km WNW of the Chernobyl reactor close to the former village Novoshepelichi. It was a pine forest of cultivated planting surrounded by grassland.

In order to provide the radioecological description of the experimental sites, the samples of soil and vegetation were collected before the experiments and analyzed in the laboratory conditions. The characteristics of the three plots and measured background values of the radionuclides airborne concentrations and deposition intensities are summarized in Tables 1–4.

TABLE 1. Characteristics of the plots used for the active fire experiments

	Plot #1	Plot #2	Plot #3
Area (m^2)	3,600	5,400	8,770
Land type	Wildland (grass)	Wildland (grass)	Forest
Vegetation species	*Elytrigia repens* (L.) *Nevski* (85%)	*Elytrigia repens* (L.) *Nevski* (85%)	*Pinus silvestris*
Vegetation height (m)	0.5	0.2	15
Vegetation density (kg m^{-2})	0.4	0.3	24
Litter density (kg m^{-2})	0.7	1.25	2.3
Dose rate range (μGy h^{-1})	10–16	6–10	2–4

TABLE 2. Background values of the airborne activity concentration A_V (Bq m^{-3}) and the deposition intensity I (Bq m^{-2} s^{-1}) of the radionuclides before the experiments

	^{137}Cs		^{90}Sr		^{238}Pu		$^{239+240}Pu$	
	$A_V \cdot 10^{-3}$	$I \cdot 10^{-4}$	$A_V \cdot 10^{-3}$	$I \cdot 10^{-5}$	$A_V \cdot 10^{-6}$	$I \cdot 10^{-8}$	$A_V \cdot 10^{-6}$	$I \cdot 10^{-7}$
Plot #1	2.1 ± 0.3	0.5 ± 0.1	0.8 ± 0.2	2.0 ± 0.3	1.5 ± 0.3	1.2 ± 0.3	2.9 ± 0.6	0.3 ± 0.1
Plot #2	4.6 ± 0.6	1.0 ± 0.1	1.4 ± 0.4	3.5 ± 0.4	4.3 ± 0.7	4.4 ± 0.7	8.5 ± 1.0	1.1 ± 0.1
Plot #3	3.2 ± 0.4	0.6 ± 0.1	4.4 ± 0.5	2.2 ± 0.2	8.3 ± 2.1	23 ± 4	22 ± 5	4.5 ± 0.7

TABLE 3. Radionuclides in the grassland plots: activity concentration C, terrestrial contamination density S and total inventory of the plots

		Soil[a]	Grass	Litter	Total	Total combustible
			Plot #1			
^{137}Cs	C, kBq kg^{-1}	80 ± 50	86 ± 50	340 ± 210		
	S, MBq m^{-2}	29 ± 18	0.026 ± 0.015	0.26 ± 0.17		
	Total, GBq	104 ± 70	0.094 ± 0.065	0.94 ± 0.7	105 ± 71	1.0 ± 0.8
^{90}Sr	C, kBq kg^{-1}	21 ± 14	120 ± 90	210 ± 150		
	S, MBq m^{-2}	7.6 ± 47	0.041 ± 0.031	0.15 ± 0.11		
	Total, GBq	27 ± 18	0.15 ± 0.12	0.54 ± 0.4	28 ± 19	0.7 ± 0.5
^{238}Pu	C, Bq kg^{-1}	140 ± 80	1.1 ± 0.7	960 ± 500		
	S, kBq m^{-2}	50 ± 30	$(4 \pm 3) \cdot 10^{-4}$	0.72 ± 0.4		
	Total, MBq	180 ± 110	$(1.4 \pm 1.1) \cdot 10^{-3}$	2.6 ± 1.5	183 ± 112	2.6 ± 1.5

$^{239+240}$Pu	C, Bq kg^{-1}	280 ± 160	2.4 ± 1.6	2,150 ± 1,500		
	S, kBq m^{-2}	100 ± 60	(8 ± 6)·10^{-4}	1.6 ± 1.2		
	Total, MBq	360 ± 200	(2.8 ± 2.1) 10^{-3}	5.8 ± 4.8	366 ± 205	5.8 ± 4.8
			Plot #2			
^{137}Cs	C, KBq kg^{-1}	14 ± 9	49 ± 21	85 ± 5		
	S, MBq m^{-2}	5.1 ± 3.1	0.014 ± 0.006	0.11 ± 0.07		
	Total, GBq	27 ± 17	0.07 ± 0.03	0.57 ± 0.36	28 ± 17	0.64 ± 0.39
^{90}Sr	C, KBq kg^{-1}	7.3 ± 6.0	62 ± 19	81 ± 35		
	S, MBq m^{-2}	2.7 ± 2.2	0.018 ± 0.005	0.087 ± 0.053		
	Total, GBq	15 ± 12	0.10 ± 0.01	0.47 ± 0.29	16 ± 12	0.57 ± 0.30
^{238}Pu	C, Bq kg^{-1}	57 ± 33	0.17 ± 0.13	130 ± 87		
	S, KBq m^{-2}	20 ± 12	(4.7 ± 4.0)·10^{-5}	0.087 ± 0.060		
	Total, MBq	110 ± 60	(2.5 ± 2.1)·10^{-4}	0.47 ± 0.33	110 ± 60	0.47 ± 0.33
$^{239+240}$Pu	C, Bq kg^{-1}	110 ± 65	0.44 ± 0.33	270 ± 190		
	S, KBq m^{-2}	40 ± 27	(1.2 ± 0.9)·10^{-4}	0.19 ± 0.14		
	Total, MBq	210 ± 150	(6.7 ± 5.3) 10^{-4}	1.1 ± 0.7	211 ± 151	1.1 ± 0.7

[a]Upper 30-cm layer.

3. Experimental results

3.1. RADIONUCLIDES AIRBORNE CONCENTRATION AND DEPOSITION DURING THE FIRES

Averaged values of the airborne radionuclide concentrations during the fires at the various distances from the source of release at 1 m height along the plumes axes are presented in Fig. 2 (background values have been subtracted). During the grassland fires, the airborne concentrations of ^{137}Cs and ^{90}Sr reached values of several Bq m^{-3} near the source of release and decreased with the distance. The concentration of plutonium was generally three orders of magnitude lower and an increase with the distance was observed in plot #1. In plot #2, rather constant values were measured at greater distances after an initial decrease. During the forest fire experiment, in general, one order of magnitude lower activity concentrations for ^{137}Cs and ^{90}Sr were measured in comparison to the grassland fires. After an initial decrease in concentration over long distances a rather constant airborne concentration was observed for all nuclides. At the most far sampling site the concentration increased again.

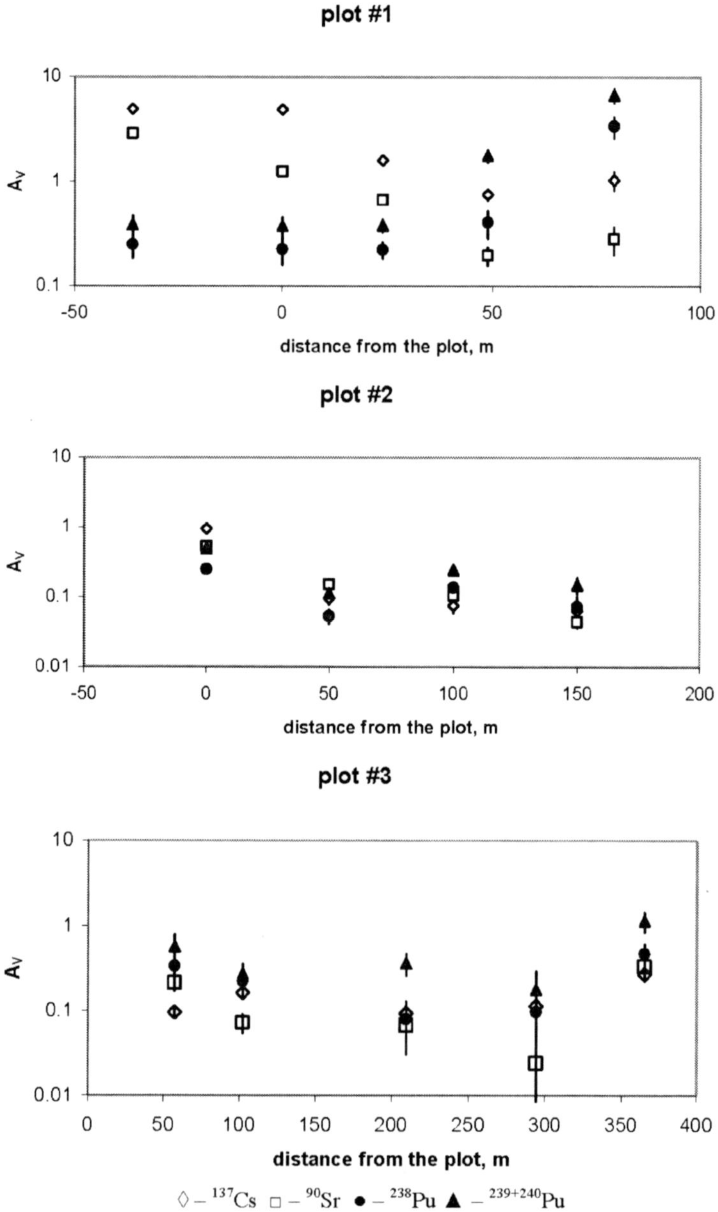

Figure 2. Airborne radionuclide concentrations during the fire experiments in Bq m^{-3} for ^{137}Cs and ^{90}Sr, and in mBq m^{-3} for plutonium. Vertical bars represent the measurement STD. The distance is calculated from the closest plot border, negative values indicate sample locations inside the plot.

TABLE 4. Biomass and radionuclides in the forest plot: biomass density, activity concentration C, terrestrial contamination density S and total inventory of the plots

		Soil[a]	Timber	Bark	1st year needles	2nd year needles	Forest litter	Fallen branches	Total	Total combustible
Biomass, $kg\ m^{-2}$		–	20.5 ± 3.5	2.6 ± 0.4	0.6 ± 0.1	0.30 ± 0.05	2.15 ± 0.28	0.11 ± 0.03		
^{137}Cs	C, $kBq\ kg^{-1}$	5.8 ± 1.1	0.83 ± 0.20	5.1 ± 1.4	57 ± 15	6.9 ± 1.9	64 ± 17	1.1 ± 0.2		
	S, $kBq\ m^{-2}$	1,700 ± 300	17 ± 7	13 ± 6	34 ± 15	2.1 ± 0.9	137 ± 53	0.12 ± 0.06		
	Total, GBq	14.9 ± 2.6	0.15 ± 0.06	0.12 ± 0.05	0.30 ± 0.13	0.018 ± 0.008	1.2 ± 0.5	0.0011 ± 0.0005	16.7 ± 3.3	1.8 ± 0.7
^{90}Sr	C, $kBq\ kg^{-1}$	3.7 ± 0.9	12.5 ± 2.5	45.8 ± 9.2	26.4 ± 5.2	70 ± 15	81 ± 18	16.6 ± 3.5		
	S, $kBq\ m^{-2}$	1,100 ± 300	260 ± 100	120 ± 40	15.8 ± 5.8	21 ± 8	173 ± 62	1.8 ± 0.9		
	Total, GBq	9.6 ± 2.6	2.25 ± 0.83	1.04 ± 0.37	0.14 ± 0.05	0.19 ± 0.07	1.52 ± 0.54	0.016 ± 0.008	14.8 ± 4.5	5.2 ± 1.9
^{238}Pu	C, $Bq\ kg^{-1}$	33 ± 8	0.24 ± 0.07	7.4 ± 2.1	0.24 ± 0.07	0.33 ± 0.10	130 ± 40	0.3 ± 0.1		
	S, $Bq\ m^{-2}$	9,800 ± 2,300	4.9 ± 2.3	19 ± 8	0.14 ± 0.07	0.10 ± 0.05	280 ± 120	0.03 ± 0.02		
	Total, MBq	86 ± 20	0.043 ± 0.027	0.17 ± 0.07	0.0013 ± 0.0008	0.0009 ± 0.0004	2.45 ± 1.07	0.0003 ± 0.0002	89 ± 21	2.7 ± 1.2
$^{239+240}Pu$	C, $Bq\ kg^{-1}$	72 ± 17	1.45 ± 0.5	18 ± 5	0.9 ± 0.3	0.55 ± 0.25	280 ± 70	1.9 ± 0.6		
	S, $Bq\ m^{-2}$	21,000 ± 5,000	30 ± 15	48 ± 21	0.5 ± 0.3	0.17 ± 0.10	600 ± 230	0.2 ± 0.1		
	Total, MBq	184 ± 44	0.26 ± 0.13	0.42 ± 0.18	0.0047 ± 0.0024	0.0015 ± 0.0009	5.3 ± 2.0	0.0018 ± 0.0011	190 ± 46	6.0 ± 2.3

[a]Upper 20-cm layer.

The deposition patterns after the grassland fires are presented in Fig. 3. In general, the fallout density of the radionuclides during the grassland fires decrease with the distance from the source of release. For ^{137}Cs and ^{90}Sr, the fallout density can reach some hundreds Bq m^{-2} near the burning plot, while for plutonium it is three to four orders of magnitude lower.

During the forest fire, the fallout density of ^{137}Cs and ^{90}Sr reaches some tens Bq m^{-2}, and for plutonium these values are two orders of magnitude lower (Fig. 4). The plutonium fallout decreases with the distance from the source of release, while the higher values of the ^{137}Cs and ^{90}Sr fallout density were measured along the dominating wind direction during the experiment.

3.2. RESUSPENSION FACTOR AND DEPOSITION PARAMETERS

During the grassland fires, the resuspension factor (RF) of ^{137}Cs and ^{90}Sr was in the range 10^{-5}–10^{-6} m^{-1}, and the resuspension factor of the plutonium nuclides in the range 10^{-6}–10^{-7} m^{-1}. The above values relate to the burning biomass material only. If the contamination density S includes also the soil contamination, the calculated resuspension factors will be two orders of magnitude lower. The grassland fires were carried out in very favourable conditions, when one could expect an intensive resuspension. Thus, the obtained values can be considered as an upper estimate for the Chernobyl zone conditions.

During the forest fire, resuspension factors of all radionuclides were in the range 10^{-6}–10^{-7} m^{-1} if they were related to the biomass contamination only, and 10^{-7}–10^{-8} m^{-1} if they were related to the total contamination. In general, they were lower than those during the grassland fires, probably because of the absence of grass and underwood in the forest. The forest litter was mainly formed by the fallen needles, while the meadow litter consisted of dry rests of grass.

The deposition intensity and deposition velocity depended on the radionuclide because of the size dependence of the deposition process. However, the observed differences were not so large because integral quantities were measured. In general, higher values of the deposition velocity were observed for ^{137}Cs and ^{90}Sr (up to 60 cm s^{-1}), while for plutonium they mainly reached several cm s^{-1} (Table 5).

Both intensity and velocity of aerosol deposition were lower during the forest fire. Probably, it can be explained by the absence of giant particles (or fragments of burning material) in the smoke plume of the forest fire.

In our previous studies (Kashparov et al., 2000), we determined for ^{137}Cs a resuspension factor of 10^{-7}–10^{-8} m^{-1} during the forest fires experiments in the contaminated territories outside the 30-km Chernobyl zone. These values are calculated for the total contamination of the sites and are in good agreement with those obtained in the present study. Thus, for the typical pine forests of the

Ukrainian Polessie we have confirmed the above values as ^{137}Cs resuspension factors during forest fires.

3.3. DISPERSAL COMPOSITION OF THE RADIOACTIVE AEROSOLS

It was observed that both the airborne concentration and the fallout density have local maxima at relatively large distances from the source of release, especially for the plutonium radionuclides. Under the assumption that the radionuclides are released during the fires in different fractions of the polydisperse aerosol, this observation may be explained. The heavier fractions of the radioactive aerosol are released at lower height and have higher deposition velocities; therefore, they are mainly deposited closer to the plot. The measured airborne concentration and fallout density decrease with the distance till the area of deposition of the finer fractions of radioactive aerosols, which lifted up to a higher altitude and deposited with a smaller deposition velocity.

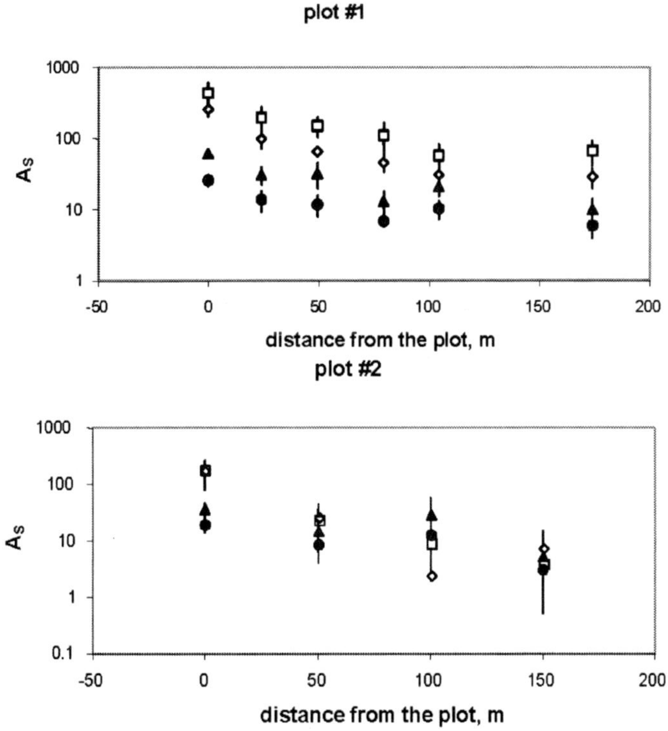

Figure 3. The deposition densities of radionuclides during the grassland fires, A_S, in Bq m^{-2} for ^{137}Cs and ^{90}Sr, and in mBq m^{-2} for plutonium. Vertical bars represent the measurement STD.

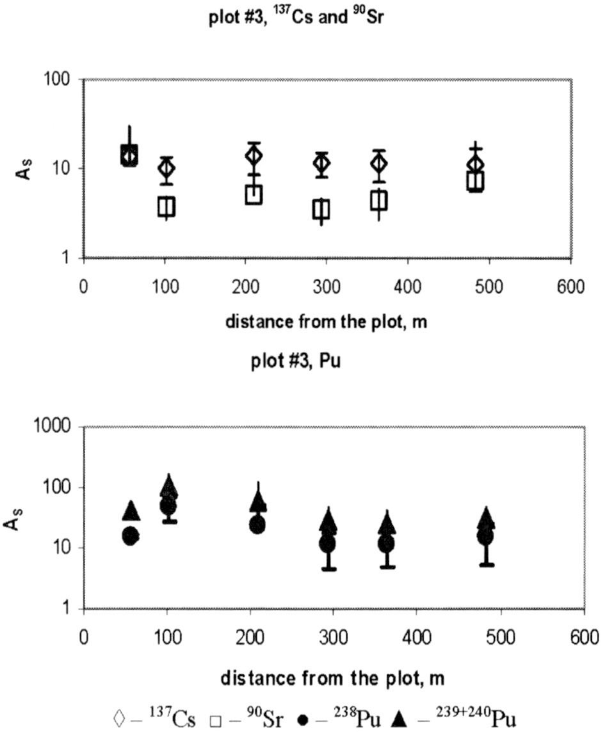

Figure 4. The deposition densities of radionuclides (averaged for each row of plates) during the forest fire, A_S, in Bq m^{-2} for ^{137}Cs and ^{90}Sr, and in mBq m^{-2} for plutonium. Vertical bars represent the variation of values (STD) in each row.

The polydisperse nature of the released radioactive aerosol is confirmed by the results of the measurements of the cascades of the deployed impactors. During the grassland fires, the contribution of the giant particles (with AMAD >25 μm) to the total airborne concentration of the radioactive aerosol is comparable high near the plot, and then it is decreasing with distance (Figs. 5 and 6). The visual analysis of the sediments on the cascades collecting large particles showed that they were mainly composite ash, while the sediments on the cascades collecting finer particles that they were mainly composite of water drops, tars, small ash particles etc. One should note that the dispersal composition of the various radionuclides was different and reflects the origin of each radionuclide. For instance, the percentage of ^{90}Sr in the giant particles was the highest in both grassland experiments. Among the other studied radionuclides, ^{90}Sr has the highest soil-to-plant transfer factor and, therefore, the biomass contamination with this radionuclide is high. In the grassland plots, the activity ratio grass/litter is the highest for ^{90}Sr (Table 3). Taking into account that grass burning produces a huge amount of large ash particles, the ^{90}Sr contamination of these large particles is reasonable. For plutonium and ^{137}Cs, the contribution of litter in the total resuspended activity is higher (because of the lower contamination of grass). These nuclides are found to a higher fraction in the fine particle range. Therefore, mainly litter burning shall be responsible for the production of the fine aerosol particles. The consequences of the differences in the contamination of the various biomass compartments (Table 4) can be seen in the forest fire data as well. In the dispersal composition of the forest fire aerosol (Fig. 7) a change can be observed with the distance from the plot. Close to the plot, the most nuclides were measured in the large particle range; in the far distance ^{90}Sr is detected in the fine and giant particles ranges and the Pu nuclides still dominated in the large particle range.

4. Dose assessment

External constituent of the total dose to the firemen is formed by the radionuclides in soil, vegetation and in the passing plume. Being inhaled from the plume, the radionuclides form the dose of internal irradiation. In our study, the external dose from the radionuclides in the biomass and the soil was determined from the measured dose rate in air. The maximum exposure for a 1-h stay at the plots was 16 μSv in plot #1, 10 μSv in plot #2 and 4.2 μSv in plot #3. The external dose of the plume was calculated from the airborne radionuclide concentration and the corresponding dose coefficients (Gusev and Beliayev, 1991).

The inhalation dose is calculated using the dosimetric model of the human respiratory tract presented in Publication 66 of the ICRP (1994). According to the regulatory documents (NCRPU, 1997), the dose absorbed in the whole body (effective equivalent dose, EED) during 50 years after the radionuclide intake is determined.

The results of the dose calculation due to 1-h stay of the firemen in the experimental sites during the fire experiments are presented in Table 6. It can be seen that beside the dose of the external irradiation from soil and vegetation, a significant contribution to the total dose can be provided by the inhalation intake of the plutonium radioisotopes and their progenies. In some conditions this contribution can reach half of the total dose. Although this estimate is conservative, the problem of protection of the firemen should be considered in more detail.

In the same time, taking into account the sharp decrease of the airborne radionuclide concentration with the distance from the source of release (Fig. 2), we can state that the inhalation constituent of the total dose (as well as the external irradiation from radionuclides in air) is not important for the personnel of the Exclusion zone which is not involved in the fire fighting and is not occasionally located in the fire area.

TABLE 5. Deposition intensity, I, and deposition velocity, vg, during the experiments for the considered radionuclides. X is the distance from the plot along the plume axis

	X (m)	^{137}Cs		^{90}Sr		^{238}Pu		$^{239+240}Pu$	
		I (mBq m^{-2} s^{-1})	v_g (cm s^{-1})	I (mBq m^{-2} s^{-1})	v_g (cm s^{-1})	I (mBq m^{-2} s^{-1})	v_g (cm s^{-1})	I (mBq m^{-2}) s^{-1}	v_g (cm s^{-1})
Grassland fire, plot #1	0	120 ± 10	2.6 ± 0.2	200 ± 10	16 ± 1	14 ± 3	6.3 ± 1.4	30 ± 5	9.5 ± 1.2
	24	60 ± 5	4.0 ± 0.5	150 ± 10	22 ± 2	12 ± 2	5.3 ± 0.9	25 ± 3	6.6 ± 0.8
	49	40 ± 3	5.3 ± 0.5	110 ± 10	57 ± 6	9.1 ± 1.5	2.2 ± 0.4	20 ± 2	1.1 ± 0.1
	79	30 ± 3	3.1 ± 0.3	60 ± 6	21 ± 4	4.2 ± 1.0	0.1 ± 0.03	10 ± 1	0.1 ± 0.01
Grassland fire, plot #2	0	340 ± 10	35 ± 3	300 ± 10	60 ± 6	31 ± 2	12 ± 2	58 ± 3	11 ± 1
	50	30 ± 3	34 ± 7	20 ± 2	15 ± 2	17 ± 2	28 ± 7	26 ± 2	21 ± 4
	100	9 ± 2	13 ± 3	10 ± 1	9 ± 1	15 ± 2	17 ± 6	38 ± 3	26 ± 7
	150	16 ± 3	22 ± 6	6 ± 1	12 ± 3	5.4 ± 1.8	7.0 ± 3.5	13 ± 2	8 ± 3
Forest fire, plot #3	57	2.0 ± 0.3	2.6 ± 0.4	1.0 ± 0.2	1.3 ± 0.3	3.1 ± 0.6	2.7 ± 1.4	9.1 ± 2.4	1.4 ± 0.6
	102	1.5 ± 0.2	1.2 ± 0.2	4.6 ± 1.8	1.0 ± 0.4	8.6 ± 1.3	4.2 ± 1.7	16 ± 2	5.2 ± 1.8
	210	4.1 ± 0.4	2.8 ± 0.5	2.8 ± 0.4	1.9 ± 0.8	14 ± 5	7.8 ± 5.3	32 ± 10	4.1 ± 1.5
	294	1.7 ± 0.2	1.9 ± 0.4	0.5 ± 0.2	2.7 ± 2.3	1.1 ± 0.4	4.7 ± 4.5	2.6 ± 0.6	3.1 ± 2.1
	365	1.3 ± 0.2	0.6 ± 0.1	0.6 ± 0.2	0.2 ± 0.07	2.6 ± 1.1	0.35 ± 0.19	7.7 ± 2.2	0.29 ± 0.14
	483	1.8 ± 0.3		0.5 ± 0.2		1.5 ± 0.8		1.8 ± 0.6	

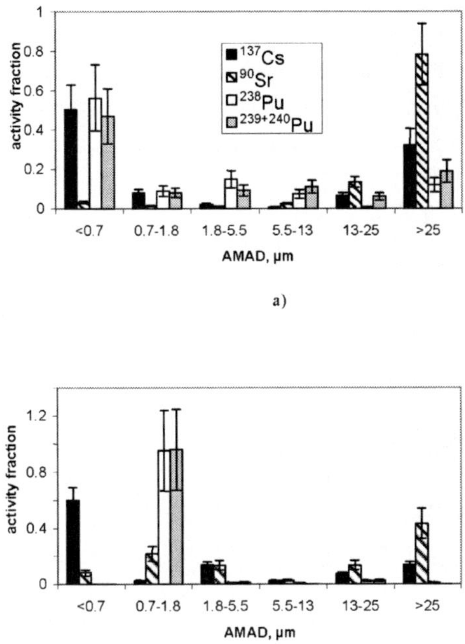

Figure 5. Dispersal composition of the radioactive aerosol during the grassland fire (plot #1): at the border of the plot (0 m) and b) 24 m far from the plot along the plume axis. Vertical bars represent the STD.

5. Modelling the radionuclides resuspension and transportation during the wildland fires

5.1. MATHEMATICAL FORMULATION OF THE MODELS

To model transportation process, the standard Gauss model for short-term source of release was applied (IAEA, 1980).

The ideology proposed by Talerko (1990) was applied to describe the initial rise of the radioactive plume from the fire zone. However, the mathematical formalization of the plume rise model was modified, as well as an approach for calculation of the initial plume parameters (temperature, velocity and radius) was created (Yoschenko et al., 2006b).

In the present formulation of the model we consider 2-d problem assuming that the particles move in the plume in two directions: vertical, along *z*-axis, with a velocity of W, m s^{-1}, and horizontal, along *x*-axis, which coincides with the direction of the wind speed U, m s^{-1}. Further, we assume that the source of release has a vertical dimension of H_{Mix}, which corresponds to the thickness of the above-ground air layer subjected to the full homogenization (mixing) above the burning fuel material during the fire. Visually, H_{Mix} might be estimated as a flame height. Within this layer, temperature of release is constant (this is an initial temperature T_i) exceeding the ambient air temperature T_e by ΔT, K. At the upper edge of the mixing layer, the fire products have only the vertical constituent of velocity, W_0, m s^{-1}. Air density is denoted as ρ_i (kg m^{-3}) in the plume and as ρ_e outside. Then the basic equations system is as follows:

$$\ddot{x} \cdot W^2 - \frac{C_W \cdot U^2 \cdot T_i}{2 \cdot \pi \cdot R \cdot T_e} + \frac{g \cdot \Delta T}{T_e} \cdot \dot{x} = 0 \quad (1)$$

$$dW / dz = \frac{g \cdot \Delta T}{T_e \cdot W} - W \cdot \varphi \cdot \frac{C \cdot T_i}{R \cdot T_e} - \frac{C_W \cdot U^2}{2 \cdot \pi \cdot R \cdot W \cdot \varphi^2} \cdot \dot{x} \quad (2)$$

$$d\Delta T / dz = -dT_e / dz - \frac{g \cdot T_i}{C_P \cdot T_e} - \frac{\Delta T \cdot C \cdot T_i \cdot \varphi}{R \cdot T_e} \quad (3)$$

$$dR / dz = \frac{R}{2} \cdot \left[C/_R \cdot \varphi \cdot (2 + \Delta T/_{T_e}) - 1/_P \cdot dP / dz - \frac{g}{C_P \cdot T_e} - \frac{g \cdot \Delta T}{T_e \cdot W^2 \cdot \varphi^2} \right] \quad (4)$$

where

$\dot{x} = dx / dz$, $\ddot{x} = d^2 x / dz^2$;
$g = 9.8$ m s^{-2} is the gravitational acceleration;
$C_W = 0.9$ is the resistance factor of the plume;
$C = 0.24$ is the entrainment parameter (Bruyatsky, 2000);

$\varphi(z) = 1 / \sin \beta(z) = \sqrt{1 + \dot{x}^2}$, and $\beta(z)$ is angle between the trajectory tangent and x-axis;
C_P is a specific heat and
P is air pressure (equal inside and outside the plume).

Assuming a linear dependence of the wind speed on elevation

$$dU / dz = U_Z$$

$$U_{(z=0)} = U_0, \quad (5)$$

we add the above system of Eqs. (1)–(4). Equation (5) represents a simplified approach, but it gives the wind speed profile in the observed plume elevation range, which is quite similar to the one calculated according to the widely used equation

$$U(z) = U_{10} \cdot (z/_{10})^P \quad (6)$$

5.2. CALCULATION ALGORITHM

The Monte–Carlo approach was applied for calculation of the values of the radionuclides airborne concentration and fallout in the points of the observation meshes during three fires. For these purpose, for each elementary cell of the fire area, the set of stochastic parameters of calculation was generated using either observed ranges of the parameters variation or the literature data. Such input parameters as wind direction, wind speed during gusts, relative duration of gusts, deposition velocity of aerosol particles, and density of contamination of the fuel material were considered as stochastic, and other necessary input data were interpreted as deterministic (Yoschenko et al., 2006b). Value of the radionuclide release during the fire was found from the best fitting the modeled values of the airborne concentration and fallout density to those measured in the experiments.

5.3. MODELING RESULTS

The model predictions are realistic and fit well the observed shapes of the plumes (also, we can state the similarity of the obtained plume trajectories to those calculated using the Briggs' equation (Briggs, 1972), especially at higher values of the wind speed).

With respect to the stochastic nature of the studied processes, one can also state a good agreement between the empirical data and calculation results. As an example, the comparison is presented for the forest fire (Fig. 8).

A summary of modeling exercises in terms of signed deviation SD and unsigned deviation UD, and released fractions α (calculated relatively to the radionuclide inventory in the fuel material) is presented in Table 7 with the following notations:

$$SD = \frac{\sum_{k=1}^{N} \frac{(\xi^{model}{}_k - \xi^{\exp}{}_k)}{\xi^{\exp}{}_k}}{N} \cdot 100\% \tag{7}$$

$$UD = \frac{\sum_{k=1}^{N} \left| \frac{\xi^{model}{}_k - \xi^{\exp}{}_k}{\xi^{\exp}{}_k} \right|}{N} \cdot 100\% \tag{8}$$

where N is a number of the observation points and ξ^{model} and ξ^{exp} are modeled and experimental values of either the radionuclide airborne concentration or fallout density in observation point k, respectively

TABLE 6. Doses by irradiation of the firemen due to a 1-hour stay in the fire zone

Radio-nuclide	Maximum airborne concentration, Bq m^{-3}			Dose origin	Dose, Sv		
	Plot #1	Plot #2	Plot #3		Plot #1	Plot #2	Plot #3
^{137}Cs	5	1	0.27	External[a]	$6.9 \cdot 10^{-10}$	$1.4 \cdot 10^{-10}$	$3.7 \cdot 10^{-11}$
				Inhalation	$6 \cdot 10^{-8}$	$1.2 \cdot 10^{-8}$	$3.2 \cdot 10^{-9}$
^{90}Sr	3	0.5	0.33	External[a]	10^{-10}	$1.7 \cdot 10^{-11}$	$1.1 \cdot 10^{-11}$
				Inhalation	$2.4 \cdot 10^{-7}$	$4.1 \cdot 10^{-8}$	$2.6 \cdot 10^{-8}$
^{238}Pu	$3.4 \cdot 10^{-3}$	$2.5 \cdot 10^{-4}$	$4.6 \cdot 10^{-4}$	Inhalation	$7.1 \cdot 10^{-6}$	$5.3 \cdot 10^{-7}$	10^{-6}
$^{239+240}Pu$	$6.7 \cdot 10^{-3}$	$5.1 \cdot 10^{-4}$	$1.1 \cdot 10^{-3}$	Inhalation	$1.7 \cdot 10^{-5}$	$1.3 \cdot 10^{-6}$	$2.8 \cdot 10^{-6}$
External irradiation from soil and vegetation					$1.6 \cdot 10^{-5}$	$1 \cdot 10^{-5}$	$4.2 \cdot 10^{-6}$
Total dose					$4.0 \cdot 10^{-5}$	$1.2 \cdot 10^{-5}$	$8 \cdot 10^{-6}$

[a]Irradiation from the radioactive cloud.

TABLE 7. Summaries of the experimental fires modelling

Parameter	Radionuclide			
	^{137}Cs	^{90}Sr	^{238}Pu	$^{239+240}Pu$
Grassland fire, plot #1				
SD, %	–30	–11	–14	
UD, %	66	46	61	
α, %	0.062	0.15	0.033	
Grassland fire, plot #2				
SD, %	67	160	2	
UD, %	109	190	72	
α	0.14	0.16	0.069	
SD, %	–10	46	9	40
UD, %	21	55	57	71
α	4.2	2.9	0.78	0.77

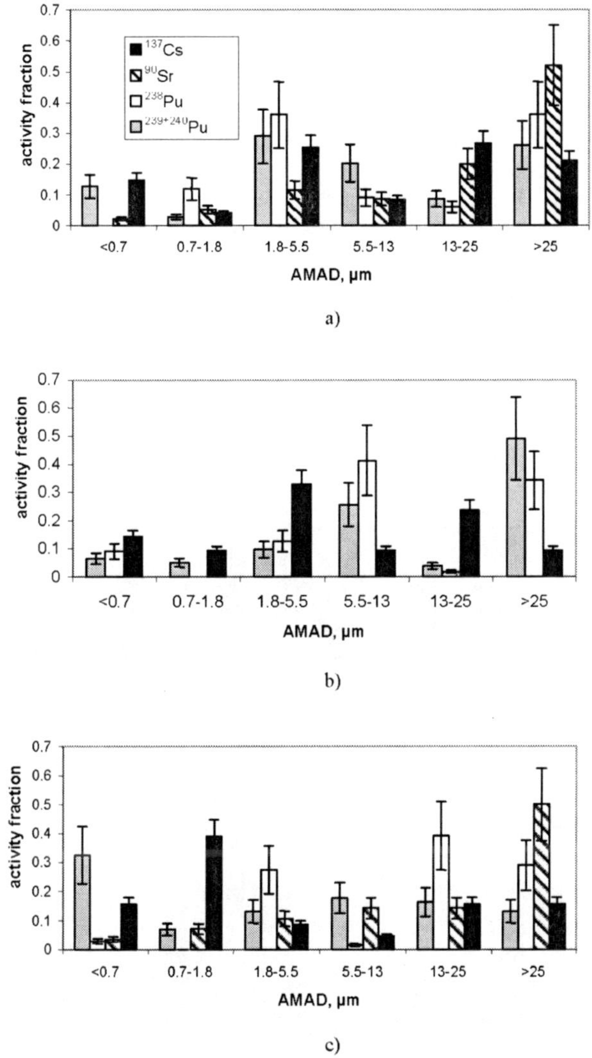

Figure 6. Dispersal composition of the radioactive aerosol during the grassland fire (plot #2): at the border of the plot (a) (0 m) and (b) 50 m far from the plot along the plume axis; (c) 100 m far from the plot along the plume axis. Vertical bars represent the STD.

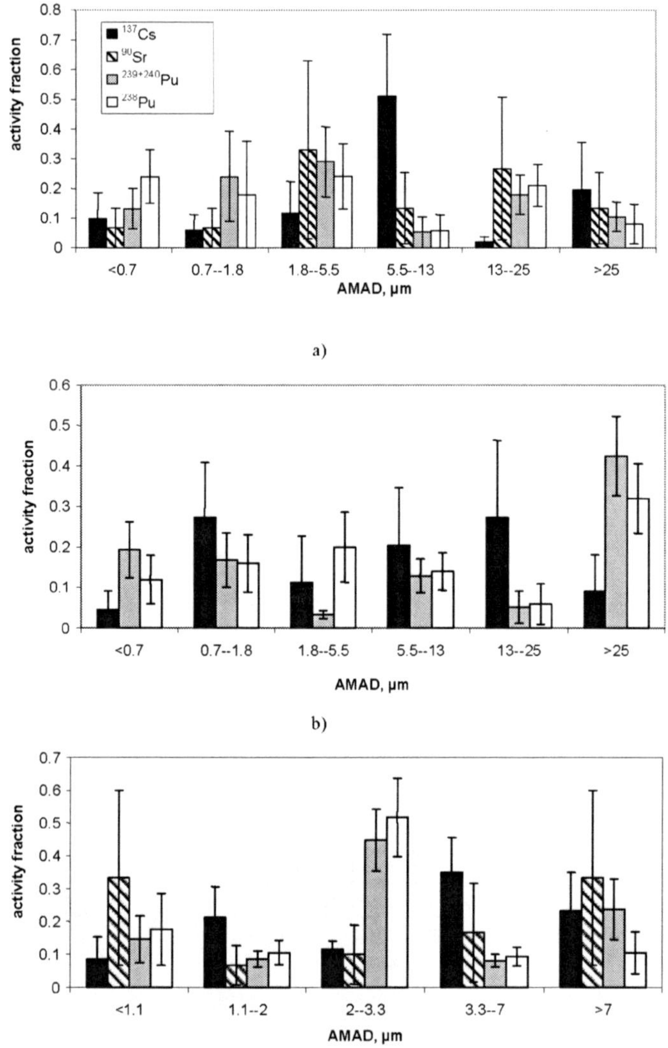

c)

Figure 7. Dispersal composition of the radioactive aerosol during the forest fire (plot #3): (a) 37 m far from the plot along the plume axis; (b) 105 m far from the plot along the plume axis and (c) 492 m far from the plot along the plume axis. Vertical bars represent the STD.

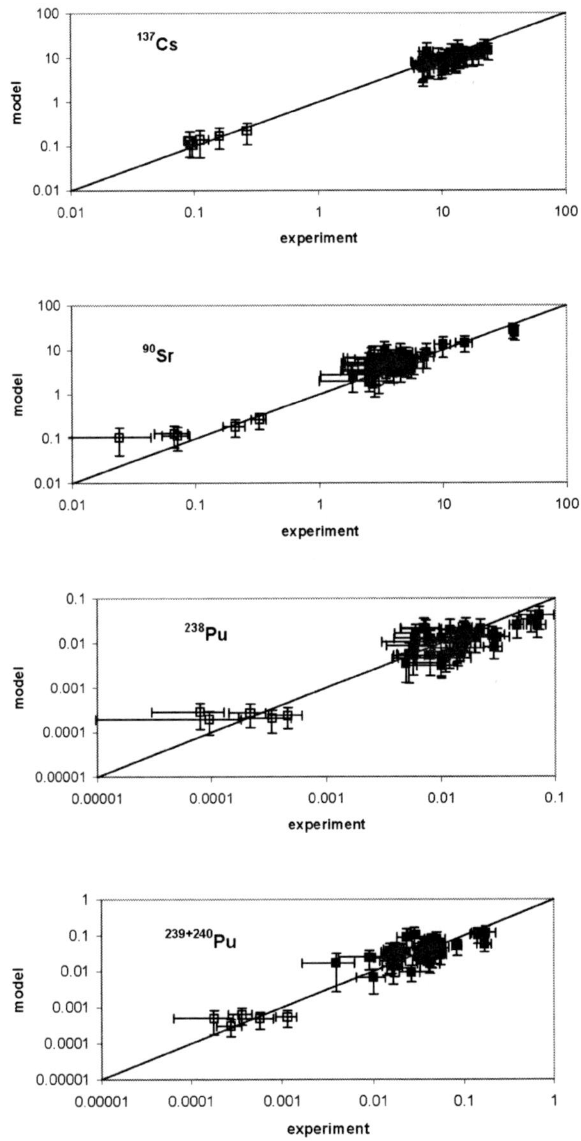

Figure 8. Model predictions vs experimental data. Forest fire (plot #3). In each graph, the left cluster of data (open points) depicts the airborne concentration values, and the right cluster (black points) depicts the fallout density data. Solid lines correspond to ideal fit (equality of modeled and empirical data in each point).

6. Conclusions

The fires lead to the significant increase of the radionuclides airborne concentrations at the close distances from the source of release.

For the grassland fires, RF of ^{137}Cs and ^{90}Sr ~ 10^{-5}–10^{-6} m^{-1}, Pu ~ 10^{-6}–10^{-7} m^{-1} relating to the fuel material contamination, and two orders of magnitude lower relating the total contamination. For the forest fire, RF ~ 10^{-6}–10^{-7} m^{-1} for the fuel material contamination, and 10^{-7}–10^{-8} m^{-1} for the total contamination;

Although the increase of the radionuclides airborne concentration, the fires do not lead to any significant redistribution of the radioactive contamination in the local scale;

However, the increase of the radionuclides airborne concentration in the surface air layer is very important, because it can provide a significant contribution to the doses to firefighters and other personnel of the Chernobyl zone. This contribution is formed by the inhalation of the radionuclides, mainly of Pu isotopes. 50-years EED from exposure in the points of the maximum airborne concentration during the fire can exceed the external irradiation dose received during the same exposure period;

Fractions of ^{137}Cs and ^{90}Sr release during the grassland fires are 0.n % of their activity in the fuel material, and can reach 3–4% during the forest fires, which is in a good accordance with the estimates of other researchers. For a first time, the corresponding values for Pu isotopes were obtained, which were up to one order of magnitude lower of the above-mentioned values for cesium and strontium;

The proposed model satisfactory describes the plume formation and transportation in the atmosphere. The model operates with the directly measured input parameters, which determines its prognostic capability and the wide application area;

The above numerical values were obtained for the limited-scale fires. One can expect that the large-scale intensive fires can lead to the formation of the high plumes, higher relative release of the radioactivity from the fuel material and bigger impact area of the fire.

Acknowledgement

Authors express their sincere gratitude to the STCU for the financial support of the researches, as well as to the Administration of the Exclusion zone, the ChornobylLis Enterprise, the Ecocentre and the firefighting divisions of the Exclusion zone for the invaluable support of the experimental works.

References

Briggs, G.A. (1972) Discussion: chimney plumes in neutral and stable surroundings, *Atmospherical Environment* **6p**: 507–510.

Bruyatsky, Ye.B. (2000) Theory of atmospheric diffusion of the radioactive releases, Institute of Hydromechanics of NAS of Ukraine, Kiev, 444 p. (in Russian).

Gusev, N.G. and Beliayev, V.A. (1991) Radioactive releases into the biosphere, book company energoatomizdat, Moscow, 256 p. (in Russian).

ICRP (1994) Publication 66. Human respiratory tract model for radiological protection, *Annals of the ICRP 24*, Pergamon Press, Oxford.

International Atomic Energy Agency (IAEA) (1980) *Atmospheric Dispersion in Nuclear Power Plant Siting: A Safety Guide*. International Atomic Energy Agency, IAEA, Safety series No 50-SG-53, Vienna.

Kashparov, V.A., Lundin, S.M., Kadygrib, A.M., Protsak, V.P., Levtchuk, S.E., Yoschenko, V.I., Kashpur, V.A., and Talerko, N.N. (2000) Forest fires in the territory contaminated as a result of the Chernobyl accident: radioactive aerosol resuspension and exposure of firefighters, *Journal of Environmental Radioactivity* **51**: 281–298.

Kashparov, V.A., Lundin, S.M., Khomtinin, Y.V., Kaminsky, S.P, Levchuk, S.E., Protsak, V.P., Kadygrib, A.M., Zvarich, S.I., Yoschenko, V.I., and Tschiersch, J. (2001) Soil contamination with ^{90}Sr in the near zone of Chernobyl, *Journal of Environmental Radioactivity* **56**: 285–298.

NCRPU (1997) Radiation Safety Norms of Ukraine, Kyiv (in Ukrainian).

Talerko N.N. (1990) Calculation of radioactive admixture ascent from Chernobyl NPP accidental unit, *Meteorology and Hydrology* **19**: 39–46.

Yoschenko, V.I., Kashparov, V.A., Protsak, V.P., Lundin, S.M., Levchuk, S.E., Kadygrib, A.M., Zvarich, S.I., Khomutinin, Yu.V., Maloshtan, I.M., Lanshin, V.P., Kovtun, M.V., and Tschiersch, J. (2006a) Resuspension and redistribution of radionuclides during grassland and forest fires in the Chernobyl exclusion zone: part I. Fire experiments, *Journal of Environmental Radioactivity* **86**(2): 143–163.

Yoschenko, V.I., Kashparov, V.A., Levchuk, S.E., Glukhovskiy, A.S., Khomutinin, Yu.V., V.A., Protsak, V.P., Lundin, S.M., and Tschiersch, J. (2006b) Resuspension and redistribution of radionuclides during grassland and forest fires in the Chernobyl exclusion zone: part II. Modeling the transport process, *Journal of Environmental Radioactivity* **87**(3): 260–278.

PART 2: ECOSYSTEM TRANSFER

THULE EXPEDITION 2003 – STUDIES ON RADIOACTIVE CONTAMINATION AND PARTICLES

SVEN P. NIELSEN[1], PER ROOS[1], HENNING DAHLGAARD[1], SVEND K. OLSEN[1], JUSSI JERNSTRÖM[1], MATS ERIKSSON[2]

[1]*Radiation Research Division, Risoe National Laboratory for Sustainable Energy, Technical University of Denmark – DTU, P.O. Box 49, 4000 Roskilde, Denmark*

[2]*IAEA Marine Environment Laboratory (IAEA-MEL), 4 Quai Antoine 1er, 98000, Monaco*

Abstract: Analyses of marine and terrestrial samples collected in August 2003 from Bylot Sound at Thule, Northwest Greenland, showed that plutonium from nuclear weapons in the American B52 plane, which crashed on the sea ice in January 1968, persists in the environment. The highest concentrations of plutonium were found in the marine sediments under the location where the plane crashed. The distribution of plutonium in the marine sediment is very inhomogeneous and associated with hot particles with activities found up to 1500 Bq ^{239,240}Pu. Sediment samples collected in Wolstenholme Fjord north of the accident site showed plutonium concentrations, which illustrate the redistribution of plutonium after the accident. Seawater and seaweed samples showed increased concentrations of plutonium in Bylot Sound. The increased concentrations are due to resuspension of plutonium-containing particles from the seabed and transport further away from the area. Plutonium concentrations in seawater, seaweed and benthic animals in Bylot Sound were low but clearly above background levels. All soil samples collected from Narssarssuk showed accident plutonium with levels above background. Plutonium was very inhomogeneously distributed and associated with particles in the surface layers. Hot particles were found in soil with activities up to 150 Bq ^{239,240}Pu. Various techniques were applied in studying and characterising hot particles, mainly those found in sediments during earlier expeditions. These techniques included electron microscopy (SEM-EDX/WDX), secondary ion mass spectrometry (SIMS), and various synchrotron radiation induced techniques including micro X-ray fluorescence spectrometry (SR-μ-XRF), micro X-ray fluorescence tomography and micro X-ray absorption near-edge structure spectrometry (SR-μ-XANES). The techniques proved successful in assessing the source of the particles and their environmental behaviour. Overall, it can be concluded that plutonium in the marine environment at Thule presents an insignificant risk to man. Most plutonium remains in the seabed under Bylot Sound far from man under relatively stable conditions and concentrations of plutonium in seawater and animals are low. However, the plutonium contamination of surface soil at Narssarssuk could constitute a small risk to humans visiting the location if radioactive particles are resuspended in the air so that they might be inhaled.

Keywords: Thule, plutonium, accident, SIMS, SR-μ-XRF, SR-μ-XANES.

1. Introduction

In January 1968 a B52 plane from the US strategic Air Command caught fire and crashed on the sea ice in Bylot Sound about 12 km west of the Thule Air Base, Northwest Greenland. The plane carried four nuclear weapons. Part of the weapons plutonium was distributed over some square kilometers of the ice in the explosive fire that followed and radioactive particles were carried by wind in the direction of Narssarssuk approximately 7 km south of the crash site. Plutonium-contaminated ice was recovered and shipped back to the U.S.A., as was the plutonium-contaminated wreck. The underlying sea sediments received a fraction of the weapons plutonium when the sea ice melted the following summer and probably also during the accident as the impact caused the ice to break up (U.S. Air Force, 1970).

Previous sample collections in 1968, 1970, 1974, 1979, 1984, 1991 (Aarkrog, 1971, 1977; Aarkrog et al., 1981, 1984, 1987, 1988, 1997) have estimated that the pollution remaining in the seabed in Bylot Sound by 1968 amounted to approximately 1.4 TBq ^{239,240}Pu (about 0.5 kg), 0.025 TBq ^{238}Pu, 4.6 TBq ^{241}Pu and 0.07 TBq ^{241}Am. Results from a sample collection in 1997 indicate, however, that previous estimates of the plutonium inventory may not have included radioactive particles in the seabed sediments in sufficient detail (Eriksson, 2002).

Single particle analysis gives additional information to the bulk analysis, especially when considering that these, also called hot particles, most probably originate from a discrete part of the fission material in the weapons. Such additional information includes isotopic compositions of uranium and plutonium of the fissile material, elemental ratios of uranium and plutonium, and heterogeneity of the fission material. In addition, it has been shown that the hot particles in Thule are very inert; radiochemical procedures applying leaching with aqua regia may not be sufficient to completely dissolve the particles (Eriksson, 2002; Moring et al., 2001; Lind et al., 2007). Therefore, radiochemical analysis based on aqua regia leaching of bulk material contains a risk of underestimating plutonium concentration in sediments. As also the particle size distribution has shown that most of the activity released from the Thule accident is bound to radioactive particles so that the dominant activity is associated to relatively few but large particles, the presence of particles should be included in the inventory calculations. The inventory of ^{239,240}Pu in the sediments taking into account the presence of hot particles was estimated in the range 5-15 TBq (Eriksson, 2002) and later in the range 1.4-6 TBq (Nielsen and Roos, 2006).

The Thule Air Base is located at 76°30'N in NW Greenland near the former main living area for the Thule culture, an indigenous Inuit society surviving mainly through hunting marine mammals and birds. In 1953 the main part of the Inuit population was forced to move about 100 km further north to Qaanaaq because of expansion of the activities at the Thule Air Base. However, they still use the area for hunting marine mammals and birds. The rich marine benthic fauna, the basis for the important walrus population was intensively studied in 1939–1941 (Vibe, 1950).

Selected data from the sampling in 1997 has been published (Dahlgaard et al., 1999a,b; Eriksson et al., 1999; McMahon et al., 2000). The present investigation was carried out during 2003–2005 and supported financially by the Danish Environmental Protection Agency, Ministry of the Environment. Further information concerning this investigation may be found in Nielsen and Roos (2006).

2. Materials and methods

2.1. SAMPLING

Sampling was carried out from vessel Adolf Jensen belonging to Pinngortitaleriffik (Nature Institute of Greenland). Sediment mapping using acoustic techniques, performed by Geological Survey of Denmark and Greenland (GEUS), was carried out in Bylot Sound in parallel with collection of seaweed and soil samples using Adolf Jensen's rubber dinghy (Figs. 1 and 2). For sampling location coordinates refer to Nielsen and Roos (2006).

Brown seaweed, *Fucus distichus*, was collected from the Thule area, six locations in Bylot Sound, four locations at Dundas, one on Wolstenholme Island and one at the Eider Islands. Sample sizes ranged from 1 to 4 kg fresh weight (fw).

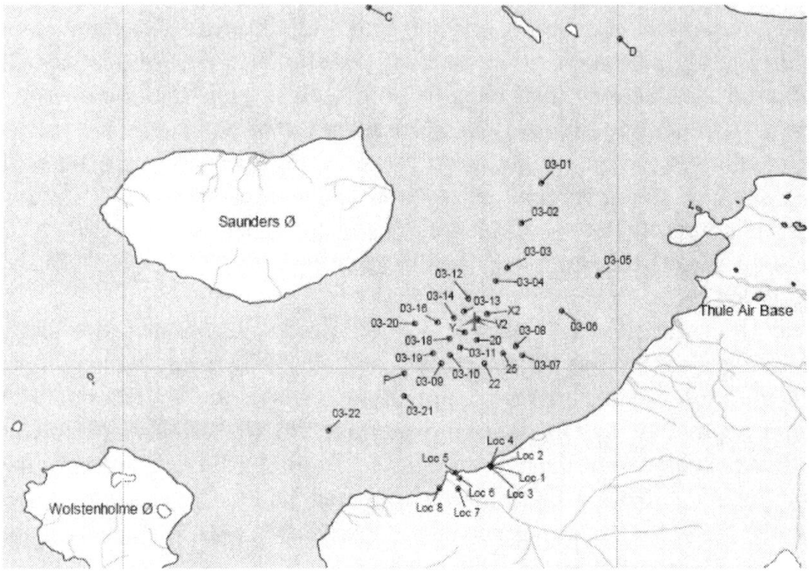

Figure 1. Sampling locations for marine sediment in Bylot Sound and Wolstenholme Fjord, and for soil at Narssarssuk.

Locations of sea water samples were chosen within the contaminated area in Bylot Sound and outside the contaminated area. A location south of Wolstenholme Island served as reference to background levels and locations north of Saunders Island served as expected low-level sites. Three samples (170–200 l) were taken at each location, a surface sample collected at 5 m below the surface and two deep samples at 15 and 35 m above the seabed. Seawater samples were passed through 0.45 μm Millipore filters to separate particles from seawater. Additional surface seawater samples were collected in Bylot Sound and at Dundas to determine the levels of the anthropogenic radionuclides ^{90}Sr, ^{137}Cs and ^{99}Tc in the Thule area and compare with levels in a sample, which was collected at Ilulissat some 1,000 km SSE of Thule.

Sediment cores were collected with a GEMAX gravity corer from 31 locations of which 10 were stations that had been sampled previously with success. All new stations were selected based on information from the sediment mapping done by GEUS. Sampling at these new stations was a great success demonstrating the usefulness of sediment mapping.

Benthic biota was collected along four 500-m transects using a Sigsbee trawl. Three of the trawl hauls were taken within a few kilometers from the accident site at 200–250 m while the fourth was taken 17 km north of the accident site at 135 m. The biota was taken from the trawl, cleaned initially and transferred to tubs where the animals were cleaned more thoroughly in seawater. Special focus was on collecting bivalves, whelks, shrimp and brittle stars; some small fish and squid were also collected. Shrimp, whelks, fish and squid were cleaned for a few hours in surface water. Bivalves and brittle stars were kept overnight in surface water to allow them to depurate. In Bylot Sound a seal was shot from which samples of flesh and liver were collected.

Low levels of plutonium contamination on land were found at Narssarssuk in 1968 (U.S. Air Force, 1970) and this area was therefore selected for collection of soil samples. Samples were collected from eight locations and at each location three soil profiles were collected down to depths of 20–30 cm within an area of one square meter. The cores were generally separated into the depth sections: 0–3, 3–6, 6–10, 10–15, 15–20, 20–25 and 25–30 cm. The resulting 145 soil samples ranged in size from a few tens of grams to about a kilogram of dry weight (dw) sieved material.

2.2. ACOUSTIC MAPPING

Sediment mapping was carried out by GEUS from Adolf Jensen and included recording of bathymetry and characterisation of the seabed. Bathymetry, sediment profiles and surface seabed mosaic were mapped by chirp and side-scan sonar. The following conclusions were drawn from the acoustic mapping, which is reported in detail by Jensen (2003):

- The bathymetry data show more details of the water depths in Bylot Sound than was known previously
- The most pronounced morphological features are a north-south going ridge in Bylot Sound with water depths less than 100 m and a southern channel partitioned in a northern and southern basin with water depths down to 250 m.
- Five characteristic seabed types reflect the sedimentation pattern in the area investigated.
- Relatively soft bottom with accumulation of sediment.
- Hard bottom with some accumulation of sediment.
- Transition zone between hard residual bottom and sedimentation bottom.
- Hard residual bottom dominated by erosion.
- Bottom with scour marks from icebergs which frequently strike the bottom and disrupt the sedimentation layers.
- The sediment accumulation areas are interesting for sampling in connection with plutonium since these areas represent sinks for fine-grained material in Bylot Sound.
- The predominant bottom type in the area near the point of impact of the B52 aircraft is a transition zone between hard residual and accumulation bottom. It is important that samples for plutonium are collected from the accumulation bottom.

- The residual bottom and the bottom with iceberg scouring marks are not interesting for sediment sampling since these bottoms do not represent recent sediments and are often disturbed.
- The screening of the side-scan data showed a nearly rectangular object of about 5 × 5 m, which apparently has a depression in the middle. The object may be a foreign body on the seabed (U.S. Air Force, 1970).

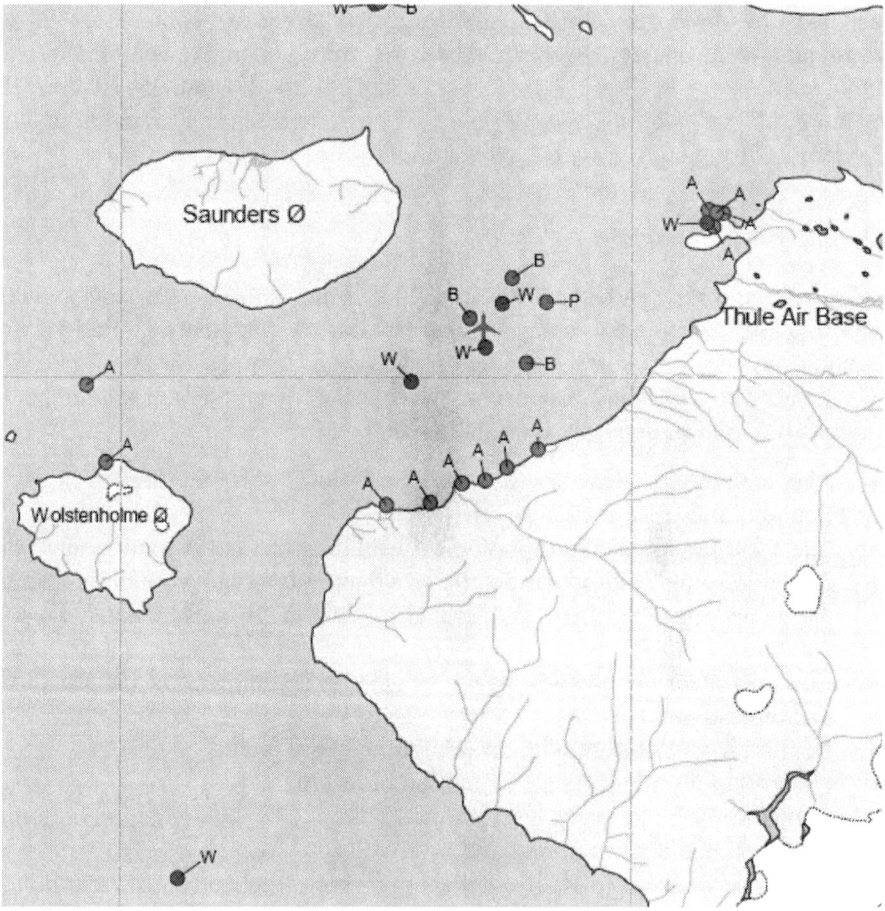

Figure 2. Sampling locations for seawater (W), seaweed (A), benthic fauna (B) and seal (P).

2.3. ANALYTICAL METHODOLOGY

2.3.1. Sample preparation

Standard preparation methods used include drying at 105°C, freeze drying and ashing at 450°C. Soil and sediment samples were sieved by removing stones larger than 1.5 mm and in addition soil samples were ashed in order to eliminate the organic content, which varied from 5% to 50%.

2.3.2. Radiochemical analyses

Gamma-emitting radionuclides (^{137}Cs and naturally occurring radionuclides) were analysed by direct measurements of samples after mass reduction. For radionuclides emitting alpha and beta particles such as plutonium isotopes, ^{241}Am, ^{90}Sr and ^{99}Tc, chemical procedures are required to separate the radionuclides in question from the sample matrix and interfering radioisotopes prior to measurement. Measurements of alpha-emitting radionuclides were made by alpha spectrometry using Si detectors and beta-emitting radionuclides were measured by low-level gas-flow Geiger–Müller counters.

The radiochemical analyses of plutonium, americium, strontium and technetium were based on procedures developed in Risø National Laboratory (Chen et al., 2001, 2002). Chemical yields were determined using radioactive tracers. Due to the low concentrations of plutonium and technetium in seawater at Thule, large samples of 175 liters were collected for determination of these radionuclides. Other sample types were collected in kilogram quantities.

2.3.3. Gamma screening

Gamma screening was carried out on selected sediment cores and all surface soil samples. Usually, analysis has previously been carried out on small aliquots selected from homogenised samples assuming that the aliquots are representative for the entire sample. The purpose of the screening was to account for all hot radioactive particles in the sample material and obtain an estimate of the amount of radioactivity associated with these particles. Plutonium isotopes do not emit gamma radiation but ^{241}Am does, and this radionuclide is a decay product of ^{241}Pu which at the time of the accident in 1968 accounted for about 2‰ by mass of the weapons plutonium. The gamma radiation from ^{241}Am is therefore an indicator of accident plutonium in sediment and soil. The screening was carried out on relatively small aliquots typically in the range of 5–15 g decreasing attenuation in the samples of the low-energy gamma radiation from ^{241}Am. The screening estimates of activity in the samples assume uniform distributions of radioactivity throughout the samples, which is clearly not the case if a sample contains one or more hot particles. Therefore, a number of sediment cores were selected for a second screening after re-packing all subsamples and thus re-distributing possible hot particles in the aliquots. Comparison of the activity estimates of the sediment cores screened twice gives an indication of the accuracy of the activity estimates associated with the screening. About 2,500 gamma spectrometric analyses were carried out covering about 43 kg dw sieved soil and 23 kg dw sieved sediment. The gamma spectrometric measurements were made using high-purity Ge detectors.

2.3.4. Sediment size distribution

In order to characterise the sediment and determine the distribution of plutonium on different grain size fractions, a number of sediment samples were sorted by wet sieving. This was done using sieves of size 0.4, 0.125 and 0.06 mm. Four different fractions were collected (>0.4, 0.125–0.4, 0.06–0.125 and <0.06 mm) and analysed for Pu isotopes. Each size fraction was analysed in five 1-g replicates.

2.3.5. Characterization of hot particles

Hot particles were separated from bulk material by sample splitting method. The method is based on the gamma-emitting radionuclide ^{241}Am present in the particles and its measurement with a real time image system. By this way the particle is localized and identified (Eriksson et al., 2002). Gamma spectrometry was applied to particles for determining ^{241}Am activity and, in some cases, for determining Pu/Am activity ratios (Eriksson et al., 2005). SEM-EDX/WDX was applied to particles in order to determine morphology, surface/sub-surface elemental composition and size. SIMS was used for isotopic ratio determinations, and synchrotron radiation (SR) based microanalytical methods μ-XRF, μ-XRF tomography and μ-XANES were applied to investigate 2D and 3D elemental distribution and oxidation states of matrix elements, respectively.

3. Results

3.1. SEDIMENTS

The radiochemical procedures and alpha-spectrometry measurements were carried out on 1-g aliquots of the samples permitting identification of ^{238}Pu, ^{239,240}Pu and ^{241}Am. Most of the profiles are based on results of the alpha-spectrometry measurements but in some cases ^{241}Am concentrations determined by gamma spectrometry have been used to estimate the plutonium profiles. The concentrations of ^{239,240}Pu varied considerably across the locations. The maximum concentration at each location ranged from 2 to 5,400 Bq kg^{-1}. At the reference site some 50 km from the contamination in Bylot Sound the concentrations of ^{239,240}Pu in the sediments were low and due to global fallout from nuclear weapons testing. At one location in Bylot Sound, 3 km from the point of impact of the B52 plane, the maximum concentration of ^{239,240}Pu in the sediment was clearly due to the presence of a hot particle in the 0–3 cm layer. The maximum concentrations across the locations occurred at varying depths in the sediments ranging from the surface layer to about 15 cm under the surface. The profiles illustrate the non-uniform distribution of plutonium with depth in the sediments and the effects of the biological mixing due to the benthic fauna.

Concentrations of ^{238}Pu were detected in 150 out of 500 of the sediment samples analysed by alpha spectrometry. The activity ratios of ^{238}Pu to ^{239,240}Pu showed clear differences between locations. The activity ratios from the sediment samples in Bylot Sound were characteristic of accident plutonium and showed little variability with an average value of 0.013 ± 0.005 (1 SD, n = 127). The average activity ratio at the reference site was 0.035 ± 0.020 (1 SD, n = 9), which is close to the typical value of the activity ratio in global fallout of 0.04 (Perkins and Thomas, 1980). The activity ratios at the locations in Wolstenholme Fjord some 17 km from the point of impact were intermediate with a mean of 0.029 ± 0.024 (1 SD, n = 14). Concentrations of ^{241}Am were detected in 37 sediment samples, and also here the activity ratios showed clear differences by location. The activity ratios of ^{241}Am to ^{239,240}Pu at the background location showed a mean value of 0.46 ± 0.13 (1 SD, n = 10) while the ratios in samples from Bylot Sound and Wolstenholme Fjord showed a mean value of 0.15 ± 0.05 (1 SD, n = 27).

The comparison of the ^{239,240}Pu depositions obtained by the different analytical methods showed considerable variation within locations thus illustrating the inhomogeneous distribution of plutonium in the sediments. In some cases hot particles were included in the 1-g aliquots analysed by chemical separation and alpha spectrometry, e.g., in the sediment core from one location where the aliquot from the 3–6 cm layer contained a ^{239,240}Pu activity of about 1 Bq. In other cases the standard gamma spectrometric analyses of typical 15-g aliquots contained hot particles, e.g., in the sediment core from a location where an aliquot from the 0–3 cm layer contained a ^{239,240}Pu activity of about 100 Bq. The systematic screening of all material from the sediment cores also in several cases revealed hot particles, which were not identified from the alpha-spectrometric analyses or the standard gamma spectrometric analyses.

3.1.1. Size fractionation

The upper 12 cm of sediments from four sediment cores were subjected to wet sieving in order to evaluate the activity distribution of plutonium among the different size fractions. Although the activity concentration does not change much over the various size classes the major part of the sediment mass is generally found in the finer fractions which therefore hold most of the Pu activity seen here. The distribution of plutonium among the sediment size classes is of importance for the resuspension of bottom sediments and their transport away from the area. Analysis of plutonium isotope signature in the suspended matter close to the bottom in Bylot Sound indicates that the elevated levels are due to resuspended sediments containing accident plutonium.

3.2. SEAWATER

The concentrations of ^{239,240}Pu in seawater at the reference station south of Wolstenholme Island were 10 mBq m^{-3} and below, and only a small fraction, 6% or less, was associated with suspended particles in the water. The concentrations in Bylot Sound were significantly higher in near-bottom waters, up to 40 mBq m^{-3} and a significant fraction of the activity, 46–78%, was associated with suspended particles. However, the samples of surface water in Bylot Sound showed low concentrations corresponding to those at the reference station and with a small fraction, 8% or lower, of the activity on suspended particles. The variability of the activity ratios of ^{238}Pu to ^{239,240}Pu on suspended particles collected on the filters was relatively small with an average value of 0.014 ± 0.001 (1 SE, n = 5), which is similar to that of accident plutonium. The variability of the isotopic ratios of ^{241}Am to ^{239,240}Pu in seawater is larger with an average value of 0.13 ± 0.02 (1 SE, n = 18). The samples from a station in Wolstenholme Fjord north of Saunders Island were influenced by accident plutonium from Bylot Sound throughout the water column with concentrations higher than at the reference site and showing higher fractions, 28–74%, of plutonium on suspended particles.

The seawater data illustrated in a consistent manner that plutonium is remobilised from the seabed in Bylot Sound on suspended particles and transported to the north out of Wolstenholme Fjord into Baffin Bay.

The concentrations of ^{90}Sr, ^{137}Cs and ^{99}Tc in seawater were low at Thule and Ilulissat and compared well with levels in samples collected in 2004 along the west coast of Greenland. The radionuclides ^{90}Sr and ^{137}Cs are well mixed in North Atlantic seawater and show little variability in contrast to ^{99}Tc, which shows a marked gradient due to the transport with sea currents from Europe.

3.3. SEAWEED

Seaweed samples were analysed for gamma-emitting radionuclides, ^{99}Tc, plutonium and americium. The concentrations of ^{239,240}Pu in seaweed at Narssarssuk in Bylot Sound (0.19–0.44 Bq kg^{-1}) were systematically higher than at locations outside Bylot Sound (0.10–0.16 Bq kg^{-1}). The plutonium levels in seaweed from Bylot Sound indicated that accident plutonium is remobilised from the contaminated sediments and transported to surface waters in the sound. This is supported by the fact that activity ratios of ^{238}Pu to ^{239,240}Pu in seaweed samples from Bylot Sound reflected accident plutonium of 0.01–0.02 while the activity ratios in samples from outside the sound of 0.02–0.03 were closer to the typical value of the activity ratio in global fallout of 0.04 (Perkins and Thomas, 1980).

The concentrations of ^{241}Am in seaweed followed the same pattern as for ^{239,240}Pu with higher concentrations at locations in Bylot Sound than outside the sound. The activity ratios of ^{241}Am to ^{239,240}Pu in seaweed showed no significant variation by location and had a mean value of 0.17 ± 0.01 (1 SE, n = 12).

The concentrations of ^{99}Tc and ^{137}Cs in seaweed showed no significant variation by location over the area. The mean concentration of ^{99}Tc in seaweed was 5.0 ± 0.1 Bq kg^{-1} dw (1 SE, n = 12) and the mean concentration of ^{137}Cs was 0.39 ± 0.02 Bq kg^{-1} dw (1 SE, n = 12).

3.4. BENTHIC BIOTA

The biota from each haul was sorted by species resulting in composite samples including from a single to over 300 individuals and sample amounts in the range 0.02–0.8 kg fw. Bivalves comprised the following species: *Macoma calcarea, Clinocardium ciliatum, Nuculana minuta, Astarte borealis, Serripes groenlandicus, Musculus niger*. Brittle stars were *Ophiura albida*, shrimp were *Crangonidae* and *Pandalidae*, and whelks were *Buccinum finmarchianum*.

The radionuclide concentrations were generally higher in bivalves than in whelks and shrimps. Geometrical mean concentrations of ^{239,240}Pu in bivalves, whelks and shrimps were between 98–760, 73–280 and 4–44 mBq kg^{-1} fw, respectively. Mean concentrations of ^{238}Pu varied between 1.5–10, 0.7–4.6 and 0.3–2.7 mBq kg^{-1} fw, respectively, and concentrations of ^{241}Am were found between 45–200, 21–150 and 2–8 mBq kg^{-1} fw, respectively.

The isotopic ratios of concentrations of ^{238}Pu to ^{239,240}Pu in bivalves did not vary across the hauls and showed a mean value of 0.016 ± 0.003 (1 SE, n = 18). The corresponding ratios for shrimp and whelks were 0.034 ± 0.005 (1 SE, n = 5) and 0.012 ± 0.002 (1 SE, n = 4), respectively. The lower ratios of 0.01–0.02 are typical for the accident plutonium while the higher ratio is typical for global fallout. This indicates that shrimp feed is not of local origin in contrast to feed for bivalves and whelks.

The variation of the isotopic ratios of concentrations of ^{241}Am to ^{239,240}Pu in biota showed no distinct pattern across species or locations. The ratios varied from 0.13 to 0.79 with a mean value of 0.40 ± 0.04 (1 SE, n = 31).

3.5. SEAL

Samples of a harp seal (Phoca groenlandica) which was shot in Bylot Sound were collected of flesh (3.3 kg fw) and liver (1.9 kg fw). Analyses were carried out of plutonium, americium, ^{137}Cs and the naturally occurring radionuclides ^{210}Po and ^{210}Pb. The concentrations of plutonium and americium were very low and detectable only for ^{239,240}Pu in liver. Concentrations of ^{238}Pu were below the limit of detection of 0.1 mBq kg^{-1} fw. The concentrations of ^{137}Cs in seal of 0.2–0.3 Bq kg^{-1} fw correspond to concentrations in fish, which have been found at 0.1–0.2 Bq kg^{-1} fw in shorthorn sculpin collected in the area. The concentrations of ^{210}Po and ^{210}Pb were typical for seals and reflect the fact that harp seals feed primarily on fish and crustaceans.

3.6. SOIL

The concentrations of ^{238}Pu, ^{239,240}Pu and ^{241}Am varied considerably across the locations and sub sites. The maximum concentration at each location ranged from 2 to 80 Bq kg^{-1} dw. In all cases the maximum concentrations occurred in the surface layers of 0–3 cm or 3–6 cm. A hot particle containing 150 Bq of ^{239,240}Pu was identified in one sample at depth 3–6 cm.

The concentrations of ^{238}Pu in the soil samples relative to ^{239,240}Pu had a mean value of 0.018 ± 0.001 (1 SE, n = 37) and showed no significant variation between locations. The ^{238}Pu/^{239,240}Pu activity ratios were all lower than the value of 0.04 typical for global fallout (Perkins and Thomas, 1980) indicating that the soil was contaminated with accident plutonium at all locations. The concentrations of ^{241}Am in the soil relative to ^{239,240}Pu had a mean value of 0.27 ± 0.02 (1 SE, n = 27) and showed no significant variation between locations. The ^{241}Am/^{239,240}Pu activity ratios were intermediate between values of 0.1–0.2 characterising accident plutonium and 0.4 characterising plutonium from global fallout (Michel et al., 2002). The ^{241}Am/^{239,240}Pu activity ratios thus also indicated that the soil sampled at Narssarssuk was contaminated with accident plutonium at all locations.

The variability of the ^{137}Cs concentrations across the locations and sub sites was less pronounced than that of the plutonium concentrations. The ^{137}Cs concentrations ranged from a low value of 14 Bq m^{-2} to a high value of 6,400 Bq m^{-2} with an average value of 1,040 ± 280 Bq m^{-2} (1 SE, n = 24). The source of the ^{137}Cs in the environment at Thule is atmospheric fallout from nuclear weapons testing, which was estimated by Hanson (1980) at a value of 777 Bq m^{-2} in 1974. Radioactive decay reduces this concentration to 400 Bq m^{-2} in 2003, which compares quite well with the concentrations found at most of the sites.

3.7. HOT PARTICLES

The analyses of the sediment samples have shown that plutonium is very inhomogeneously distributed in the sediments due to the presence of hot particles. The systematic gamma spectrometry screening of ^{241}Am in 5–15 g aliquots of the sediment samples have indicated a geometric mean ^{239,240}Pu activity of a few Bq with an approximate 95% confidence interval of from 0.1 to 50 Bq. The maximum ^{239,240}Pu activity determined in a hot particle in the sediments was 1,500 Bq.

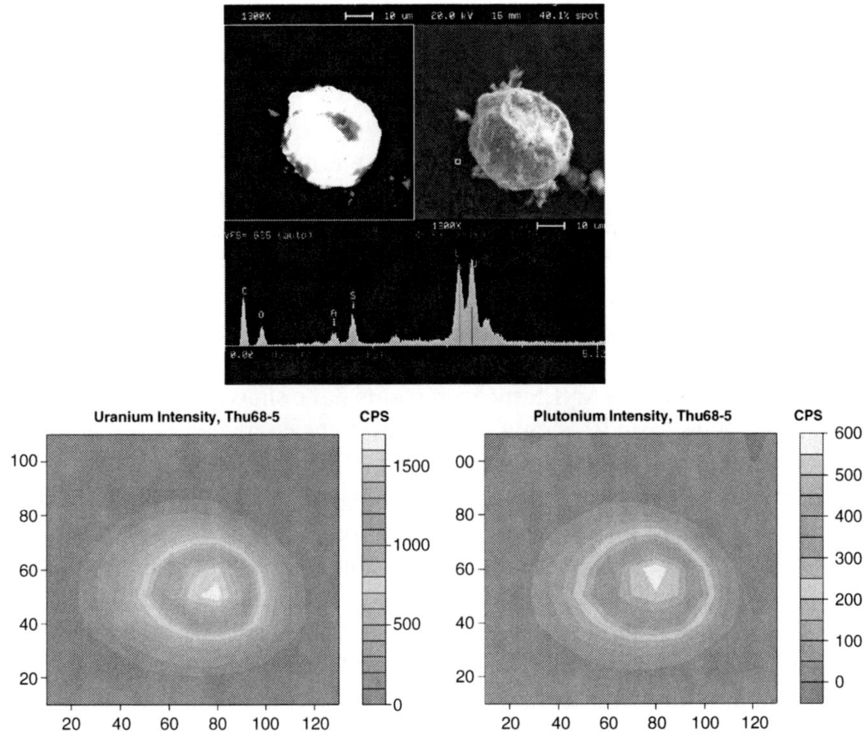

Figure 3. A hot particle from Thule sediments visualized by SEM-EDX (upper image).The lower images present intensity distribution of U and Pu in the same particle mapped with SR-μ-XRF (Eriksson et al., 2005).

A number of hot particles were identified in the soil samples. One of these samples, collected at depth 3–6 cm, had a mass of 54 g dw for which the gamma spectrometric analysis yielded an estimated concentration of 100 Bq kg^{-1} of ^{241}Am corresponding to about 5 Bq of ^{241}Am in the sample. Further investigations showed, however, that the activity was due mainly to a single hot particle, which was isolated by successive subdivisions of the sample material combined with gamma spectrometric measurements. This particle was subject to leaching experiments to identify chemical procedures that would completely dissolve the plutonium and americium in the particle and to determine the isotopic ratios. The particle was analysed by chemical separation and alpha spectrometry and found to contain 3.6 Bq ^{241}Am and 23 Bq ^{239,240}Pu with the activity ratios of 0.015 for ^{238}Pu/^{239,240}Pu and 0.158 for ^{241}Am/^{239,240}Pu. The latter value is in good agreement with activity ratios of ^{241}Am/^{239,240}Pu from five hot particles from Thule sediment samples showing a mean value of 0.169 ± 0.002 (1 SE, n = 5). These hot particles were from sediment samples collected from earlier expeditions and had $^{239,240}Pu$ activities ranging from 6 to 84 Bq (Eriksson, 2002).

3.7.1. Characterisation of hot particles

An extensive study made by using SEM-EDX/WDX has revealed mainly three different morphological structures of the marine hot particles from Thule (Ranebo et al., 2007). One group of the particles has a "spongy" matrix, one other has a "popcorn" structure, and the third group consists of particles where uranium and plutonium is present as inclusions.

It has been revealed that the particles have rather low density, about half of the density of solid UO_2/PuO_2 matrix. In addition, based on the tomography studies the 3D elemental distribution of U and Pu in some of the particles is heterogeneous. The μ-XANES measurements have shown that the Pu oxidation state in the Thule particles shows larger variations compared to uranium, where the most dominant oxidation state is +4 (Eriksson et al., 2005). ICP-MS measurements (Eriksson et al., 2007; Dahlgaard et al., 1999b) on dissolved particles in order to identify the isotopic ratios of U and Pu have revealed that the isotopic composition of uranium and plutonium show variations ranging from 0.01–0.07 for $^{240}Pu/^{239}Pu$ (atom ratio) and 0.96–1.44 for $^{235}U/^{238}U$ (atom ratio). These results are in good agreement with the isotopic ratios obtained by SIMS on single hot particles. The SIMS results indicated that there exist up to four different $^{240}Pu/^{239}Pu$ isotopic groups in the fission material of the Thule debris. Using SIMS in analyzing U isotopic compositions has been noticed to be advantageous as it reduces the risk of cross-contamination from the surrounding environment (laboratory equipment, etc.), other particles, and reagents used during radiochemical processes. The low $^{235}U/^{238}U$ ratios measured with ICP-MS (Moring et al., 2001; Lind et al., 2007) may be as a result of the above mentioned reasons.

4. Conclusions

Analyses of marine and terrestrial samples collected in August 2003 from Bylot Sound have shown that plutonium from nuclear weapons in the American B52 plane, which in January 1968 crashed on the sea ice and caught fire, persists in the environment at Thule.

Successful sampling of marine sediments was accomplished by selecting sampling locations based on information obtained from acoustic mapping of the seabed. The highest concentrations of plutonium were found in the marine sediments under the location where the plane crashed. The distribution of plutonium in the marine sediment is very inhomogeneous and associated with hot particles with activities found up to 1,500 Bq $^{239,240}Pu$. Sediment samples collected in Wolstenholme Fjord north of the accident site show plutonium concentrations more than an order of magnitude above background levels, which illustrates the redistribution of plutonium after the accident.

Seawater samples showed increased concentrations of particle-associated plutonium in near-bottom water in Bylot Sound and seaweed samples showed increased concentrations in Bylot Sound compared to locations outside Bylot Sound. These increased concentrations are due to resuspension of plutonium-containing particles from the seabed and transport further away from the area. The continuous mixing of the sediments by the benthic fauna has the effect that plutonium concentrations in the surface sediment layers are high in general and not buried under uncontaminated sediment.

Plutonium concentrations in seawater, seaweed and benthic animals in Bylot Sound were low but clearly above background levels.

All soil samples collected from Narssarssuk showed accident plutonium with levels above background. Plutonium was very inhomogeneously distributed and associated with particles in the surface layers. Hot particles were found in soil with activities up to 150 Bq ^{239,240}Pu.

Different analytical and microanalytical methods have been successfully applied to study radioactive (hot) particles. The various kinds of information related to particle structure, elemental composition and distribution, activity, isotopic ratios, and oxidation state of matrix elements make it possible to assess the origin of a particle, and its behaviour in the environment.

Plutonium in the marine environment at Thule presents an insignificant risk to man. Most plutonium remains in the seabed under Bylot Sound far from man under relatively stable conditions. Concentrations of plutonium in seawater and animals are low and can not present any risks to man even by consumption of shellfish at the highest concentrations. However, the plutonium contamination of surface soil at Narssarssuk could constitute a small risk to humans visiting the location if radioactive particles are resuspended in the air so that they might be inhaled.

Acknowledgements

The authors wish to thank the staff of the Radioecology Programme in the Radiation Research Division for their conscientious work on samples and analyses. The investigation vessel Adolf Jensen belonging to Pinngortitaleriffik collected the marine samples at Thule, and GEUS carried out the acoustic mapping of sediments in Bylot Sound. The authors gratefully acknowledge the financial support from the Ministry of Environment of Denmark to this investigation. The conclusions in this report do not necessarily reflect the views of the Ministry. The IAEA is grateful for the support provided to its Marine Environment Laboratory by the Government of the Principality of Monaco.

References

Aarkrog, A. (1971) Radioecological investigation of plutonium in an arctic marine environment, *Health Phys.* **20**:31–47.

Aarkrog, A. (1977) Environmental behavior of plutonium accidentally released at Thule, Greenland, *Health Phys.* **32**:271–284.

Aarkrog, A., Buch, E., Chen, Q.J., Christensen, G.C., Dahlgaard, H., Hansen, H., Holm, E., and Nielsen, S.P. (1988) Environmental Radioactivity in the North Atlantic Region Including the Faroe Islands and Greenland. 1986, Risø-R-550, Risø National Laboratory, Roskilde, Denmark.

Aarkrog, A., Chen, Q.J., Clausen, J., Christensen, G.C., Dahlgaard, H., Ellis, K., Hansen, H., Holm, E., Joensen, H.P., Nielsen, S.P., and Strandberg, M. (1997) Environmental Radioactivity in the North Atlantic Region Including the Faroe Islands and Greenland. 1992 and 1993, Risø-R-757, Risø National Laboratory, Roskilde, Denmark.

Aarkrog, A., Boelskifte, S., Dahlgaard, H., Duniec, S., Holm, E., and Smith, J.N. (1987) Studies of transuranics in an arctic marine environment, *J. Radioanal. Nucl. Chem. Art.* **115**:39–50.

Aarkrog, A., Dahlgaard, H., and Nilsson, K. (1984) Further studies of Plutonium and Americium at Thule, Greenland, *Health Phys.* **46**:29–44.

Aarkrog, A., Dahlgaard, H., Holm, E., Hansen, H., Lippert, J., and Nilsson, K. (1981) Environmental Radioactivity in Greenland in 1980, Risø-R-449, Risø National Laboratory, Roskilde, Denmark.

Chen, Q.J., Aarkrog, A., Nielsen, S.P., Dahlgaard, H., Lind, B., Kolstad, A.K., and Yu, Yixuan (2001) Procedures for Determination of $^{239,240}Pu$, ^{241}Am, ^{237}Np, $^{234,238}U$, $^{228,230,232}Th$, ^{99}Tc and ^{210}Pb-^{210}Po in Environmental Materials, Risø-R-1263(EN), Risø National Laboratory, Roskilde, Denmark.

Chen, Q.J., Hou, X.L., Yu, Y.X., Dahlgaard, H., and Nielsen, S.P. (2002) Separation of Sr from Ca, Ba and Ra by means of $Ca(OH)_2$ and $Ba(Ra)C_{12}$ or $Ba(Ra)SO_4$ for the determination of radiostrontium, *Anal. Chim. Acta* **466**:109–116.

Dahlgaard, H., Chen, Q.J., Stiirup, S., Eriksson, M., Nielsen, S.P., and Aarkrog, A. (1999a) Plutonium Isotope Ratios in Environmental Samples from Thule (Greenland) and the Techa River (Russia) Measured by ICP-MS and a-Spectrometry, IAEA-SM-3541109, Vienna, Austria, pp. 254–259.

Dahlgaard, H., Eriksson, M., Ilus, E., Ryan, T., McMahon, C.A., and Nielsen, S.P. (1999b) Plutonium in an Arctic Marine Environment 29 years After the Thule Accident, Nordic Society for Radiation Protection, 12th ordinary meeting, Skagen, Denmark.

Eriksson, M. (2002) On Weapons Plutonium in the Arctic Environment (Thule, Greenland), Risø-R-1321(EN), Risø National Laboratory, Roskilde, Denmark

Eriksson, M., Dahlgaard, H., Ilus, E., Ryan, T., Chen, Q.J., Holm, E., and Nielsen, S.P. (1999) Plutonium in the marine environment of Thule Air Base, N.W. Greenland. Inventories and distribution in sediments 29 years after the accident, in: *Extended Abstracts. 4. International Conference on Environmental Radioactivity in the Arctic*, P. Strand, T. Jølle, eds., Norwegian Radiation Protection Authority, Østerås, pp. 60–62.

Eriksson, M., Lindahl, P., Roos, P., Dahlgaard, H., and Holm, E. (2007) The U, Pu and Am nuclear signatures of the Thule hydrogen bomb debris, submitted to *Environ. Sci. Tech.*

Eriksson, M., Ljunggren, K., and Hindorf, C. (2002) Plutonium hot particle separation techniques using real-time digital image systems, *Nucl. Instrum. Meth. A* **488**:375–380.

Eriksson, M., Osan, J., Jernström, J., Wegrzynek, D., Simon, R., Chinea-Cano, E., Markowicz, A., Bamford, S., Tamborini, G., Török, S., Falkenberg, G., Alsecz, A., Dahlgaard, H., Wobrauschek, P., Streli, C., Zoeger, N., and Betti, M. (2005) Source term identification of environmental radioactive Pu/U particles by their characterization with non-destructive spectrochemical analytical techniques, *Spectrochim. Acta Part B* **60**:455–469.

Hanson, W.C. (1980) Transuranic elements in arctic tundra ecosystems, in: *Transuranic Elements in the Environment*, W. C. Hanson, ed., DOE/TIC-22800, Springfield, VA.

Jensen, J.B. (2003) Akustisk Kortlægning af Havbundssedimenterne i Bylot Sund, Thule. GEUS Rapport 2003/103, Danmarks og Grønlands Geologiske Undersøgelse, Miljøministeriet.

Lind, O.C., Salbu, B., Janssens, K., Proost, K., García-León, M., and García-Tenorio, R. (2007) Characterization of U/Pu particles originating from the nuclear weapon accidents at Palomares, Spain, 1966 and Thule, Greenland, 1968, *Sci. Total Environ.* **376**:294–305.

McMahon, C.A., Leon Vintro, L., Mitchell, P.I., and Dahlgaard, H. (2000) Oxidation-state distribution of plutonium in surface and subsurface waters at Thule, northwest Greenland, *Appl. Radiat. Isot.* **52**:697–703.

Michel, H., Barci-Funel, G., Dalmasso, J., Ardisson, G., Appelby, P.G., Haworth, E., and El-Daoushy, F. (2002) Plutonium and americium inventories in atmospheric fallout and sediment cores from Blelham tarn, Cumbria (UK), *J. Environ. Radioact.* **59**:127–137.

Moring, M., Ikäheimonen, T.K., Pöllänen, R., Ilus, E., Klemola, S., Juhanoja, J., and Eriksson, M. (2001) Uranium and plutonium containing particles in a sea sediment sample from Thule, Greenland, *J. Radioanal. Nucl. Chem.* **248**:623–627.

Nielsen, S.P. and Roos, P. (2006) Thule-2003 – Investigation of Radioactive Contamination, Risø-R-1549(EN), Risø National Laboratory, Roskilde, Denmark.

Perkins, R.W. and Thomas, C.W. (1980) World wide fallout, in: *Transuranic Elements in the Environment*, W. C. Hanson, ed., DOE/TIC-22800, Springfield, VA.

Ranebo, Y., Eriksson, M., Tamborini, G., Niagolova, N., Bildstein, O., and Betti, M. (2007) The use of SIMS and SEM for the characterization of individual particles with a matrix originating from a nuclear weapon, *Microsc. Microanal.* **13**:179–190.

U.S. Air Force (1970) Project Crested Ice, *USAF Nucl. Safety* **65**:1–97.

Vibe, C. (1950) The marine mammals and the marine fauna in the Thule district (Northwest Greenland) with observations on ice conditions in 1939-41, *Meddelelser om Grønland* **150**: 1–117.

HOT PARTICLES IN THE FLOODPLAIN OF THE YENISEI RIVER

ALEXANDER BOLSUNOVSKY
Institute of Biophysics, Siberian Branch of the Russian Academy of Sciences, Akademgorodok, Krasnoyarsk 660036, Russia
phone: ??????? e-mail: radecol@ibp.ru

Abstract: The purpose of this study was to investigate radionuclide composition of the hot particles found in the floodplain of the Yenisei River and to discuss their possible sources. Since 1995, researchers of the Institute of Biophysics have found a considerable number of hot particles that contain up to 29,000 kBq/particle of ^{137}Cs in the Yenisei River floodplain. Investigations of the particles in Krasnoyarsk, Moscow, and Novosibirsk have confirmed their reactor origin and made it possible to roughly estimate their age. Based on comparative analysis of ^{137}Cs/^{134}Cs ratios, all the particles can be divided into two or three major groups, suggesting that over the 50-year period of the MCC operation, there have been two or three emergency situations at the MCC reactors, with nuclear fuel microparticles released into the Yenisei. Microparticles containing up to 36 Bq/particle of ^{241}Am were detected in soil and sediment samples collected at the Yenisei River. In some of these particles the ^{137}Cs/^{241}Am ratio is high but in the others – low. It remains unknown how these particles have been formed. However, these microparticles occur in floodplain soils and sediments very frequently and, thus, can be considered to be the main form in which radionuclides are present in the Yenisei River floodplain. The presence of a large number of hot particles of different origin in the floodplain of the Yenisei River from the MCC production area down to the town of Yeniseisk (and, probably, farther downstream) is a source of potential health hazard to people.

Keywords: hot particles, Yenisei River, floodplain soils and sediments, plutonium complex, cesium and plutonium isotopes, ^{241}Am, dating of particles.

1. Introduction

The Yenisei is one of the largest rivers in the world. The Mining-and-Chemical Combine (MCC) of Rosatom is situated on the bank of the Yenisei River, 60 km downstream of the city of Krasnoyarsk (Fig. 1). The Combine, which produces weapons-grade plutonium, consists of a reactor plant and a radiochemical plant. Two reactors that used the Yenisei water as coolant for the reactor core have been shut down. The third reactor is still working, using the Yenisei water as coolant for some reactor parts. The Rosatom officials have repeatedly declared that, unlike the other similar plants (the "Mayak" Production Association in the Chelyabinsk Province and the Siberian Chemical Combine in the Tomsk Province), the MCC in the Krasnoyarsk Territory has been accident-free. In 1994, Timofeev (1995), a geologist, for the first time found hot particles in soil layers of a radioactive anomaly on the riverside of the town of Yeniseisk, 330 km downstream of the MCC. The γ-spectrometric analysis of those particles showed the presence of a large number of radionuclides, both short-lived (^{51}Cr, ^{54}Mn, ^{59}Fe, Ru and Ce isotopes, etc.) and long-lived ones, such as ^{137}Cs and Eu isotopes (Timofeev, 1995). Investigations conducted by Timofeev allowed him to assume that there could be only one source of those particles – the Mining-and-Chemical Combine at Zheleznogorsk. The MCC officials and specialists, though, rejected this assumption on the grounds that, first, Timofeev found the particles in the Yenisei flood plain at a considerable distance from the MCC, and, second, the radionuclide analysis of the particles was not complete. Indeed, at that time the particles did not undergo complete radionuclide analysis, and their reactor origin was not proven.

In August 1995, during the joint expedition of the Institute of Biophysics SB RAS and the MCC to the Yenisei, the author of this paper found several hot particles (Bolsunovsky et al., 1998; Bolsunovsky and Tcherkezian, 2001). These particles were found on islands of the Yenisei River near the village of Atamanovo, in the immediate vicinity of the MCC and the place where the Combine releases the radionuclide-contaminated water (Fig. 1). Later, in 1997–2007, during expeditions to the Yenisei River the author and other researchers detected new hot particles every year (Sukhorukov et al., 2004).

In recent years, a number of researchers, trying to find out the reasons of inhomogeneous distribution of radionuclides in floodplain soils and sediments of the Yenisei River, detected hot particles of smaller sizes and, hence, lower (several orders of magnitude) activities (Gritchenko et al., 2001; Sukhorukov et al., 2004). They were termed "secondary" hot particles (Gritchenko et al., 2001).

The purpose of this study was to investigate radionuclide composition of the hot particles found in the floodplain of the Yenisei River and to discuss their possible sources.

2. Materials and methods

In 1995 hot particles were detected on islands in the Yenisei River near the village of Atamanovo. Two particles (denoted as 1542 and 1543) were found in the soil layer at 20–25 cm below the soil surface and the particle denoted as

1544 was found at a depth of 5–10 cm. During 1997–2007 more hot particles were found in the islands of the Yenisei River near the village of Atamanovo and at the riverside near the village of B. Balchug, which is situated downstream of the MCC production area. Some of the particles were found in the top soil layer, the others – in the 10–15 cm layer (Tables 1 and 2).

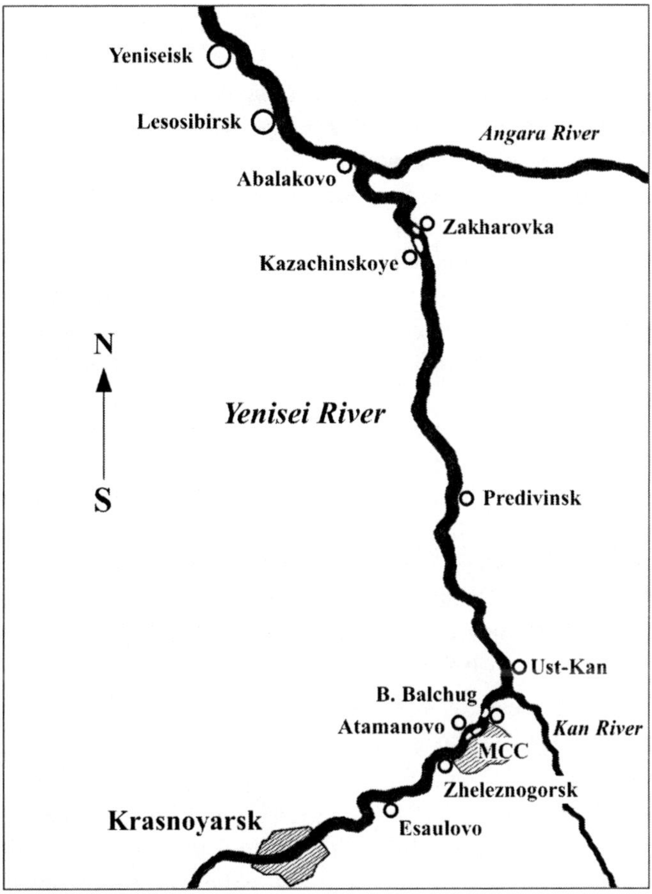

Figure 1. Diagrammatic map of the south of the Krasnoyarsk Territory (Russia), showing settlements near which the particles were found. Scale 1 : 2,000,000.

The hot particles were investigated in the Institute of Biophysics SB RAS (IBP) (Krasnoyarsk), the Analytical Center at the Institute of Geology and Mineralogy SB RAS (IGM) (Novosibirsk), and the Radiochemistry Laboratory at the Vernadsky Institute of Geochemistry and Analytical Chemistry RAS (IGAC) (Moscow). Specialists of the IGAC are experienced in the investigation of hot particles, including those from Chernobyl. The hot particles were cleaned in isopropanol to remove soil adhering to them (Tcherkezian et al., 1994). Some of the separated and washed particles were examined under a microscope. They were black or dark-brown, non-spherical, of linear dimension up to 2 mm, and up to 1 mg in mass. Outwardly they resembled the fuel hot particles found at the Chernobyl NPP after the accident.

At the IBP, γ-spectrometric analysis of the particles was carried out on a "Canberra" (USA) γ-spectrometer coupled to a hyperpure germanium (HPGe) detector. Spectra were analyzed using GENIE PC software ("Canberra"). At the IGAC, the γ-spectra of the hot particles were measured using spectrometers with DGDK-80 and DGDK-100 Ge(Li)-detectors as well as an "Ortec" detector (HPGe) linked to an MCA 8100 analyzer. Spectra were processed using "ASPRO" software. To determine isotopes of plutonium and americium as well as strontium, the particles were dissolved in boiling 8 M HNO_3 + 0.2 M $KBrO_3$. The radiochemical methods for concentration and refinement of the elements are described in detail elsewhere (Tcherkezian et al., 1994; Bolsunovsky, Tcherkezian, 2001). Plutonium and americium were determined using "Canberra" α-spectrometers and an ionization chamber with a grid and an LP4050 analyzer. Strontium activity was measured with a "Tesla" low-background flow-proportional counter.

Gamma-spectrometric measurements showed abnormally high concentrations of such radionuclides as ^{137}Cs or ^{241}Am in some soil and sediment samples collected from the floodplain of the Yenisei River near the villages of Atamanovo and B. Balchug. Having split the samples into sub-samples of increasingly smaller size, we found that radioactivity was almost entirely concentrated in one of the small portions of the sample. That was indicative of the presence of microparticles; being too small to be extracted, they were investigated together with the surrounding soils.

3. Results and discussion

Results of the analysis of the high-activity particles are listed in Table 1 and demonstrate that the radionuclide composition of the particles is essentially the same: cesium, plutonium, and strontium isotopes. Some particles contained traces of americium. It is noteworthy that the ^{137}Cs activity of the particles was very high – up to 29,200 kBq/particle. The hot particles were all found in the Yenisei River floodplain, the fact suggesting that they had entered the environment by water, during accidents at the MCC plutonium complex in the Krasnoyarsk Territory. The presence of plutonium isotopes (^{238}Pu and ^{239,240}Pu) (Table 1) is indicative of the reactor origin of these particles. Hot particles were detected not only on the islands of the Yenisei River near the village of Atamanovo but also on the river bank, near the village of B. Balchug, situated

TABLE 1. Radionuclide composition of the hot particles of the Yenisei River floodplain

Particle label	Sampling position and date	^{137}Cs kBq/particle	^{137}Cs/^{134}Cs	^{239,240}Pu Bq/particle	^{238}Pu Bq/particle	^{90}Sr Bq/particle
1542	Atamanovskii Island, August 1995	29,200 ± 900	4,000 ± 300	n.d.	n.d.	n.d.
1543	Atamanovskii Island, August 1995	1,880 ± 60	4,000 ± 500	0.01 ± 0.005	0.03 ± 0.01	840 ± 10
1544	Atamanovskaya Spit, August 1995	6,850 ± 200	4,000 ± 400	0.05 ± 0.01	0.09 ± 0.02	1,350 ± 30
HP8/1	Atamanovskaya Spit, September 1997	182 ± 4 198 ± 6	1,300 ± 100 700 ± 100	0.80 ± 0.06	0.50 ± 0.06	475 ± 5
HP8/2	Atamanovskaya Spit, September 1997	n.d. 778 ± 24	n.d. 900 ± 100	n.d.	n.d.	n.d.
HP4	Atamanovskaya Spit, September 1998	489 ± 8 516 ± 16	2,400 ± 300 2,600 ± 200	0.78 ± 0.07	0.09 ± 0.01	34 ± 1
HP13	Atamanovskaya Spit, September 2002	n.d. 754 ± 30	n.d. 420 ± 50	n.d.	n.d.	n.d.
HP5	Atamanovskii Island, September 1998	n.d. 1,860 ± 56	n.d. 2,600 ± 200	n.d.	n.d.	n.d.
HP6	Atamanovskii Island, September 1998	n.d. 18,000 ± 550	n.d. 4,500 ± 400	n.d.	n.d.	n.d.
HP9	Village of B. Balchug, September 1997	n.d. 249 ± 8	n.d. 1,000 ± 100	n.d.	n.d.	n.d.
HP7	Village of B. Balchug, September 1998	744 ± 2 700 ± 21	41,000 ± 12,000 >10,000	3.80 ± 0.14	2.0 ± 0.1	54 ± 1
HP10	Village of B. Balchug, July 1999	n.d. 27 ± 1	n.d. 3,400 ± 600	n.d.	n.d.	n.d.
HP11	Village of B. Balchug, September 2000	n.d 2,050 ± 60	n.d. 2,800 ± 300	n.d.	n.d.	n.d.
	Town of Yeniseisk*, 1994	159 ± 16*	600 ± 100*			

Measurements of the IGAC RAS (underlined values); Measurements of the IBP SB RAS;
n.d. – not determined. All activities are decay-corrected to sampling date. * – Timofeev's (1995) data.

TABLE 2. Radionuclide composition of hot particles investigated at the Institute of Geology and Mineralogy SB RAS (Novosibirsk) (Sukhorukov et al., 2004)

Particle label	Sampling position and date	Activity, Bq/particle								$^{137}Cs/^{134}Cs$	$^{238}Pu/^{239,240}Pu$
		^{137}Cs	^{134}Cs	^{154}Eu	^{155}Eu	^{60}Co	$^{239,240}Pu$	^{238}Pu	^{241}Am		
10A1	Taivan Island, August 1998	648,500	384	1,050	300	286	n.d.	n.d.	60.3	1,689	
E2/97	Atamanovskaya Spit, September 1997	737,000	–	–	–	–	n.d.	n.d.	–	–	
E3/97-1	Atamanovskaya Spit, September 1997	196,000	225	–	–	–	n.d.	n.d.	n.d.	871	0.54
E3/97-2		747,000	859	–	–	–	7.0	n.d.	n.d.	870	
E3/97-3		416,000	468–	–	–	–	n.d.	3.8	n.d.	889	
E3/97-4		240,000	272–	–	–	–	–	n.d.	–	882	
E505/01	Atamanovskaya Spit, September 2001	40,100	188	3,100	1,130	136	n.d.	n.d.	65.9	213	–
E1/98	Atamanovskii Island, October 1998	105,900	1,578	4,811	2,000	22	2.3	119	71	67	51.7
E8/97	Atamanovskii Island, September 1997	74,700	53.4	–	–	–	–	–	–	1,399	
E001*	Atamanovskii Island, September 1997	24,480	524	1,598	685	65.2	0.77	44.7	22.9	46.7	58.1
E002		33,250	705	1,635	710	81.5			35	47.2	
E2/98	Village of B. Balchug, October 1998	89,200	47	–	–	–	0.02	1.2	n.d.	1,898	60.0
E22/98	Village of B. Balchug, September 1997	238,700	170.7	–	–	–	–	n.d.	n.d.	1,398	

* – particles E3/97-1–E3/97-4, E001 and E002 are fragments of two particles that broke apart when being prepared for measurements.
–, below detection limit; n.d. – not determined.

downstream of the MCC production area (Fig. 1). The ^{137}Cs activity concentration of the new particles found near the village of B. Balchug is much lower than that of the other particles. However, the ^{239,240}Pu activity concentration in particle HP7 is 3.8 Bq/particle, which is one or two orders of magnitude higher than in other, previously measured, particles (Table 1). The ^{238}Pu/^{239,240}Pu ratio varies considerably, from 2.0 in particles 1543 and 1544 to 0.11 in particle HP4. These ratios are higher than those typical of weapons-grade plutonium, which can be accounted for by a longer exposure of the particles to neutron flux. This can be the case if the spent fuel was not removed from the reactor in proper time. Table 1 also gives the maximal ^{137}Cs activity in the particle found earlier by Timofeev near the town of Yeniseisk. The Yeniseisk particles clearly have a significantly lower ^{137}Cs activity than the particles found near the MCC production area. The reason is that the particles split into small fragments while being transported over long distances down the river.

Researchers of the IGM SB RAS (Novosibirsk) also found several particles in the floodplain of the Yenisei River (Sukhorukov et al., 2004). The radionuclide composition of those particles is listed in Table 2. It is noteworthy that they contain lower ^{137}Cs concentrations (up to 737 000 Bq/particle) than the particles found by the IBP researchers (Table 1). Moreover, analysis of the IGM particles revealed the presence of ^{60}Co, ^{154}Eu, ^{155}Eu and ^{241}Am. As already noted, the ^{238}Pu/^{239,240}Pu ratio in the IBP particles varies from 0.11 to 2.0 (Table 1), while in the IGM particles it varies from 0.54 to 60 (Table 2). Long-duration exposure of the particles to neutron flux cannot be the only factor accounting for the high ^{238}Pu/^{239,240}Pu ratios in the IGM particles. Other factors can be involved too.

We have compared the ^{137}Cs/^{134}Cs ratios for the particles and have thereby divided them into two major groups:

- Particles for which the ^{137}Cs/^{134}Cs ratio is more than 3,000. These are the particles found earlier, in 1995, in which the ^{137}Cs/^{134}Cs ratio equals 4,000 and which are 28–33 years old (as of the sampling date – 1995).
- Particles for which the ^{137}Cs/^{134}Cs ratio is less than or equal to 1,000. The particle found by Timofeev at the town of Yeniseisk can be considered as belonging to this group. Evidently, the ratio of the cesium isotopes suggests a much more recent time of formation of these particles as compared to the particles of Group 1. Statistical analysis has shown that for the first group the mean value of the ^{137}Cs/^{134}Cs ratio is $3,600 \pm 700$ and for the second group – 900 ± 500.

The Novosibirsk researchers (Sukhorukov et al., 2004) suggest adding a third group of particles:

- Particles for which the ^{137}Cs/^{134}Cs ratio is less than or equal to 200 ($50 \div 200$), according to the data of the Institute of Geology and Mineralogy SB

RAS. These are young particles and their estimated age is 10–13 years (Sukhorukov et al., 2004).

In our earlier papers we wrote that hot particles had been released from the reactor into the Yenisei River just once, about 30 years ago (Bolsunovsky et al., 1998). However, more recent investigations of high-activity particles in Krasnoyarsk and Novosibirsk suggest that over the 50-year period of the MCC operation there have been two or three emergency situations at the MCC reactors, with nuclear fuel microparticles released into the Yenisei. As particles were found both near the MCC and 330 km downstream, we can infer that the accidents were quite serious (Fig. 1). The presence of a large number of hot particles in the floodplain of the Yenisei River from the MCC production area down to the town of Yeniseisk (and, probably, farther downstream) is a source of potential health hazard to people.

It should be noted that the calculated values of fuel burn-up and the age of the particles are just estimates, because both the reactor power and the cesium isotope ratio in the spent fuel are MCC closed data.

In addition to reactor fuel particles, one would expect that in the Yenisei floodplain there would also be high-activity microparticles originating from corrosion and deformation of metal pipes in the reactor systems, which become radioactive while passing through the reactor core with coolant. However, activation radionuclides usually have shorter half-lives than fission radionuclides. Particles earlier found by Timofeev (1995) near Yeniseisk could probably have been activation products.

As mentioned above, a number of researchers, trying to find out the reasons of inhomogeneous distribution of radionuclides in floodplain soils and sediments, detected hot particles of smaller sizes and, hence, lower (several orders of magnitude) activities. They were termed "secondary" hot particles (Gritchenko et al., 2001). However, these microparticles occur in floodplain soils and sediments very frequently and, thus, can be considered to be the main form in which radionuclides are present in the Yenisei River floodplain (Gritchenko et al., 2001; Sukhorukov et al., 2004).

Radiochemical investigations of sediment samples collected from the Yenisei River showed that in some parts of the riverbed, levels of transuranium elements (^{238,239,240}Pu, ^{241}Am) were at least 10 times higher than earlier reported and at least 100 times higher than global fallout levels. Results of our measurements of transuranium elements in sediments show that their levels remain high even at a distance of 200 km downstream of the MCC (Bolsunovsky and Bondareva, 2007). ^{241}Am levels in sediment samples were determined by three methods: using γ-spectrometry, X-ray-spectrometry results and results of α-spectrometry, for which the radionuclide was separated radiochemically. The advantage of radiochemical method is higher accuracy of ^{241}Am determination, but the sub-sample used in it is very small (not more than 10 grams biomass). In γ-spectrometric determination of ^{241}Am, a whole sample can be used (several dozen and hundred grams biomass), although the accuracy of ^{241}Am determination is somewhat lower. For some of the sediment layers of the Yenisei γ-spectrometry showed up to 600 Bq/kg ^{241}Am and X-ray-spectrometry – 470 Bq/kg ^{241}Am. Sub-samples of these layers were used for radiochemical separation of ^{241}Am and no anomalous amounts of either transuranic elements or ^{241}Am were recorded. The results of the investigation of these samples suggest that ^{241}Am may be present in the samples mostly as microparticles. However, the nature and origin of these microparticles is unknown. Anomalous ^{241}Am content was revealed not only in the sediment samples of the Yenisei River collected in close proximity to the MCC but also at a distance of 200 and even 500 km downstream of the MCC.

We conducted detailed investigations of samples of the Yenisei sediment for the presence of Am particles, based on measurements of sub-samples of different masses. A nearly ten-fold decrease in the mass of the initial sample, from 145–155 to 19 g, did not cause any significant changes in the measured activity concentrations of almost all radionuclides except ^{241}Am (Table 3). As the mass of the measured samples was decreased, ^{241}Am activity concentration increased dramatically, from 400 to 1,860 Bq/kg, indicating the presence of a particle with ^{241}Am activity of 36 Bq/particle. ^{137}Cs concentration in the particle is insignificant, which is surprising, as ^{137}Cs is the principal element of fuel particles.

TABLE 3. Activity concentrations of artificial radionuclides in samples of the Yenisei River sediments collected near the village of B. Balchug, based on measurements of sub-samples of different masses

Sample label	Sub-sample mass, g/geometry	^{60}Co, Bq/kg	^{137}Cs, Bq/kg	^{152}Eu, Bq/kg	^{154}Eu, Bq/kg	^{241}Am, Bq/kg	^{40}K, Bq/kg
1302-1-2-1	155/120	1,200 ± 50	640 ± 35	800 ± 20	180 ± 11	400 ± 23	470 ± 47
1302-1-2-2	145/120	610 ± 26	550 ± 30	835 ± 20	210 ± 11	7 ± 3	460 ± 57
1302-1-2-1-1	39/35	2,300 ± 140	630 ± 80	930 ± 50	270 ± 40	19 ± 5	690 ± 250
1302-1-2-1-2	39/35	510 ± 25	690 ± 60	870 ± 30	230 ± 20	<50	580 ± 100
1302-1-2-1-3	39/35	1,050 ± 50	760 ± 70	710 ± 30	210 ± 20	<50	520 ± 110
1302-1-2-1-4	39/35	560 ± 26	800 ± 70	940 ± 33	280 ± 21	765 ± 50	700 ± 100
1302-1-2-1-4-1	19/15	490 ± 25 (9.4)*	660 ± 50 (12.7)*	810 ± 30 (15.6)*	230 ± 20 (4.4)*	143 ± 11 (2.8)*	560 ± 100
1302-1-2-1-4-2	19/15	670 ± 30 (12.9)*	930 ± 70 (17.9)*	1,090 ± 35 (21.0)*	300 ± 20 (5.8)*	1,860 ± 100 (35.8)*	470 ± 85

* – radionuclide activity per sample (Bq/sample).

The Novosibirsk researchers investigated a large sediment sample (about 17 kg) collected from the Yenisei at the village of Atamanovo and found abnormally high ^{241}Am concentrations in some of the analyzed 5-g sub-samples – up to 3,570 Bq/kg, or 17.8 Bq/sample (Sukhorukov et al., 2004). Measurements of 5-g sub-samples revealed a high correlation between ^{241}Am and ^{137}Cs concentrations in anomalous samples. Studies performed in the IBP (Krasnoyarsk) showed that increased concentration of ^{241}Am in the sample containing a microparticle is not correlated to ^{137}Cs concentration, while the IGM (Novosibirsk) researchers revealed a high correlation between ^{241}Am and ^{137}Cs concentrations in anomalous samples. Thus, sediment samples could contain two types of ^{241}Am particles: with high ^{137}Cs concentration and a high ^{137}Cs/^{241}Am ratio and with low ^{137}Cs concentration. However, the nature and origin of both types of microparticles with ^{241}Am has not been determined yet. Microparticles with enhanced ^{241}Am concentration were also found in floodplain soils and their nuclide composition was similar to that of sediment particles.

The presence of a large number of high-activity particles on the Yenisei riverside, from the MCC to Yeniseisk (and, probably, farther downstream), is a source of potential hazard to the health of residents. The dose rate of γ-radiation at a distance of 1 m from some of the particles amounted to 0.1 mSv/h, and the laboratory investigations had to be conducted in a glove box. The particles were found at the riverside in places frequented by the local population, and by staying near a particle for some hours a person is sure to receive a yearly irradiation dose (by the Russian State standards) – 1 mSv. Further investigations of active particles will provide grounds for an objective estimate of radioactive contamination of the Yenisei River and for uncovering accidents at the MCC that are now unknown to the public.

4. Conclusion

- Since 1995, researchers of the Institute of Biophysics have found a considerable number of hot particles that contain up to 29,000 kBq/particle of ^{137}Cs in the Yenisei River floodplain. Investigations of the particles in Krasnoyarsk, Moscow, and Novosibirsk have confirmed their reactor origin and made it possible to roughly estimate their age. Based on comparative analysis of ^{137}Cs/^{134}Cs ratios, all the particles can be divided into two or three major groups, suggesting that over the 50-year period of the MCC operation, there have been two or three emergency situations at the MCC reactors, with nuclear fuel microparticles released into the Yenisei. The Krasnoyarsk and Novosibirsk researchers have found a total of 50 hot particles in the floodplain of the Yenisei River.
- Microparticles containing up to 36 Bq/particle of ^{241}Am were detected in soil and sediment samples collected at the Yenisei River. In some of these

particles the $^{137}Cs/^{241}Am$ ratio is high but in the others – low. It remains unknown how these particles have been formed. However, these microparticles occur in floodplain soils and sediments very frequently and, thus, can be considered to be the main form in which radionuclides are present in the Yenisei River floodplain.

- In addition to reactor fuel particles, there may also be high-activity microparticles originating from corrosion and deformation of metal pipes in the reactor systems. They become radioactive while passing through the reactor core with coolant. However, activation radionuclides in these metal particles usually have shorter half-lives than fission radionuclides.
- The presence of a large number of hot particles of different origin in the floodplain of the Yenisei River from the MCC production area down to the town of Yeniseisk (and, probably, farther downstream) is a source of potential health hazard to people.

Acknowledgements

The work was partly supported by the Siberian Branch of the Russian Academy of Sciences within the framework of Integration Project No. 30.

References

Bolsunovsky, A. and Bondareva, L. (2007), Actinides and other radionuclides in sediments and submerged plants of the Yenisei River, *Journal of Alloys and Compounds* **444–445**: 495–499.

Bolsunovsky, A.Ya. and Tcherkezian, V.O. (2001), Hot particles of the Yenisei River flood plain, Russia, *Journal Environmental Radioactivity* **57**(3): 167–174.

Bolsunovsky, A.Ya., Goryachenkova, T.A., Cherkezyan, V.O., and Myasoedov, B.F. (1998), Hot particles in the Krasnoyarsk Territory, *Radiokhimiya*, **40**(3): 271–274.

Gritchenko, Z.G., Kuznetsov, Yu.V., Legin, V.N., et al. (2001), Secondary hot particles in the floodplain soils of the Yenisei River, *Radiokhimiya*, **43**(6): 863–865.

Sukhorukov, F.V., Degermendzhy, A.G., Bolsunovsky, A.Ya., Belolipetskii, V.M., Kosolapova, L.G., et al. (2004), Zakonomernosti raspredeleniya i migratsii radionuklidov v doline reki Yenisei (Distribution and Migration Behaviours of Radionuclides in the Yenisei River Floodplain), SB RAS Publishers "Geo" Branch, Novosibirsk (in Russian).

Tcherkezian (Cherkezyan), V., Shkinev, V., Khitrov, L., and Kolesov, G. (1994), Experimental approach to Chernobyl hot particles, *Journal Environmental Radioactivity* **22**: 127–139.

Tcherkezian, V., Galushkin, B., Goryachenkova, T., Kashkarov, L., Liul, A., Roschina, A., and Rumiantsev, O. (1995), Forms of contamination of the environment by radionuclides after the Tomsk accident (Russia, 1993), *Journal Environmental Radioactivity* **27**: 133–139.

Timofeev, V.A. (1995), Technogenic radioactive contamination of alluvial deposits of the Yenisei, in: *Sbornik dokladov II Mezhdunarodnoi radioekologicheskoi konferentsii, Krasnoyarsk (Proceedings of the II International Radioecological Conference,* Krasnoyarsk) pp. 165–171 (in Russian).

MIGRATION OF FUEL PARTICLES OF ChNPP FALLOUT AND LEACHED RADIONUCLIDES IN SOILS AND SOIL-TO-PLANT SYSTEM

YURIY IVANOV
International Radioecology Laboratory of Chernobyl Center for Nuclear Safety, Radioactive Waste and Radioecology, P.O. Box 151, Slavutych 07100, Ukraine
phone: (380 4479) 61562, e-mail: ivanov@chornobyl.net, yai@scienrad.kiev.ua

Abstract: The behaviour of the Chernobyl fuel particles in soil and soil-to-plant system has been studied in model experiments, by natural observation as well as by mathematical modeling. Parameters of vertical migration in soils of ^{90}Sr, ^{137}Cs and ^{239}Pu in initial water-soluble form, fuel particles and leached radionuclides, dynamics of radionuclides mobile forms in soils as well as dynamics of radionuclides transfer in soil-to-plant system have been estimated.

Keywords: Chernobyl fuel particles, radionuclide, vertical transfer in soil, radionuclide mobile forms, bioavailability.

1. Introduction

The radionuclide behaviour in the biogeochemical chains of migration significantly depends on physical–chemical forms of radioactive substances ingressed into the soil: water-soluble forms of global fallout after the ground nuclear explosion on the carbonate rock, emissions and faulting of nuclear fuel cycle enterprises; sparingly soluble fallout after the ground nuclear explosion on the silicate rock, fallout of Chernobyl NPP accidental release etc. (Ivanov et al., 1991). The fuel component is represented by sparingly soluble particles of irradiated nuclear fuel with heterogeneously dispersed composition, in which the matrix is comprised predominantly of Uranium oxides and chemical elements from construction materials (Begichev et al., 1990; Kuriny et al., 1993). These characteristics of fallout predetermine the peculiarities of Chernobyl radionuclides in the environment.

2. Methods

In order to analyze the fuel particle behaviour in soil and soil-to-plant systems, complex perennial field observations, laboratory experiments, and mathematical modeling have been carried out.

2.1. FIELD OBSERVATIONS OF VERTICAL TRANSFER OF RADIONUCLIDES IN SOIL

Radionuclide redistribution in soils profile has been studied at different experimental sites of observation network organized at 1986–1987 with taking into account of spatial distribution of fuel and condensation components of fallout. Two soil columns have been sampled in two to three places of each experimental site, columns have been sliced in layers 0–2; 2–5; 5–10;...; 35–40 cm (Ivanov et al., 1996a).

2.2. STUDY OF VERTICAL TRANSFER OF RADIONUCLIDES AND FUEL PARTICLES IN MODELLING EXPERIMENTS

Diffusion transfer of ^{90}Sr, ^{137}Cs and ^{239}Pu was studied in column experiment with some typical for Ukrainian Polesye soils. Size of columns ∅8 × 120 mm. Soils are incubated during 14 days before radionuclide introducing. Radionuclide activity 7,4 kBq per column (specific activity 370 kBq·ml^{-1}, without carrier). After incubation during 210–600 days at 18–20°C columns are dissected at the temperature of liquid nitrogen and are divided into layers 5 mm. Repeatability – three to six multiple (Ivanov et al., 1996b).

Vertical transfer of fuel particles was studied in column experiment with mentioned above soils. Size of columns ∅60 × 200 mm. Soils are incubated during 14 days before radionuclide introducing. Sandy soil sampled in 1.5 km from ChNPP site at 1988 (soil layer 0–2 cm) are mineralized (<300°C) and are used as initial samples of fuel particles with real distribution on dispersed composition. Main part of Caesium radioisotopes fallen out as condensation component (85–95%) was removed from soil with the use of isotopic exchange technique. Total activity of gamma-emitting radionuclide was about 70 MBq per column. Repeatability – three multiple. At the end of experiment soil columns are divided into layers 5 mm (Ivanov et al., 1996b).

2.3. MATHEMATICAL MODELLING

Mathematical modeling of radionuclide vertical transfer in soils and radionuclide speciation dynamics has been carried out with the use of modified convective-diffusion model permitting to take into account dynamics different forms of radionuclide transfer in soil layer (Ivanov et al., 1996c).

3. Experimental results and discussion

3.1. VERTICAL TRANSFER OF RADIONUCLIDES IN SOIL

Using data from perennial field observations, laboratory experiments, and mathematical modeling, analyses of radionuclide vertical migration in soil after the Chernobyl accident have revealed a number of peculiarities in radionuclide behaviour within the contaminated territory.

Analysis of radionuclide diffusion coefficients in representative types of soils has shown that ability to migrate decreases in the following order: ^{90}Sr > ^{137}Cs > plutonium isotopes (Ivanov, 1997). When soil moisture is increased, the mobility of radionuclide increases with preservation of mentioned sequence (Table 1).

TABLE 1. Diffusion coefficients of ^{137}Cs, ^{90}Sr and ^{239}Pu in various soils depending on its moisture

Soil	Radionuclide	D, $\cdot 10^{-8} \cdot cm^2 \cdot s^{-1}$ at soil moisture, % TMC		
		30	60	100
WHS	^{90}Sr	20.0 ± 1.0	26.0 ± 3.0	35.0 ± 5.0
SPSLS		9.0 ± 1.0	16.0 ± 3.0	15.0 ± 1.0
PBS		0.6 ± 0.1	3.7 ± 0.2	13.0 ± 1.0
WHS	^{137}Cs	0.50 ± 0.04	0.70 ± 0.15	0.9 ± 0.4
SPSLS		0.8 ± 0.1	1.0 ± 0.1	1.7 ± 0.5
SPRLS		Not determined	Not determined	0.28 ± 0.08
CRL		0.27 ± 0.10	0.63 ± 0.20	Not determined
PBS		1.7 ± 0.3	7.9 ± 3.0	31.6 ± 5.0
SPSLS	^{239}Pu	0.081 ± 0.020	0.077 ± 0.023	0.19 ± 0.03
CRL		0.067 ± 0.022	0.069 ± 0.07	Not determined
SPRLS		Not determined	Not determined	0.086 ± 0.020

D – diffusion coefficients; TMC – total moisture capacity of soil; WHS – weakly humous sand; SPSLS – soddy-podsolic sandy-loam soil; SPRLS – soddy-podsolic residual loamy soil; PBS – peaty-boggy soil; CRL – chernozem (black earth) regraded loamy.

The migration of radionuclide of accidental release in soils in environmental conditions has the specificity stipulated by different forms of radioactive fallout as well as peculiarities of soil-climatic conditions of the territory. The migratory ability of "Chernobyl" radionuclide in soil differing from that of radionuclide of global fallout is inherent first of all for radionuclide which has fallen out in the matrix of fuel particles, to a lesser degree – to radionuclide which have fallen out predominantly in a composition of condensed component (Fig. 1).

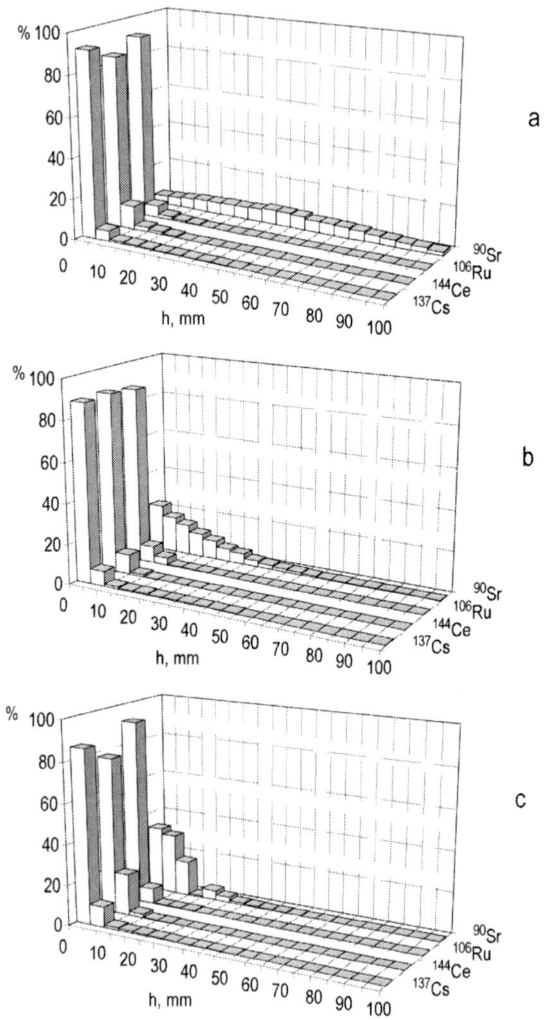

Figure 1. Distribution of ^{137}Cs, ^{144}Ce and ^{106}Ru, contained in the matrix of fuel particles, and ^{90}Sr in initial water-soluble form in the profile ofsoil columnes: A – soddy-podsolic sandy soil, B – soddy-podsolic loamy-sand soil, C – peaty soil, h – depth of soil profile.

Deposition of radionuclide in a structure of fuel particles predetermines the absence of differences in distribution in the soil profile during first 2–3 years after deposition of radioisotopes Cs, Ce, Ru and Sr, which are characterized by different physical–chemical properties. Hereinafter, as a result of fuel particles destruction, the differentiation of radioisotopes distribution profiles of mentioned elements in soil is observed. As a whole for mineral soils the higher movability of ^{90}Sr as contrasted to ^{137}Cs and Pu isotopes is observed, higher migratory ability of ^{137}Cs is observed in organic soils.

Results of model experiment on study of fuel particles vertical transfer in soil are presented in. The character of ^{90}Sr, ^{137}Cs and ^{239,240}Pu distribution in a soil profile at the sites characterized by different ratio of fuel and condensed fallout components as well as results of mentioned above model experiment indicates a capability of radionuclides migration in a structure of fuel particles.

Effective parameters of transfer of fuel particles in soil have been estimated (Table 2). Mean value of effective diffusion coefficient of fuel particles in soil has estimated as $4.5 \cdot 10^{-10}$ cm$^2 \cdot$s^{-1}.

TABLE 2. Values of effective diffusion coefficient of ^{137}Cs, ^{144}Ce and ^{106}Ru, contained in the matrix of fuel particles, and ^{90}Sr in initial water-soluble form in the profile of soil columns, cm$^2 \cdot$s^{-1}

Radionuclide	Soddy-podsolic sandy soil	Soddy-podsolic loamy-sand soil	Peaty soil
^{137}Cs	$3.8 \cdot 10^{-10}$	$5.0 \cdot 10^{-10}$	$5.0 \cdot 10^{-10}$
^{144}Ce	$5.2 \cdot 10^{-10}$	$4.2 \cdot 10^{-10}$	$7.0 \cdot 10^{-10}$
^{106}Ru	$3.8 \cdot 10^{-10}$	$4.6 \cdot 10^{-10}$	$3.7 \cdot 10^{-10}$
Mean ±SD	$(4.3 \pm 0.8) \cdot 10^{-10}$	$(4.6 \pm 0.4) \cdot 10^{-10}$	$(4.4 \pm 1.4) \cdot 10^{-10}$
^{90}Sr	$2.2 \cdot 10^{-7}$	$1.4 \cdot 10^{-8}$	$6.1 \cdot 10^{-9}$

Effective parameters of radionuclide vertical transfer in soils contaminated with fuel component of fallout calculated on the base of natural observation varies in a wide limits (D_{eff} (radionuclide) = $n \cdot 10^{-9}$–10^{-8} см$^2 \cdot$с$^{-1}$). These data together with an autoradiography data about vertical distribution of fuel particles with the different size in a soil profile (Fig. 2) allowes to make a conclusion about vertical migration of radionuclide contained in a matrix of fuel particles, predominantly by mechanical transfer of fuel particles in soil profile (Ivanov et al., 1996a,b).

Rate of radionuclide migration also depends on the velocity of destruction of fuel-particle and radionuclide leaching from its matrix. As the intensity of ^{90}Sr sorption by a solid phase of soil is less on an order of magnitude than these ones of ^{137}Cs and Plutonium isotopes, this radionuclide leached from a matrix of fuel particles, is characterized by higher migratory mobility in soils.

Radionuclide fallen out in a matrix of fuel particles are characterized by significantly less velocity of vertical migration in comparison with these ones in initial water-soluble forms.

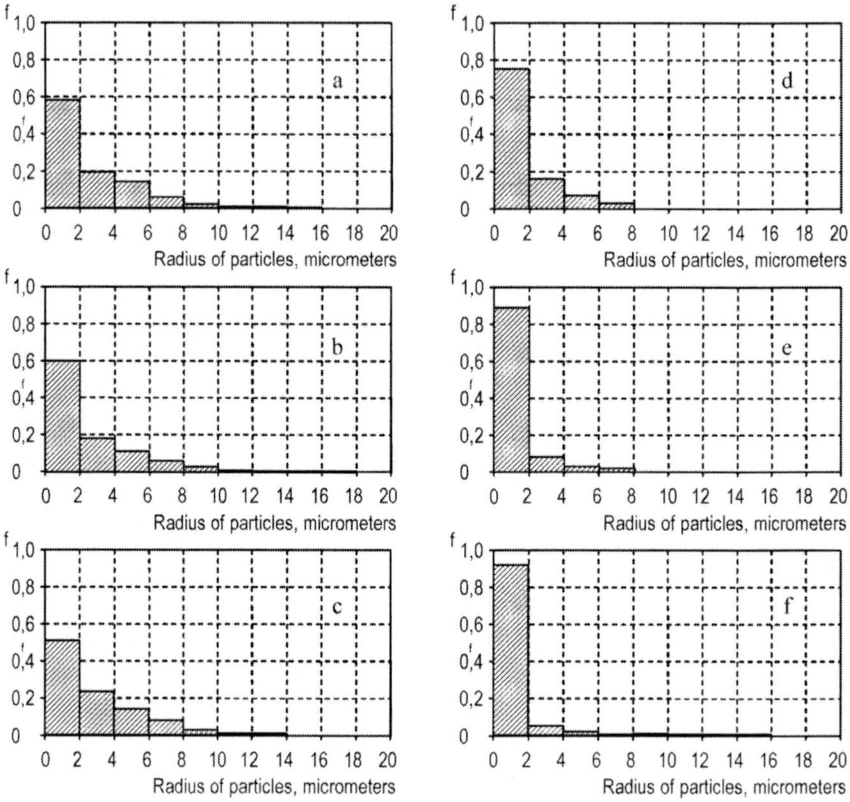

Figure 2. Change of fuel particles dispersal composition as a result of its vertical migration in the profile of peaty soil: (a) initial sample of particles; (b) initial sample of particles on the soil surface after water elution; (c) soil horizon 0–5 mm; (d) soil horizon 5–10 mm; (e) soil horizon 10–15 mm; (f) soil horizon 15–20 mm. Abscissa – size of particles, mm; ordinate – percentage of particles with defined size.

Due to leaching of ^{90}Sr from fuel particles matrix intensity of radionuclide migration increases with time, on the contrary, because significant part of ^{137}Cs has fallen out in condensation component its movability decreases with time. Example of differences of radionuclide movability in soils 15 years after deposition period predetermined both fuel particles presence and differences of physical–chemical properties of radionuclide as isotopes of different chemical elements is presented in Fig. 3.

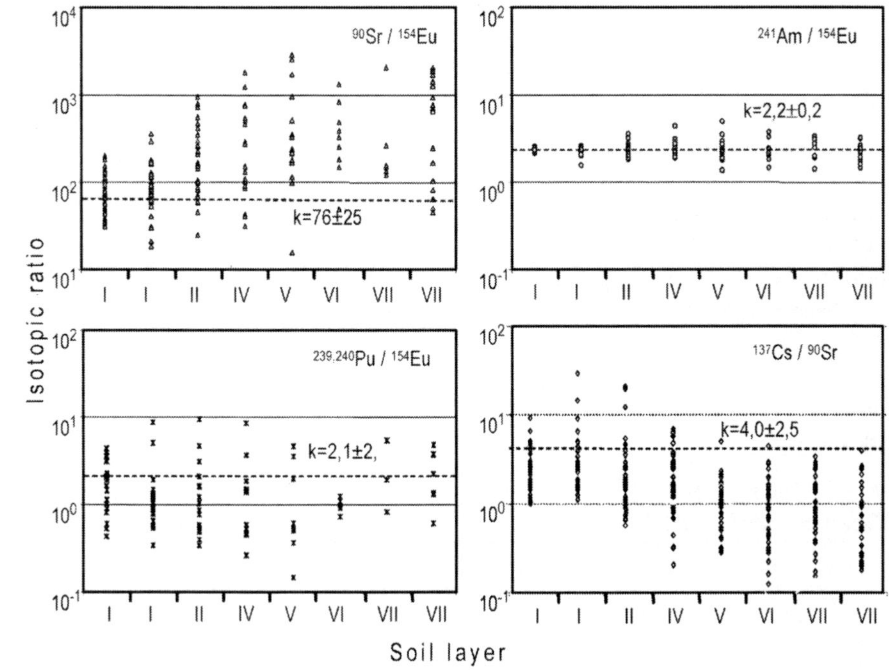

Figure 3. Values of radionuclides isotopic ratio in profile of different soils on the fuel track of fallout (experimental site "Red forest"): I, 0–2 cm; II, 2–4 cm; III, 4–7 cm; IV, 7–10 cm; V, 10–15 cm; VI, 15–20 cm; VII, 20–25 cm; VIII, 25–30 cm; k, ratio of radionuclide total activities in full soil profile (Bondarkov et al., 2003).

For hypothetical situations when fuel particles are not destructed calculated values of ecological half-life of cleaning of root-inhabited horizons of meadows, formed on typical for Ukrainian Polesye mineral and organic soils from fuel particles is about 2,050–2,250 years. Values of half-life of cleaning of root-inhabited horizons of meadows and arable lands from ^{90}Sr and ^{137}Cs at various tracks of Chernobyl fallout are presented in Table 3 (Ivanov, 1997).

Comparison of these estimations indicates that in natural conditions in fuel tracks of fallout values of ecological half-life of cleaning of root-inhabited horizons of soils from radionuclide are 1–2 order of magnitude less in comparison with mentioned hypothetical situation. These estimations are indicated significant role both mechanical transfer of fuel particles due to activity of soil animals, and fuel particles destruction and following transfer of leached radionuclides.

TABLE 3. Ecological half-life of cleaning of root-inhabited horizons of meadows and arable lands from ^{90}Sr and ^{137}Cs

Soil	^{90}Sr		^{137}Cs	
	Meadow (0–5 cm)	Arable land (20 cm)	Meadow (0–5 cm)	Arable land (20 cm)
Automorphous mineral soils with light granulometric composition	11–19	21–32	60–150	26–45
Automorphous mineral soils with a heavy granulometric composition	15–32	37–129	150–400	60–150
Hydromorphic organogenic soils	100–160	103–233	11–20	30–60
Drained organogenic soils			17–83	
Hydromorphic mineral soils			45	
Weakly humous sand	2.2–6	2.2–7.6		

3.2. MOBILE FORMS OF RADIONUCLIDES IN SOIL

Contamination of the territory with radionuclide represented by different physical–chemical forms of fallout of ChNPP release, predetermines a number of peculiarities of radionuclide behaviour in a soil-plant cover (Ivanov, 1997; Kashparov, 2000). First of all it is connected with fuel particles presence in fallout.

Behaviour in soils of radionuclide, represented by condensation component of fallout is very similar to behaviour of radionuclide of global fallout. Content of radionuclide mobile forms in soil is predetermined mainly by sorption-desorption mechanisms by solid phase of soil, and decreases during the time with velocity, depending on specific landscape-geochemical condition – acidity and composition of soil solution, granulometric and mineralogical composition of soil, water regime of soil etc. (Ivanov and Kashparov, 2003).

Territory of "near-field" tracks of fallout is characterized by superposition of two components of fallout – condensation and fuel one. Radionuclide mobility in these conditions depends firstly on ratio of mentioned forms of fallout for each radionuclide.

The dynamics of the content of radionuclide mobile forms in the soil in the mentioned territories depends on both physical–chemical properties of fuel particles, and characteristics of soils. Properties of fuel particles predetermine different velocity of particles destruction and leaching velocity of radionuclide from particles matrix in various tracks of fallout. Soils characteristics predetermine both destruction rate of fuel particles and the sorption of leached radionuclide by solid phase of soil. In any case the fuel particles are the additional to solid phase of soil long-term source of the mobile forms of radionuclide in soil (Ivanov, 1997).

During first years after the accident, the quota of exchangeable forms of ^{137}Cs in similar soils on the territories contaminated with condensation component of fallout, was usually higher than that in the same soils contaminated by fuel component. Fallout of radionuclide in a structure of fuel particles modifies essentially dynamics of radionuclide fastening by solid phase of soils (presence of the maximum of the radionuclide exchangeable forms content during the time, less intensive decline of these forms content) as contrasted to condensed traces of fallout. On the territories, contaminated with fuel component of fallout, this characteristic increased for the first 3–4 years after the accident - (Fig. 4, Ivanov, 1997). Using a biogeochemical version of isotopic dilution method, we determined that the different rate of decline of the contents of radionuclide exchangeable forms is concerned with both differences of the fallout physical–chemical forms, and differences of soil-chemical conditions of the territory (Ivanov, 1997).

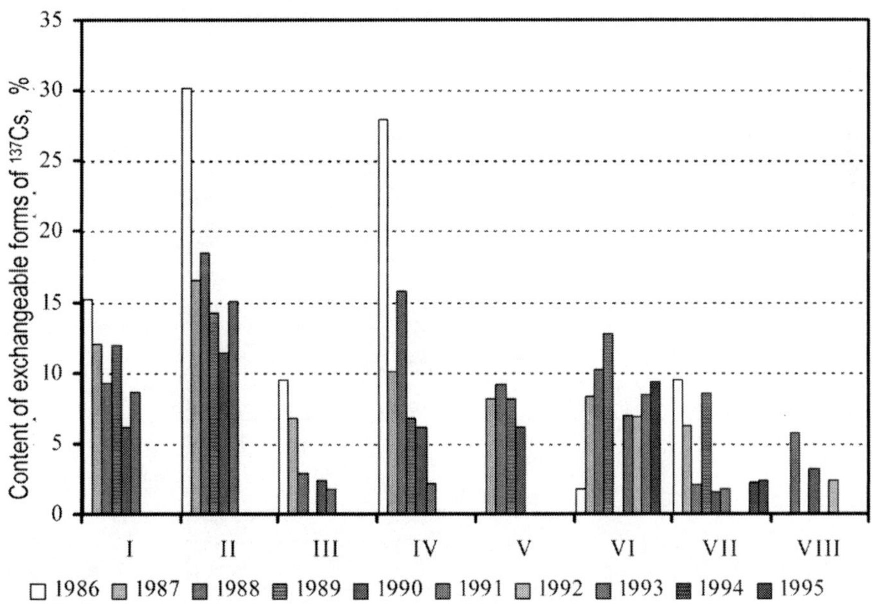

Figure 4. Dynamics of ion-exchangeable forms of ^{137}Cs in soils on various tracks of fallout: I – automorphous soils, R* = 2–15 km; II – automorphous soils, R* = 15–50 km; III – hydromorphous soils, R* = 2–15 km; IV – hydromorphous soils, R* = 15–50 km; V – peaty-boggy soils, R* = 15–50 km; VI – soddy-podsolic sandy soil, R** = 3–4 km; VII – peaty-podsolic soil, R** = 3–4 km; VIII – soddy-podsolic sandy soil, R** = 4–5 km; * Northern track; ** Western track.

Based on experimental data of the half-life of decreasing of ^{137}Cs exchangeable forms after radionuclide introduction into the soil in an initial water-soluble form have been estimated. For typical in Ukrainian Polessye soddy-podsolic soils the value of $T_{1/2}$ varied between 0.8–1.4 years during first 5 years after soil contamination (Ivanov, 1997). Subsequently, the decline rate of the content of radionuclide exchangeable forms in the soil has been essentially reduced.

The content of ^{90}Sr exchangeable forms in the soil after its deposition in initial water-soluble form is characterized by a very slow decrease. Analysis of experimental data obtained during "pre-Chernobyl" period (Yudintseva et al., 1983, 1985) permit to calculate half-life of decreasing of ^{90}Sr water-soluble and exchangeable form in soddy-podsolic loamy-sand soil after its introducing in water-soluble form (~55 years). On the fuel tracks of fallout content of ^{90}Sr exchangeable forms in soil depending on level of physical–chemical transformation of fuel particles matrix as well as soil condition was characterized by increasing during period from 3–7 to 10–15 years after deposition (Ivanov, 1997; Kashparov, 2000). Examples of forecast estimation of dynamics of ^{90}Sr different forms of transfer in 10-cm layer of typical for Ukrainian Polesye soddy-podsolic sandy and loamy-sand soils are presented in Fig. 5. Such data permit estimates of radionuclide re-distribution in soil-to-plant systems.

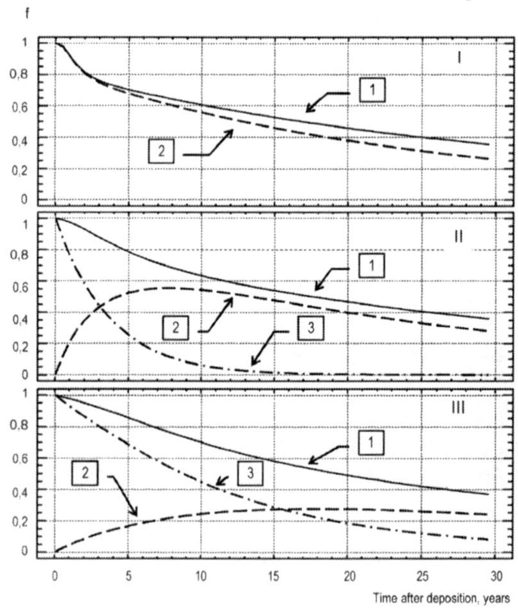

Figure 5. Dynamics of distribution of different transfer forms of ^{90}Sr in 10-cm soil layer of natural meadows, formed on soddy-podsolic sandy and sandy-loam soils: f – quota from the total content of radionuclide in profile; I – ^{90}Sr in initial water soluble form; II, III – ^{90}Sr in a structure of fuel particles with transformation constant, respectively, 0,25 $year^{-1}$ (II) and 0,05 $year^{-1}$ (III); 1 – total content of radionuclide; 2 – content of bioavailable forms; 3 – content of radionuclide in fuel particles.

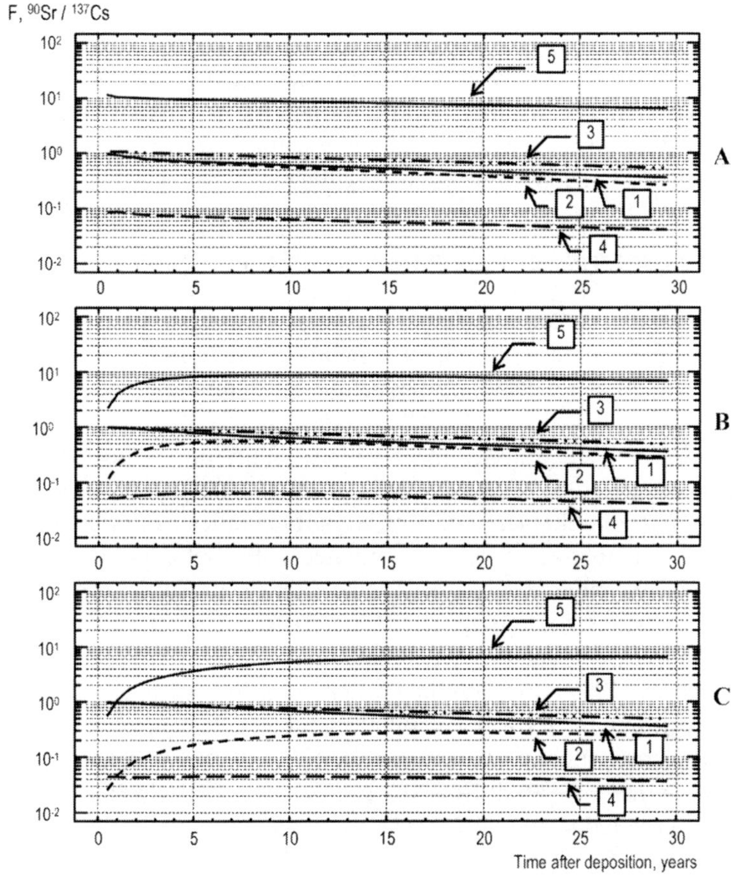

Figure 6. Dynamics of ^{90}Sr and ^{137}Cs total content, exchangeable forms content and ratio of the content of radionuclide exchangeable forms in root-inhabited layer of soil of natural meadow, formed on soddy-podsolic loamy-sand soil at various tracks of fallout: F – quota of total content; A – initial water-soluble form of radionuclide; B – fuel fallout $k_{transform}$= 0,25 year^{-1}, quota of ^{137}Cs in fuel particles – 0.5; C – fuel fallout, $k_{transform}$ = 0,05 year^{-1}, quota of ^{137}Cs in fuel particles – 0.5; 1,3 – total content of ^{90}Sr (1) and ^{137}Cs (3); 2,4 – content of exchangeable forms of ^{90}Sr (2) and ^{137}Cs (4); 5 – ratio of the content of radionuclide exchangeable forms of ^{90}Sr and ^{137}Cs.

Predicted estimates of the content of ^{137}Cs and ^{90}Sr different forms in soddy-podsolic loamy-sand soil show principal differences of dynamics of the content of these radionuclide exchangeable forms at condensation and fuel tracks of fallout Fig. 6 and, respectively, principally different dynamics of its biological availability as well as radiological significance (Ivanov, 1997).

On the whole, at the fuel tracks of fallout specific dynamics of the content of ^{137}Cs exchangeable forms in soils (the existence of the maximum of the content of radionuclide exchangeable forms, less intensive decrease of the content in comparison with condensation tracks) is revealed during first 3–4 years after deposition period. Specific dynamics of the content of ^{90}Sr exchangeable forms in soil, which is expressed as increase of the content of exchangeable forms with time, occurs during first 3–15 years after deposition period depending on level of physical–chemical transformation of fuel particles matrix as well as soil condition (Ivanov, 1997).

3.3. RADIONUCLIDE TRANSFER IN A SOIL-TO-PLANT SYSTEM

Dynamics of radionuclide transfer in soil-to-plant system depends on physical–chemical properties of soils, characteristics of radionuclides as isotopes of concrete chemical elements as well as ratio of different components of fallout at various tracks.

Dynamics of ^{137}Cs transfer from soil to plants on condensation tracks of fallout is characterized by decreasing in time; intensity of this process significantly depends on sorption characteristics of soils. Dynamics of radionuclide transfer factor (TF) to plants from soils contaminated with fuel component of fallout is characterized by more complicated character – in some cases more slow decreasing in time of intensity of ^{137}Cs transfer to plant in comparison with condensation tracks; in another cases maximum of ^{137}Cs TF to plants is observed (Fig. 7 – Ivanov, 1997). In 1988 the comparative biological availability (CBA[1]) in the territories characterized by contamination with superposition of fuel and condensation component of fallout with a quota of fuel component in the total contamination 0.3–0.6 varied from 0.4 to 0.7, some years later at the same plots CBA of ^{137}Cs ranged from 0.9 to 1.0 (Bondar and Ivanov, 2001).

Contrast to ^{137}Cs dynamics of available to plants forms of ^{90}Sr in soils is predetermined more significantly by radionuclide outflow from root-inhabited layer of soil and by its physical decay, because processes of ^{90}Sr fixation by solid phase of soil run with significantly less velocity. Content of available to plants forms of ^{90}Sr in soils is predetermined by intensity of fuel particles destruction, last one depends on soil-chemical condition and characteristics of fuel particles. Dynamics of ^{90}Sr TF to plants on fuel traks of fallout is characterized by increase of radionuclide TF to plants during certain time period. Period of the maximum coming of radionuclide transfer to plants depends on intensity of fuel particles transformation (destruction and dissolution) in soils as well as intensity of radionuclide vertical transfer. Example of off-line dynamics of redistribution of ^{90}Sr transfer forms in 10-cm soil layer of natural

---

[1] CBA – is evaluated as the ratio of biological availability of a concrete radionuclide fallen in soil in the unknown physical–chemical form, to biological availability of the same radionuclide fallen in soil in the initial water-soluble form. As a measure of biological availability the transfer factor of a radionuclide from soil to plants is used.

meadow as well as appropriate experimental data on dynamics of ^{90}Sr TF to plants of this meadow are presented in Fig. 8.

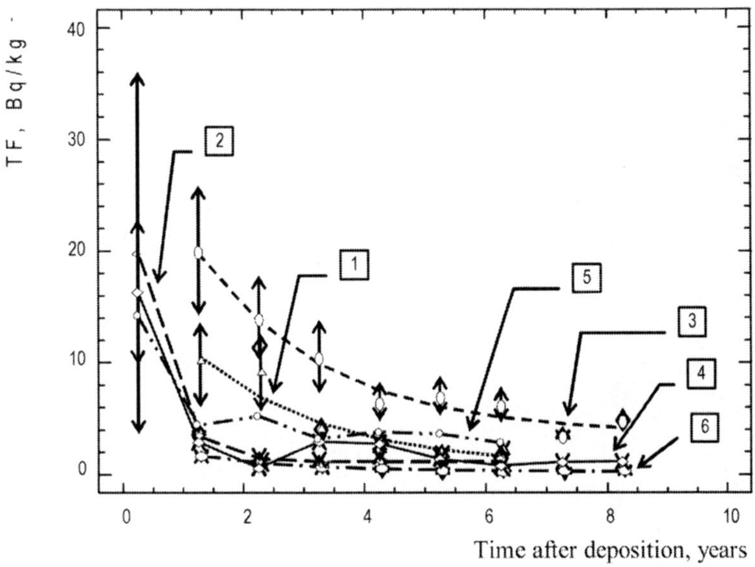

Figure 7. Dynamics of ^{137}Cs TF to plants of natural meadows formed on automorphous mineral soils: 1 – soddy-podsolic sandy; 2 – soddy-podsolic sandy; 3 – soddy-podsolic sandy and sandy-loam; 4 – soddy-podsolic sandy soil, fuel track of fallout; 5 – automorphous mineral soils, fuel tracks of fallout; 6 – heavy clay and clay soils.

Dynamics of ^{90}Sr transfer to plants after radionuclide introducing in soil in initial water-soluble form is characterized by very slow decrease in time. Since ^{90}Sr TF to plants is linearly depends on the content of radionuclide water-soluble and exchangeable forms in soil, dynamics of ^{90}Sr TF changes by the same parameters as dynamics of radionuclide water-soluble and exchangeable forms. Thus, with taking into account of mentioned in paragraph 3.2 data half-life of ^{90}Sr TF decreasing in soddy-podsolic loamy-sand soil after its introducing in water-soluble form is equal about 55 years.

4. Conclusions

Injection to the environment of particles of fine-dispersed nuclear fuel with various degree of physical–chemical characteristics of its matrix causes significant modification of regularities of behaviour of radiologically significant radionuclide (^{90}Sr, ^{137}Cs, actinides) in soils and soil-to-plant system studied during pre-Chernobyl time. This modification is appeared during first units-tens years after accidental release, and is consisted in decreasing in time of radionuclide of fuel component inclusion in migration chains. Characteristics of migratory media exerts prevailing influence on dynamics of radionuclide inclusion in migration chains on condensation tracks of fallout, whereas migrants properties (fallout in form of fuel particles) – on fuel tracks.

Deposition of radionuclide in composition of fuel particles is predetermined principally different dynamics of ^{90}Sr and ^{137}Cs radiological significance on fuel tracks of fallout. Though that ^{90}Sr in whole is characterized by higher movability in biological chains in comparison with ^{137}Cs formation of radiation exposure to population on the territories contaminated by fuel component of fallout is predetermined by Caesium radioisotopes during first decades after the accident.

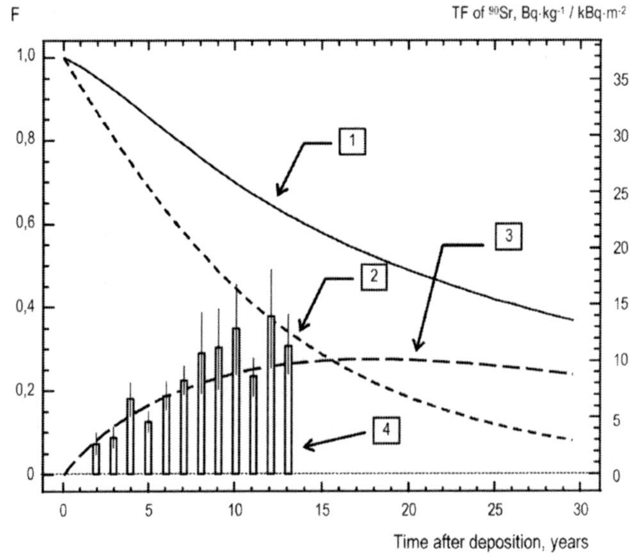

Figure 8. Off-line dynamics of redistribution of ^{90}Sr transfer forms in 10-cm soil layer of natural meadow formed on soddy-podsolic loamy-sand soil and experimental data on dynamics of ^{90}Sr TF to plants of this meadow: F – quota of ^{90}Sr from total content in the profile; 1 – total content of ^{90}Sr in soil layer; 2 – content of ^{90}Sr in the matrix of fuel particles; 3 – content of ^{90}Sr ion-exchangeable forms; 4 – TF of ^{90}Sr to meadows plant.

References

Begichev S.N., Borovoy A.A., Burlakov E.V., Gavrilov S.L., Dovbenko A.A., Levina L.A., Markushev V.M., Marchenko A.E., Stroganov A.A., Tataurov A.L., 1990, Fuel of 4 Unit reactor of ChNPP (Brief Hand-book): Preprint of the Institute of Atomic Energy; 5268/3, Moscow, 21 p. (In Russian)

Bondar P.F., Ivanov Yu.A., 2001, Estimation of biological availability of Chernobyl release radionuclides Proc. of theoretical and practical Conf. "Role of creative legacy of academician V.M. Klechkovsky in solution of actual problems of agricultural radiology", XXIX Radioecological hearings dedicated to Scientist memory (Moscow, 5–6 Dec. 2001), Alexakhin R.M., Sychev V.G. Eds., Moscow, pp. 140–150 (In Russian)

Bondarkov M.D., Gaschak S.P., Ivanov Yu.A., Maksimenko A.M., Ryabushkin A.N., Zheltonozhsky V.A., Sadovnikov L.V., Chesser R.K., Baker R.G., 2003, Parameters of radiation situation on the territory of the Red Forest site in the Chornobyl exclusion zone as impact factors for wild non-human species, *Contributed Papers of International Conference on the Protection of the Environment from the Effects of Ionizing Radiation.* 6–10 Oct. 2003, Stockholm, Sweden. IAEA-CN-109/100, 196–199.

Ivanov Yu.A., 1997, Radioecological substantiation of long-term forecast of radiation situation on agricultural lands as a result of major nuclear accidents (on an example of the accident on Chernobyl NPP): Abstract of Dr.Sc. Theses, Obninsk, 50 p. (In Russian).

Ivanov Yu.A., Alexakhin R.M., Zhigareva T.L., Ratnikov A.N., Popova G.I., 1991, Soil chemistry of radionuclides, in *Agricultuak Radioecology*, Alexakhin, R.M. Korneev N.A. Eds., Book Company Ecology, Moscow, pp. 12–54 (In Russian).

Ivanov Yu.A., Kashparov V.A., Levchuk S.E., Zvarich S.I., 1996a, Vertical transfer of radionuclides of ChNPP release in soils. 1. Long-term dynamics of radionuclides re-distribution in soils profile. *In situ Radiokhimiya (Radiochemistry)*, **38 (3)**, 264–271 (In Russian).

Ivanov Yu.A., Kashparov V.A., Levchuk S.E., Zvarich S.I., 1996b, Vertical transfer of radionuclides of ChNPP release in soils. 2. Experimental modelling of vertical transfer of radionuclides in soils profile. *Radiokhimiya (Radiochemistry)*, **38 (3)**, 272–277 (In Russian).

Ivanov Yu.A., Kashparov V.A., Khomutinin Yu.V., Levchuk S.E., 1996c, Vertical transfer of radionuclides of ChNPP release in soils. 3. Mathematical modelling of radionuclides vertical transfer in soil profile. *Radiokhimiya (Radiochemistry)*, **38 (3)**, 278–284 (In Russian).

Ivanov Yu.A., Kashparov V.A., 2003, Long-term dynamics of radioecological situation in terrestrial ecosystems on the territory of exclusion zone. *Environmental Science and Pollution Research*, 2003, **1**, 13–20.

Kashparov V.A., 2000, Estimation and forecasting of radioecological situation at radiation accidents with release of particles of the irradiated nuclear fuel (on an example of the accident on Chernobyl accident). Abstract of Dr.Sc. Theses, Obninsk, 48 p. (In Russian).

Kuriny V.D., Ivanov Yu.A., Kashparov V.A., Loshchilov N.A., Protsak V.P., Yudin E.B., Zhyrba M.A., Parshakov A.E., 1993, Particle-associated Chernobyl fall-out in the local and intermediate zones. *Annals of Nuclear Energy*, **20(6)**, 415–420.

Yudintseva E.V., Zhigareva T.L., Pavlenko L.I., 1983, Speciation of ^{137}Cs and ^{90}Sr" in soddy-podsolic soil as a result of liming and fertilization. *Pochvovedeniye (Soil Science)*, **9** (In Russian).

Yudintseva E.V., Popova G.I., Zhigareva T.L., 1985, Regularities of artificial radionuclides behaviour in soil and soil-to-plant system as a result of continuous interaction with soil. Report of RIAR 0287.0 040715, Task OSH.106. Obninsk (In Russian).

DISSOLUTION OF PARTICLES OF IRRADIATED NUCLEAR FUEL IN THE TEMPORARY STORAGES OF RADIOACTIVE WASTE IN CHERNOBYL ZONE: SOURCES FOR RADIONUCLIDES MIGRATION

VALERY KASHPAROV*[1], NOUREDDINE AHAMDACH[2], SVYATOSLAV LEVCHUK[1], VASYL YOSCHENKO[1], SERGEY FESENKO[3], IGORE MALOSHTAN[1]

[1]Ukrainian Institute of Agricultural Radiology NAU, Mashinostroitelej str., 7, Chabany, Kiev' Region, 08162, Ukraine

[2]Institute for Radioprotection and Nuclear Safety (IRSN), BP6, 92266 Fontenay aux Roses cedex, France

[3]International Atomic Energy Agency (IAEA), Department of Nuclear Science and Applications, Agency's Laboratories

**phone: (+38044) 5261246, e-mail: vak@uiar.kiev.ua*

Abstract: Results of the ongoing international study (i.e., Chernobyl Pilot Site Project) on characterization of the surface waste disposal site in the Chernobyl zone are presented. The reported experimental work includes determination of physico-chemical forms of radionuclides inside the waste storage (trench) and the data on kinetics of the fuel particles (FP) dissolution in natural conditions. In order to characterize the trench source term, the parameters of the FP dissolution kinetics in the trench were estimated and the model of the ^{90}Sr mobile forms dynamics was created.

Keywords: radionuclides, radioactive waste, hot particles, nuclear fuel, Chernobyl accident.

1. Introduction

In 1986, the near zone (30-km) of Chernobyl accident was contaminated with the radionuclides, which were deposited mainly in a matrix of fine-dispersed irradiated nuclear fuel (Kashparov et al., 2001, 2003). Among other radionuclides, it concerns one of the most-mobile long-living radionuclide ^{90}Sr ($T_{1/2}$=29 y).

D.H. Oughton and V. Kashparov (eds.), *Radioactive Particles in the Environment*,

More than 90% of ^{90}Sr activity in the initial fallout was presented in well-fixed non-exchangeable form Kashparov et al. (2001), while the condensed form mounted some percents. Nearest to ChNPP territories were contaminated for the greatest extent. The extensive works for the soil decontamination have been started just after the accident. These works included a removal and burying the upper 10-cm soil layer. For this purpose, the temporary storages of radioactive waste (TSRW) – the near-surface radioactive waste burial (trenches) had been utilized. In first post-accidental years, during very short time more than 800 trenches in the 30-km zone were created without proper hydro-isolation. Some trenches in present time are periodically flooded by the ground water, which can promote the radionuclides migration from these trenches. It determines the necessity of development of the strategy of management for TSRW in the Chernobyl zone.

Present paper reports the results obtained within the frameworks of international research project on evaluation of one of such the storages as the source of radionuclides migration in the Environment (Kashparov et al., 2001, 2003). The Chernobyl Pilot Site Project is organized through an international cooperative agreement between four organizations. From Ukraine, there are the Institute of Geological Sciences (IGS) and the Ukrainian Institute of Agricultural Radiology (UIAR), and from France, the partners are the Institute for Radiation Protection and Nuclear Safety (IRSN/DPRE/SERGD/LESTS) and the Tracer Application Office (CEA/SAT-French Atomic Energy Commission).

The study site is located 2.5 km South-West of the ChNPP in the so-called Red Forest radioactive waste dump site (Bugai et al.,). In 1987 in the course of clean-up works the radioactive materials, such as topsoil layer (>90% activity) and contaminated tree trunks from the dead pines killed by extreme radiation levels in 1986, have been bulldozed in-situ in a few meters deep trenches. Afterwards, the whole territory of Red Forest was covered with clean sand layer of about 30–50 cm. Young pine and birch trees, bushes and grass were planted to prevent the wind resuspension of sand.

A particular waste burial (trench #22), namely Chernobyl Pilot Site (CPS), was selected since 1998 by the joint French-Ukrainian scientist team for radiological characterization and monitoring studies. The burial represents ~70 m long, 8–10 m wide and 2–2.5 m deep unlined trench. The source term of radionuclide migration to the geo-environment is a heterogeneous mixture of contaminated organic materials and soil containing the micron-size reactor fuel particles. The depth to groundwater table is 2–3 m. For 15 years following disposal, the radionuclides have been leached from the trench by precipitations (average annual rainfall is 550–650 mm), and have been penetrating the underlying unsaturated soil and the aquifer. As a result, ^{90}Sr concentration in groundwater in the upper part of the aquifer in the vicinity of the trench varies between n·100 and n·10,000 Bq/l, and the radiostrontium plume has spread some 10 m downstream from the source (Bugai et al.,; Kashparov et al., 2001, 2003).

Method of radiometric measurements in boreholes drilled at the regular mesh can be considered as a reliable method to determine the trench location and its shape. After the calibration, the method can be applied for assessment of ^{137}Cs specific activity in waste material (Bugai et al., 2005). Obtained radiometric results made it possible to determine spatial distribution of ^{137}Cs activity in the trench and to estimate its total content ($A_0{}^{137}$Cs = 600 ± 240 GBq). ^{90}Sr content in the trench was estimated as $A_0{}^{90}$Sr =290 ± 120 GBq in 1999 (400 ± 160 GBq in 1986). Contamination of soil with radiostrontium under the trench can reach a significant level, while other radionuclides (^{137}Cs, TUE, ^{154}Eu) are not found to have high migration abilities. Strontium activity was estimated on the base of activities of low mobile ^{137}Cs and ^{154}Eu. Hence, distribution of ^{90}Sr activity in the trench obtained in 1999–2000 corresponds to its distribution at the moment of the trench creation (t = 0) accounting ^{90}Sr radioactive decay (Bugai et al., 2005). To predict the radionuclides migration from the trench into groundwater and vegetation, it is not enough to know how the radionuclides are distributed inside the storages and what are their total amounts. We have to know the long-term dynamics of the radionuclides mobile forms, which is determined by the fuel particles dissolution. Thus, it is necessary to develop and to parameterize a model of the FP dissolution. Only after the FP dissolution the radionuclides are transferred into a soil solution, sorbed by the soil particles and involved into abiogenic and biogenic migration. Ratio of the specific activity of the radionuclide in absorbed by soil phase (Bq/g) to the specific activity of the radionuclide in dissolved in soil solution radionuclide phase (Bq/ml) in equilibrium state is defined as distribution coefficient (Kd, ml/g). The equilibrium state in soil for strontium and cesium radioisotopes is reached in some seconds and minutes. Thus, the sorption/desorption processes are much faster than the FP dissolution and the radionuclides migration processes.

The great ^{90}Sr mass in the initial fallout in 1986 was presented by a low bioavailable form of the nuclide (Bobovnikova et al., 1991) associated with the matrix of fine-dispersed nuclear fuel particles of various level of transformation, i.e. with the fuel particles. Ratio between activities of low mobile radionuclides (^{95}Zr, ^{99}Mo, ^{89,90}Sr, ^{103,106}Ru, ^{141,144}Ce, ^{154,155}Eu, $^{238\text{-}241}$Pu, etc.) in the fuel component of Chernobyl radioactive fallout is corresponded with the same one in the nuclear fuel exactly at the moment of the accident (Bobovnikova et al., 1991; Kashparov et al., 2001, 2003). The FP matrix is subjected to dissolution in natural conditions, and the radionuclides are leached into the soil solution and involved into the migration processes. The FP dissolution rate in soils of the near zone is determined by the physico-chemical characteristics of the particles (such as ratio of uranium oxides UO_2/U_3O_8, presence of admixtures, etc.) and by an acidity of the soil solution and presence of oxygen (Bobovnikova et al., 1991; Kashparov et al., 2001, 2003).

Fuel particles (separate grains and crystallites of uranium oxide of a median diameter of 4–6 μm) and their conglomerates in the radioactive fallout can be divided into 3 types (Kashparov et al., 2001, 2004).

Chemically extra-stable particles (hypothetically U-Zr-O, formed as a result of high-temperature annealing of UO_2 in presence of zirconium in building materials (Ahamdach, 2002)). These particles were formed in the first moment of the accident on 26.04.86 and then were deposited within the narrow western trace mainly passed through the experimental site Red forest.

Non-oxidized chemically stable fuel particles (UO_2) of the first release (26.04.86), formed as a result of mechanical destruction of nuclear fuel (Kashparov et al., 2001). These particles had created narrow western trace of fallout, passed through the experimental site Red forest. Leaching the fission products from the particles of the certain type during the accident was minimal. This fact is confirmed by constant ratio of fission products.

Chemically low stable particles (UO_{2+x}), formed as a result of oxidization of nuclear fuel in a period 26.04.86-5.05.86 (Kashparov et al., 2001). These particles had created Northern and Southern traces of fuel fallout predominately.

A superposition of all three types of the fuel particles is observed in the initial fallout in various points of 30-km ChNPP zone. Their partial contributions vary depending on direction and distance from ChNPP. Classification of the fuel particles into three types is hypothetical because of difficulties in determination of the borders between various types of the particles (nuclear fuel oxidization degree and zirconium content can vary in a wide range) (Ahamdach, 2002; Salbu et al., 2001). Based on the fact that the narrow western fuel trace of the radioactive fallout passing through Red forest site was formed at the initial moment of the accident 26.04.86 mainly, it can be assumed that the trace is presented by UO_2 and U-Zr-O fuel particles to the great extent. As well as wind direction changed significantly during 10 days of the accident, and Red forest is located at a short distance from the source of release, particles of UO_{2+x} and condensed ^{90}Sr can be presented in the fallout in this area.

This paper presents the part of results obtained within the frameworks of international Chernobyl Pilot Site Project. The paper is dedicated to the studies of the FP dissolution in the trench #22 and to development and parameterization of the dynamical model of the radionuclides leaching into the soil solution.

2. Materials and methods

Soil sampling in the trench had been carried out in 2000 and 2001 at the depths up to 3.5 m using the standard agrochemical sampler with a cylindrical working part (∅ = 6 cm, h = 10 cm). Taking into account an essential heterogeneity of the radionuclides spatial distribution in the trench (Bugai et al., 2005), the samples were collected in seven points along the trench axis from different depths (82 samples).

2.1. RADIOCHEMICAL DETERMINATION OF PHYSICO-CHEMICAL FORMS OF THE RADIONUCLIDES

Gamma-spectrometry of the samples was performed using a low-background gamma-spectrometer ADCAM-300 equipped with passive shield and a semi-conducting detector of pure germanium GEM-30185, (EG&G ORTEC, USA).

Standard radiochemical method was used to measure the total soluble fraction of ^{90}Sr activity (A_s) in the collected samples. The sample is treated with 6 M HNO_3 during 2 h at permanent mixing and heating. After filtration of the solution, hydroxides of high-valence metals (Fe, Al, Ti, Mn, Th, U) are extracted by adding of ammonia without carbon into the solution. Then, after acidification, the solution is left for 3 weeks in order to reach a balance between ^{90}Sr and ^{90}Y. At the end of this period, a stable isotope of yttrium is added into solution and extracted by ammonia without carbon. Obtained precipitate is incinerated to Y_2O_3 and 1 day later is analyzed at the low-background beta-radiometer (CANBERRA-2400, USA) or beta-spectrometer (SEB-01-70, Ukraine). Chemical yield of radiostrontium is estimated using the data on ^{85}Sr.

For estimation of the fraction of chemically extra-stable particles, in part of the samples a completeness of the particles dissolution after 2-h heating in 6 M HNO_3 was evaluated. For this purpose, before the dissolution, the sample (200 g) was charred in air at the temperature of 500 C and then was divided into 10 sub-samples (20 g) and each sub-sample was analyzed by means of γ-spectrometry (^{154}Eu) and β-spectrometry (^{90}Sr). Each sub-sample was measured five times and mixed after each measurement in order to reduce the effect of heterogeneity of contamination. The sediments obtained after the 2-h dissolution of the sub-samples in 6 M HNO_3 were analyzed again. Sub-samples were merged again after that and redissolved in 6 M HNO_3. Prolongation of the dissolution period up to 4 h did not result a significant increase of the radionuclides leaching into solution. Thus, in the experiments the fraction of ^{90}Sr and ^{154}Eu was estimated that cannot be leached from soil to solution by means of standard radiochemical methods (A_{stable}). Insoluble sediment obtained after two stages of dissolution in 6 M HNO_3 can be related to chemically extra-stable fuel particles. Therefore, activity fraction of extra-stable particles ΔFP_{stable} is expressed as $\Delta FP_{stable} = A_{stable}/A_{total}*100\%$, where $A_{total} = A_s + A_{stable}$, and A_s is activity of the soluble fraction.

To estimate the fraction of non-dissolved fuel particles in waste material in the moment of analysis, the method was used, which was developed by UIAR and applied earlier for contaminated upper layer of soil (Kashparov et al., 1999,

2004). Fraction of non-dissolved fuel particles (ΔFP_s) in waste was determined using the data on contents of exchangeable (A_e) and the total soluble part of activity ($A_s = A_e+A_p$) of Chernobyl ^{90}Sr and ^{154}Eu in samples collected at the various depths. The method comparing the contents in soil of exchangeable forms of ^{90}Sr (^{154}Eu) of Chernobyl fallout and injected in water-soluble form ^{85}Sr (^{152}Eu) was used for determination of the activity fraction associated with the matrix of non-dissolved fuel particles (Kashparov et al., 1999). 5 cm^3 of ^{85}Sr and ^{152}Eu (activity about 100 Bq) and 25 cm^3 of water had been added into each sub-sample of 100 cm^3 volume. The time of incubation of radionuclides in samples was 6–10 days. Afterwards, a salt extraction of radionuclides in 2M NH_4Ac had been done. Each sample with added solution (1:10) was mixed intensively during 1 h and then the mixtures were left for 1 day. Then the standard radiochemical method had been applied for determining the radionuclides activity in solution (A_e) and in precipitation (A_p) after dissolution in 6 M HNO_3. The fractions of ^{85}Sr (^{152}Eu) and ^{90}Sr (^{154}Eu) activities, extracted from the soil, and the fraction of activity associated with the matrix of non-dissolved particles (which can be solved in natural conditions) are defined as

$$\Delta^{85}Sr=[A_e(^{85}Sr)/(A_e(^{85}Sr)+A_p(^{85}Sr))];$$

$$\Delta^{90}Sr=[A_e(^{90}Sr)/(A_e(^{90}Sr)+A_p(^{90}Sr))];$$

$$\Delta FP_s(Sr)= [1-\Delta^{90}Sr/\Delta^{85}Sr]*100\%.$$

The total soluble fraction of ^{90}Sr and ^{154}Eu activity (A_s) is measured by standard radiochemical method with sample dissolution in 6 M HNO_3. It makes up only a part of the total sample activity (about 80%) as activity (about 20%) of chemically extra-stable particles wasn't taken into account.

This method of ΔFP_s(Sr) evaluation gives the good results in case of the soil surface contamination in absence of intensive migration of ^{90}Sr (Kashparov et al., 1999, 2004), when the sampling of leached radiostrontium is representative. However, the method can mislead in case of very heterogeneous spatial distribution of contamination (Bugai et al., 2005). In this case, exchangeable ^{90}Sr content in the sample can be related not only to the dissolution of the fuel particles in the sample. Some activities of exchangeable radiostrontium could migrate from or into the sample. For waste in the trench, estimation of ΔFP_s(Eu) is more reliable because of low migration ability of ^{154}Eu in soil (waste). This estimate is based on a fact that the activity ratio A(^{90}Sr)/A(^{154}Eu) = c(t) in initial Chernobyl fuel fallout is the same in various parts of the near zone and is changed in time due to the radioactive decay (Kashparov et al., 2003). Total initial FP-associated radiostrontium content in each sample had been calculated using the values of activity of low-mobile ^{154}Eu ($A_s(^{154}$Eu)):

$$(A_s(Sr\text{-}Eu)=c(t)\cdot A_s(^{154}Eu)$$

Then the FP-associated fraction of activity in moment t can be calculated as

$$\Delta FP_s(Sr\text{-}Eu)=[A_p(^{90}Sr)\text{-}(1\text{-}\Delta^{85}Sr)*A_e(Sr)]/\cdot A_s(Sr\text{-}Eu)))\cdot 100\%.$$

Both methods gave close results.

2.2. DETERMINATION OF FP-ASSOCIATED ACTIVITY FRACTION BY MEANS OF SEDIMENTATION IN HEAVY LIQUID

We used a carefully homogenized soil sample of about 100 g. The sample was taken from the depth of 210–220 cm in borehole 42–231. Part of the sample was used to determine $\Delta FP_s(Sr)$ by above-described radiochemical method, and 2-3 g of the sample were dispersed in 30 ml of bromoform (bulk density of 2.88 g/cm^3) in plastic tubes (Nalgen, FEP Oak Ridge tube). After centrifugation the majority of soil components (density < 2.88) remains on the air/bromoform interface while heavy fraction (fuel particles and heavy minerals of soil) sinks. In order to quantify the part of activity (^{134,137}Cs, ^{154,155}Eu and ^{241}Am) associated with the fuel particles in the trench, gamma activity of sample was determined before (A_{total}) and after ($A_{FP} = A_p+A_{stable}$) separation of the fuel particles. Then the total FP-associated activity fraction in the moment of analysis was calculated as $\Delta FP=A_{FP}/A_{total}*100\%$.

2.3. CHARACTERIZATION OF CHEMICALLY EXTRA-STABLE FUEL PARTICLES

In order to study chemically extra-stable fuel particles, the precipitation obtained as a result of dissolution of the waste samples by 2-h heating in 6 M HNO_3 was used. The fuel particle characterization technique was based on direct observ-ation of the fuel particles morphology using a scanning electron microscopy (SEM) (Ahamdach, 2002; Bugai et al., 2005; Kashparov et al., 2000).

2.4. DETERMINATION OF DISSOLUTION RATE OF THE FP FROM THE TRENCH IN MODEL EXPERIMENTS IN FLUIDS A pH = 3 – 4

In order to study the fuel particles dissolution kinetics at pH = 3 – 4, three waste samples of several kilograms mass had been collected in 2001. The samples were collected in various parts of the trench at the depth of 1.5 m in boreholes 41/233, 12/241 and 32/23, and then they were carefully mixed. Two sub-samples each of 100 g were taken from each sample for studies at pH ≈3 and pH ≈4. 400 ml of 2M NH_4Ac were added to each sub-sample, then the suspensions were mixed during 10 min with mechanical mixer and filtered through the "blue ribbon" filters. Then the solid precipitates passed through the same leaching procedure again. Radionuclides activities in solutions were measured by the standard methods. Thus, the mobile fraction of ^{90}Sr was washed out from the sub-samples. 18 g of NaCl (for reduction of K_d of Sr) and 10 ml of ^{85}Sr solution were added into the solid precipitates. Volume of each solution was adjusted to 400 ml by adding the distilled water, and 0.1 M HCl was added to reach the necessary value of pH (3 or 4). To avoid evaporation, the glasses were firmly covered with Parafilm films. Once a week the solutions were carefully mixed and the pH values were controlled or adjusted at necessity with 0.1 M HCl. Three times (at the beginning of the experiment, 42 days later and at the end of the experiment – on 216th day) ^{90}Sr and ^{85}Sr activities in solutions were measured. At the end of the experiment, radionuclides activities in precipitates were measured too. The activity fraction of ^{90}Sr leached into the solution during 216 days (fraction of dissolved labile fuel particles, $1 - \Delta FP_s$) and the fuel particles transformation rates at pH = 3 – 4 were determined.

3. Obtained results

3.1. CONDENSED ^{90}Sr FRACTION IN THE INITIAL FALLOUT IN THE ChNPP ZONE

At the time of study (13–15 years after the Chernobyl accident) it was not possible to measure in experiment the initial fraction of ^{90}Sr condensed component in the radioactive fallout (i.e., the initial activity fraction non-associated with the FP matrix). In April-May 1986 the fraction of condensed ^{90}Sr component in the initial Chernobyl radioactive fallout was measured in Chernobyl town only (south-east direction at the 15 km distance from ChNPP) (Bobovnikova et al., 1991). During an active phase of release, it mounted 5% in relation to the total activity of deposited ^{90}Sr. As nearer to ChNPP, parts of condensed ^{137}Cs and ^{90}Sr in the radioactive fallout decrease, thus, at 3–4 km distance from the source the part of condensed ^{137}Cs component is two to three times less in comparison with the part at 15 km distance (Kashparov et al., 2001, 2003).

At the mechanical destruction of the real Chernobyl irradiated nuclear fuel (mechanism of UO_2 particles formation deposited near the experimental site) in model experiments, fraction of mobile ^{90}Sr (easily washed in light acids solutions) was equal to 2.3% (Kashparov et al., 2000). At the experimental modeling of oxidized UO_{2+x} particles formation during nuclear fuel annealing in air at temperature 670 K in 1–21 h, the fraction of mobile ^{90}Sr was 3–7%.

Activity ratios of refractory radionuclides (^{154}Eu, $^{238\text{-}240}Pu$, etc.) and ^{90}Sr in radioactive fallout in the near zone of Chernobyl accident correspond to those in Chernobyl nuclear fuel and in FP selected from soil (Bobovnikova et al., 1991; Kashparov et al., 2001). It means that the condensed fraction of ^{90}Sr in radioactive fallout in this territory could not exceed a relative uncertainty of the radionuclides activity measurement in the same samples (<5%).

The experimental site Red forest is located along the axis of the western trace of radioactive fallout with prevailing amount of non-oxidized UO_2 fuel particles, and the fraction of UO_{2+x} is significantly lower. Therefore, a realistic estimate of the condensed component fraction in the radioactive fallout in the experimental site Red forest is about 2% with a maximal upper level not exceeding 5% in relation to the total ^{90}Sr activity in the radioactive fallout.

3.2. FRACTION OF CHEMICALLY EXTRA-STABLE PARTICLES IN THE TRENCH

In the selected samples a completeness of the dissolution after 1–2-h heating in 6–8 M HNO_3 was estimated. Prolongation of the dissolution period up to 4 h did not result a significant increase of the radionuclides leaching into solution. Thus, in the experiments the fraction of ^{90}Sr was estimated that cannot be leached from soil to solution by means of standard radiochemical methods.

Values of ΔFP_{stable} widely vary because of very heterogeneous distribution of FP-associated activity in the trench, while the distribution of the mobile strontium is more homogeneous due to its migration. Insoluble sediment obtained after two stages of dissolution in 6–8 M HNO_3 can be related to chemically extra-stable fuel particles. Fraction of ^{154}Eu in the chemically extra-stable particles varies in a wide range (8–26%), which corresponds to the estimates derived from ^{90}Sr data. In all samples collected beneath the trench, where FP are absent and ^{90}Sr activity can be caused only by its vertical migration, we obtained $\Delta FP_{stable} = 0$.

On the base of results it can be concluded that the value of ^{90}Sr activity fraction in composition of chemically extra-stable fuel particles mounts to about $\Delta FP_{stable} = 20\% \pm 10\%$. This estimate is in good accordance with the results of our earlier studies of dependence of fraction of chemically extra-stable fuel particles in soil in the 30-km zone on the distance and direction from the source of release (Kashparov et al., 2004). It was found that along the southern and northern traces of radioactive fallout, ^{90}Sr activity in soil that is associated with acid-resistant fuel particles varies from 2% to 10% and from 3% to 13% respectively. Along the western trace this fraction varies from 6% to 21% with the maximum near to ChNPP, where the studied trench is located. In order to clarify the nature of chemically extra-stable particles a completeness of dissolution of a sample of irradiated Chernobyl nuclear fuel of the burn-up of 6.5 MW*d/kg was estimated. Detail description of physico-chemical characteristics of the UO_2 sample is presented in Kashparov et al. (2000) and Kashparov et al. (1996). The sample had been dissolved in heated to 98°C 6M HNO_3 in accordance with the standard radiochemical method (see above). More than 99.98% of initial activity of ^{90}Sr, $^{134,137}Cs$, $^{154,155}Eu$ and ^{241}Am had been leached into solution. Taking into account that oxidized nuclear fuel is dissolved much faster than UO_2 (Kashparov et al., 2000), on can state that chemically extra-stable particles are not the particles of pure uranium oxides and that they were formed in accidental conditions at high temperature.

We suspect that these stable particles are made of UZrO. X-ray micro-analysis of such particles should confirm this assumption. It seems that such the stable fuel particles can hardly be ever dissolved in soil in natural conditions and the radionuclides associated with the matrix of these particles will not influence the radiological situation. However, a presence of these particles is to be taken into account when we measure the total radionuclides activity in sample or when we develop the technology to deal with the radioactive waste.

3.3. ESTIMATE OF THE LABILE PART OF ^{90}Sr ACTIVITY IN COMPOSITION OF THE FUEL PARTICLES AT THE MOMENT OF EXPERIMENTAL STUDY

Fraction of non-dissolved fuel particles (ΔFP_s) in waste was determined using the data on contents of exchangeable (A_e) ant the total soluble part of activity ($A_s = A_e + A_p$) of Chernobyl ^{90}Sr and ^{154}Eu in samples collected at the various depths. The total soluble part of ^{90}Sr and ^{154}Eu activity (A_s) is measured by standard radiochemical method with sample dissolution in 6–8 M HNO_3. It makes up only a part of the total sample activity (about 80%) as activity (about 20%) of chemically extra-stable particles (see Section 3.2) was not taken into account.

Fraction of non-dissolved fuel particles (ΔFP_s) for various points of the trench varies in a wide range and needs the following specification. Value of ^{90}Sr and ^{154}Eu activity part in composition of fuel particles related to a labile part for all points of sampling (at the identical methods of sampling with 10 cm step and samples measurements) in 1999–2000 was $\Delta FP_s(Sr) = 30–70\%$.

More correct estimates of non-dissolved fuel particles fraction (ΔFP_s) were obtained in 2001 for three homogeneous samples collected at the 1.5 m depth and in 2002 for additional samples collected at various depths and in various parts of the trench. Both two methods of estimation described above, and a method based on ratio of sequentially displaced exchangeable ^{154}Eu part to its total content were used for given samples. Activities of ^{85}Sr, ^{152}Eu, ^{154}Eu, ^{241}Am and ^{137}Cs in obtained solutions were measured by γ-spectrometry. ^{90}Sr activities were determined 2 weeks later by means of β-radiometry.

Besides the radiochemical method of determination of FP-associated activity fraction, the method of sedimentation in heavy liquid was applied. FP were separated from the soil sample taken from the depth 210–220 cm in borehole 42–231. Sedimentation method results show that 52% of ^{137}Cs, 70% of ^{154}Eu and 65% of ^{241}Am are still associated with the fuel particles. It means that the source of radionuclides in the trench is mainly represented by the fuel particles.

The fraction of the total ^{137}Cs activity associated with the fuel particles is lower than those of ^{154}Eu and ^{241}Am because of its higher volatility. ^{90}Sr activity associated with the fuel particles is calculated using measured activity of ^{154}Eu and activity ratio $^{90}Sr/^{154}Eu$ in Chernobyl nuclear fuel ($^{90}Sr/^{154}Eu$ = 72 in 2000). Results show that 69% of ^{90}Sr still are associated with fuel particles. This result is in good agreement with radiochemistry analysis for this sample: $\Delta FP_s(Sr)$ = 65 ± 8 of ^{90}Sr and $\Delta FP_s(Eu)$ = 73 ± 7 of ^{154}Eu.

Analysis of physico-chemical forms of the radionuclides in various parts of the trench has shown that in 2001–2002 activity fractions of ^{90}Sr and ^{154}Eu in the fuel particles (ΔFP_s) from the total soluble part of activity UO_2 and UO_{2+x} particles were:

$\Delta FP_s(Sr)$ = 40 ÷ 65% (from exchangeable part of ^{90}Sr)

$\Delta FP_s(Eu)$ = 50 ÷ 70% (from exchangeable part of ^{154}Eu)

$\Delta FP_s(Eu\text{-}Sr)$ = 40 ÷ 50% (from non-exchangeable part of ^{90}Sr activity and total activity of ^{90}Sr calculated from ^{154}Eu)

- Average by activity value ΔFP_s is 55 ± 15% of soluble part of ^{90}Sr activity (in 2001–2002) or ΔFP=(0.55 ± 0,15)*(80 ± 10)% + (20 ± 10)% = 64 ± 13% of total of ^{90}Sr activity in the fuel particles (from U-Zr-O, UO_2 and UO_{2+x} particles in 2001). About 37 ± 13% of total of ^{90}Sr activity (in 2001–2002) is found beyond fuel particles.
- Obtained results correspond well to direct measurements of $^{154,155}Eu$ and ^{241}Am activity fraction ΔFP = 65 ÷ 70% in composition of the fuel particles, which were isolated directly from a sample collected in the trench (borehole 42–231, 210–220 cm) by means of sedimentation in heavy liquid.

3.4. FUEL PARTICLES DISSOLUTION KINETICS IN THE TRENCH

Process of dissolution of the fuel particles of various genesis (U-Zr-O, UO_2 and UO_{2+x} particles) can be described as a sum of the first-order kinetic equations (Kashparov et al., 1999, 2000, 2004):

$$\Delta FP(t) = \sum_{i=1}^{3} \Delta FP_i(0) \cdot \exp(-k_i \cdot t) \quad \sum_{i=1}^{3} \Delta FP_i(0) = 1 \quad \Delta FP_s(0) = \sum_{i=2}^{3} \Delta FP_i(0) = 0{,}8 \pm 0{,}1 \qquad (1)$$

where: $\Delta FP_i(0)$ – initial fraction of the particles of i-type (genesis);
k_i – transformation constant of the particles of i-type (per year);
t – time after the beginning of the dissolution process (years).

As it was already shown, fraction of the very stable particles (U-Zr-O, $k_1 = 0$) $\Delta FP_1(0)$=(0.2 ± 0.1). Transformation constant of UO_{2+x} particles is one order of magnitude higher than of UO_2 particles (Fig. 1) (Ahamdach, 2002). Half-dissolution period of UO_{2+x} particles in the trench conditions, at pH of the soil solution of about 5, is equal 2.5 years only. Therefore, to the moment, 14–16 years after the trench construction, practically all UO_{2+x} particles have been dissolved, and the particles in the trench are presented only by U-Zr-O and UO_2 particles. Dissolution of UO_2 particles is comparable slow, and their dissolution rate in the natural conditions can be hardly estimated during the observation period (2–3 years), especially taking into account the apparent deviations of this value. However, we can dissolve the waste samples in more aggressive conditions (at pH = 3 – 4) accelerating the fuel particles transformation. Then obtained dissolution parameters can be extrapolated to the natural (trench) conditions.

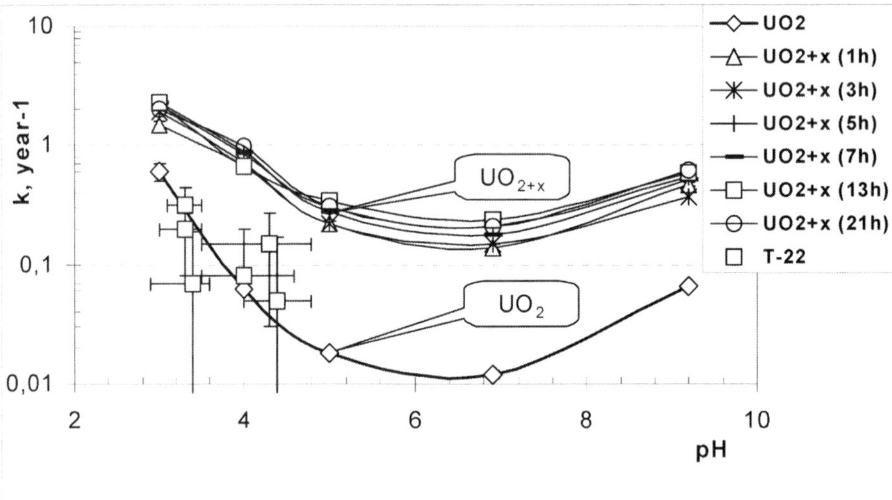

Figure 1. Results of model experiments on dissolution of nuclear fuel particles (UO_2 and UO_{2+x}, obtained by annealing in air of UO_2) and of waste material from the trench at pH = 3 and pH = 4.

3.5. DISSOLUTION KINETICS OF THE WASTE SAMPLES FROM THE TRENCH IN FLUIDS AT pH = 3 – 4

The activity fraction of ^{90}Sr leached into the solution during 216 days (fraction of dissolved labile fuel particles, $1\text{-}\Delta FP_s$) and the fuel particles transformation rates at pH = 3 – 4 were determined, (Table 1). The higher values of the dissolution velocity in the 2nd and in 3rd samples as compared to the 1st sample are related to the higher initial contents of ^{90}Sr outside the FP matrix in these samples and to the non-complete washing out the radionuclide from the samples before the beginning of the dissolution process.

TABLE 1. Activity fraction of ^{90}Sr in labile FP after the dissolution during 216 days, and FP transformation constants

Sample	pH of solution	ΔFPs, %	k, per year
41/233	2.9–3.6	96.2 ± 6.7	0.07 ± 0.12
(Sampling 2001)	4–4.8	96.8 ± 6.8	0.05 ± 0.12
12/241	3.1–3.5	82.7 ± 5.8	0.32 ± 0.12
(Sampling 2001)	3.5–4.8	91.5 ± 6.4	0.15 ± 0.12
32/23	3.0–3.5	89.1 ± 6.3	0.20 ± 0.12
(Sampling 2001)	3.5–4.6	95.1 ± 6.7	0.08 ± 0.12

It was shown in this model experiment that the dissolution rate of the fuel particles from the trench in solutions at pH = 3 and 4 corresponds to the dissolution rate of UO_2 fuel particles (real particles of irradiated Chernobyl nuclear fuel) in simulated conditions (Table 1, Fig. 2) (Ahamdach, 2002). Obtained transformation constants of UO_2 fuel particles in the trench (in assumption that 25% of activity were associated with dissolved in the first years UO_{2+x} particles) at various soil acidity (Fig. 2) also are in good accordance with the earlier results (Fig. 2). A correlation was found between ΔFP_s and acidity of media (Fig. 2). Therefore, the earlier obtained transformation constants (k, per year) of the model fuel particles of different oxidation state (UO_2 and UO_{2+x})

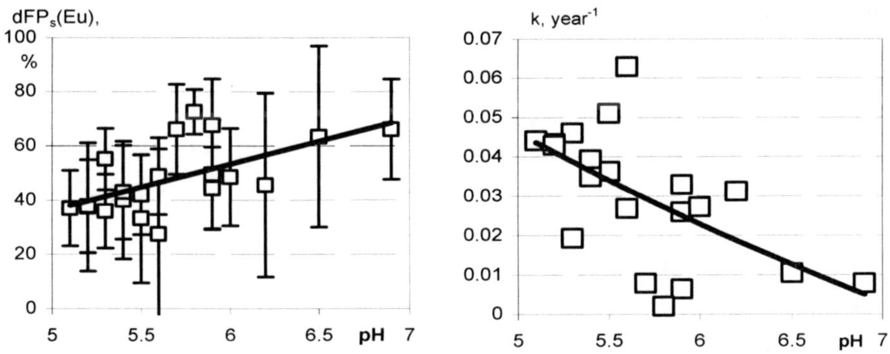

Figure 2. Dependence of ^{154}Eu FP-associated activity fraction and of the derived value of the transformation constant of UO_2 particles on the soil acidity (in 2002).

(Ahamdach, 2002) can be also applied to the fuel particles in the trench. For this purpose, we have to take the constants for the real acidity of the soil solution in the trench (acidity of the soil solution sampled by the ceramic samplers is equal to 5):

For UO_2 particles: $k_2 = 0{,}018 \pm 0.001$/year

For UO_{2+x} particles: $k_3 = 0.28 \pm 0.05$/year

3.6. ACIDITY OF THE SOIL SOLUTION IN THE TRENCH

Data on acidity of the soil solution are necessary for estimation of the FP dissolution rate. Acidity was determined in course of monitoring works after sampling the soil solution at various depths with vacuum ceramic samplers. Average pH value is 5 at its variation in small range 4.8–5.1. However, the solution in samplers is in contact with air during the long period, and its acidity can be changed with time. In this reason, acidity of soil solution was also determined by standard agrochemical method, which is based on measurement of pH at ratio distilled water/soil = 2.5. Soil samples were collected in 2002 in five points at various depths in the trench. Average pH was estimated as 5.8, which exceeds the above reported value. However, it's well known that decreasing of ratio distilled water/soil leads to decreasing of pH of water extract. At real humidity of soil (5–10%) pH will correspond to the values measured in soil solutions. Thus, we can assume that the value pH = 5 in soil solution in the trench is proved.

Before the creation of TSRW Red Forest in 1987, the fuel particles could be in more aggressive conditions in the forest litter. In 2002 pH of upper 0–5 cm layer of soil in pine forest near the TSRW (3 points) was measured by standard agrochemical method. Obtained value pH = 5 is in good accordance with the results of our previous measurements for the forests of the 30-km zone (pH = 4.8 ± 0.5) and only slightly differs comparing the value in the trench. Difference between the fuel particles transformation constants at pH = 4.5 and pH = 5.5 (Fig. 2) does not exceed uncertainty of estimation of the transformation constant and thus can be neglected.

3.7. ESTIMATION OF UO_2 AND UO_{2+X} PARTICLES FRACTIONS IN THE INITIAL MOMENT OF THEIR DISSOLUTION IN THE TRENCH

Using values of the fuel particles transformation constants ($k_2 = 0.018 \pm 0.001$/year and $k_3 = 0.28 \pm 0.05$/year at pH = 5) and the value of ^{90}Sr activity fraction in fuel particles ($\Delta FP(t) = 0.64 \pm 0.13$ for $t = 15$ years), we can calculate fractions of activity of UO_2 and UO_{2+x} particles for $t = 0$:

$$\Delta FP(t) = \sum_{i=1}^{3} \Delta FP_i(0) \cdot \exp(-k_i \cdot t)$$

$$\Delta FP(t) - \Delta FP_1(0) = \Delta FP_2(0) \cdot \exp(-k_2 \cdot t) + \Delta FP_3(0) \cdot \exp(-k_3 \cdot t)$$
$$\Delta FP_2(0) + \Delta FP_3(0) = \Delta FP(0) - \Delta FP_1(0) = 1 - (0{,}2 \pm 0{,}1) = 0{,}8 \pm 0{,}1$$
$$\Delta FP_2(0) = [(\Delta FP(t) - \Delta FP_1(0)) - (0{,}8 \pm 0{,}1) \cdot \exp(-k_3 \cdot t)] \Big/ [\exp(-k_2 \cdot t) - \exp(-k_3 \cdot t)]$$
$$\Delta FP_3(0) = (0{,}8 \pm 0{,}1) - \Delta FP_2(0)$$

where:

$$\Delta FP(t) = 0{,}64 \pm 0{,}13$$
$$\Delta FP_1(0) = 0{,}2 \pm 0{,}1$$
$$t = 15 \text{year}$$

In this way we obtained that ^{90}Sr in the initial radioactive contamination of the trench (at t = 0) was presented in four different physico-chemical forms:

- U-Zr-O particles: $\Delta FP_1(0) = 0.20 \pm 0.10$ of ^{90}Sr total activity ($k_0 \approx 0$)
- UO_2 particles: $\Delta FP_2(0)$ = *0.57 ± 0.15 of* ^{90}Sr *total activity* ($k_1 = 0.018 \pm 0.001$/*year for pH = 5*)
- UO_{2+x} particles: $\Delta FP_3(0)$ = *0.21 ± 0.15 of* ^{90}Sr *total activity* ($k_2 = 0.28 \pm 0.05$/*year for pH = 5*)
- Condensed form of ^{90}Sr outside particles: $A_k(0)$ = *0.02 ± 0.01 of* ^{90}Sr *total activity* ($A_{Sr\text{-}total}(0)$) ($k_4 = \infty$).

3.8. DYNAMICS OF ^{90}Sr LEACHING INTO MOBILE FORMS IN THE TRENCH

Thus, we can calculate the relative dynamics of the ^{90}Sr FP-associated fraction $\Delta FP(t)$ and leached from FP into mobile forms fraction of strontium activity $A_{Sr}(t)$ in the trench (Fig. 3):

$$\Delta FP(t) = (0{,}20 \pm 0{,}10) + +(0{,}57 \pm 0{,}15) * \exp(-(0{,}018 \pm 0{,}001) * t) + (0{,}21 \pm 0{,}15) * \exp(-(0{,}28 \pm 0{,}05) * t);$$

$$A_{Sr}(t) / (A_{Sr\text{-}total}(0) * \exp(-\lambda * t)) = (1 - \Delta FP(t))$$

Obtained data show that now the activity of the mobile forms of ^{90}Sr in the trench had reached the maximal value.

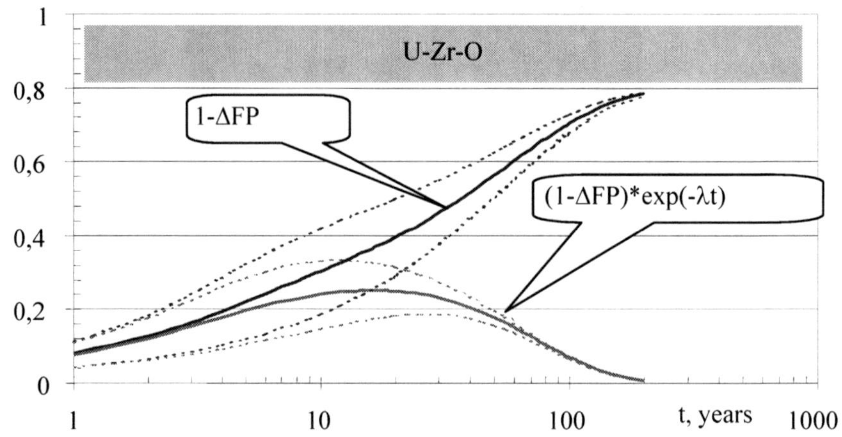

Figure 3. Relative dynamics of ^{90}Sr contents in the mobile forms in the trench (without release due the migrations from the trench).

4. Conclusion

As a result of studies of the fuel particles dissolution kinetics in natural and simulated conditions, the activity fractions of various FP types in Chernobyl radioactive fallout were estimated for a first time. It was found that in TSRW Red forest (2.5 km South-West of the ChNPP) ^{90}Sr in initial radioactive contamination of the trench (at t = 0) was presented in four different physico-chemical forms:

U-Zr-O particles: $\Delta FP_1(0) = 0.20 \pm 0.10$ of ^{90}Sr total activity ($k_0 \approx 0$)

UO_2 particles: $\Delta FP_2(0) = 0.57 \pm 0.15$ *of* ^{90}Sr *total activity* ($k_1 = 0.018 \pm 0.001$/*year for pH=5*)

UO_{2+x} particles: $\Delta FP_3(0)=0.21 \pm 0.15$ *of* ^{90}Sr *total activity* ($k_2 = 0.28 \pm 0.05$/*year for pH = 5*)

Condensed form of ^{90}Sr outside particles: $A_k(0) = 0.02 \pm 0.01$ *of* ^{90}Sr *total activity* ($A_{Sr\text{-}total}(0)$) ($k_4 = \infty$).

In order to characterize the trench source term, the parameters of the FP dissolution kinetics in the trench were estimated and the model of the ^{90}Sr mobile forms dynamics was created.

Obtained data show that now the activity of the mobile form of ^{90}Sr in the trench had reached the maximal value.

Acknowledgements

The present work was supported by funding in the frame of Chernobyl Pilot Site Project, a collaborative French-Ukrainian program of the Institute for Radioprotection and Nuclear Safety (France).

References

Ahamdach, N. (2002) The Chernobyl pilot site project: Isolation and microscopic characterisation of fuel particles. *Radioprotection-Colloques*, **37**(C1) 1055–1060.

Bobovnikova, T.I. et al. (1991) Phisical chemical forms of radionuclides in atmospheric follout after the Chernobyl accident and their transformation in soil. *Atomnaya Energiya*, **69**(1), 449 (Russ.).

Bugai, D., Kashparov, V., Dewiére L., Khomutinin, Yu., Levchuk, S., and Yoschenko, V. (2005) Characterization of subsurface geometry and radioactivity distribution in the trench containing Chernobyl clean-up wastes. *Environmental Geology*, **47**, 869–881.

Bugai, D.A., Dewiere, L., Kashparov, V.A., and Ahamdach, N., Strontium-90 transport parameters from source term to aquifer in Chernobyl Pilot Site. *Radioprotection – Colloques*, **37**(C1) 11–16.

Dewiere, L., Bugai, D., Grenier, C., Kashparov, V., and Ahamdach, N. (2004) ^{90}Sr migration to the geosphere from a waste burial in the Chernobyl exclusion zone. *Journal of Environmental Radioactivity*, **74**(1–3), 139–150.

Dewiere, L., Bugai, D., Kashparov, V., and Barthès V. (2005) Validation of the global model for ^{90}Sr migration from the waste burial in the Chernobyl exclusion zone. *Radioprotection* **40**(1) (May 2005), S245–S251.

Kashparov, V.A., Ahamdach, N., Zvarich, S.I., Yoschenko, V.I., Maloshtan, I.N., and Dewiere, L. (2004) Kinetics of dissolution of Chernobyl fuel particles in soil in natural conditions. *Journal of Environmental Radioactivity*, **72**(3), 335–353.

Kashparov, V.A., Ivanov, Yu.A., Zvarich, S.I., Protsak, V.P., Khomutinin, Yu.V., Kurepin, A.D., and Pazukhin, E.M. (1996) Formation of hot particles during the chernobyl nuclear power plant accident. *Nuclear Technology*, **114**(1), 246–253.

Kashparov, V.A., Lundin, S.M., Khomutinin, Yu.V., Kaminsky, S.P., Levtchuk, S.E., Protsak, V.P., Kadygrib, A.M., Zvarich, S.I., Yoschenko, V.I., and Tschiersch, J. (2001) Soil contamination with ^{90}Sr in the near zone of the Chernobyl accident. *Journal of Environment Radioactivity*, **56**(3), 285–298.

Kashparov, V.A., Lundin, S.M., Zvarich, S.I., Yoschenko, V.I., Levtchuk, S.E., Khomutinin, Yu.V., Maloshtan, I.N., and Protsak, V.P. (2003) Territory contamination with the radionuclides representing the fuel component of Chernobyl fallout. *The Science of The Total Environment*, **317**(1–3), 105–119.

Kashparov, V.A., Oughton, D.H., Zvarich, S.I., Protsak, V.P., and Levchuk, S.E. (1999) Kinetics of fuel particle weathering and ^{90}Sr mobility in the Chernobyl 30-km exclusion zone. *Health Physics*, **76**(3), 251–259.

Kashparov, V.A., Protsak, V.P., Ahamdach, N., Stammose, D., Peres, J.M., Yoschenko, V.I., and Zvarich, S.I. (2000) Dissolution kinetics of particles of irradiated Chernobyl nuclear fuel: influence of pH and oxidation state on the release of radionuclides in contaminated soil of Chernobyl. *Journal of Nuclear Materials*, **279**, 225–233.

Salbu, B., Krekling, T., Lind, O.C., Oughton, D.H., Drakopoulas, M., Simionovichi, A., Snigireva, I., Snigirev, A., Weitkamp, T., Adams, F., Janssens, K., and Kashparov, V. (2001) High energy X-ray microscopy for characterization of fuel particles. *Nuclear Instruments and Methods in Physics Research A*, **467–468**, 1249–1252.

PHENOMENON OF A FAST MIGRATION OF PLUTONIUM RADIOISOTOPS IN GROUND WATER: COLLOIDS OR SOLUBLE FORM?

SVYATOSLAV LEVCHUK*[1], VASYL YOSCHENKO[1], VALERY KASHPAROV[1], NATHALIE VAN MEIR[2], CHRISTOPHE ARDOIS[2], DMITRIY BUGAÏ[3]
[1]*Ukrainian Institute of Agricultural Radiology, Mashinobudivnykiv str.7, Chabany, Kyiv-Svjatoshin distr., Kyiv reg., 08162, Ukraine*
[2]*Institute of Radioprotection and Nuclear Safety, BP 17, 92262, Fontenay aux Roses cedex, France*
[3]*Institute of Geological Science, Gonchara str.55b, Kyiv 01054, Ukraine*
**phone: (380 44) 526 2444, e-mail: slava@uiar.kiev.ua*

Abstract: Field experiments and laboratory studies were performed to investigate migration processes of plutonium isotopes in the aquifer at the experimental site. The objectives of these experiments were to characterize the spatial distribution and possible migration mechanisms of plutonium in the aquifer, and to obtain the parameters that determine radionuclide retardation. During 2005–2006 experimental investigations were carried out and spatial distributions of plutonium isotopes (^{239,240}Pu, ^{238}Pu), ^{90}Sr and main dissolved ions in aquifer along the groundwater flow were obtained. Radionuclide specific activities in groundwater depended on the location of the piezometer and varied in the range of 1 ÷ 200 mBq/L for ^{239,240}Pu, 0.5 ÷ 100 mBq/L for ^{238}Pu and n·100 ÷ n·104 Bq/L for ^{90}Sr. It was found that the spatial features of the specific activities distributions of plutonium and strontium in groundwater were similar, i.e., there was a correlation between the positions of the radionuclides activity maximums. The Pu isotopes' plume in the aquifer spreads about 10 m downstream of the radionuclides source. Characterization of the initial radionuclide composition of waste showed that all plutonium in the aquifer originated from the trench. Ratio of plutonium isotopes (^{239,240}Pu/^{238}Pu) was the same in waste material and in groundwater samples. In-situ ultrafiltration of several groundwater samples was carried out. Obtained size fractionation data on plutonium activities in collected colloids suggest that significant part of Pu (50–80%) was associated with low molecular weight fraction (<5 kDa).

D.H. Oughton and V. Kashparov (eds.), *Radioactive Particles in the Environment*,

Keywords: plutonium, groundwater, aquifer, migration, distribution coefficient.

1. Introduction

The safety of the radioactive wastes in the nuclear fuel storages is related to their capacity to confine radioactivity and isolate it from the biosphere. The most likely process that can result in the radionuclides release from a storage is their migration with ground water. There have been many studies of migration of transuranium radionuclides from the radioactive waste storages (Buddemeier and Hunt, 1988; Penrose et al., 1990; Kaplan et al., 1994; Kersting et al., 1999; Dai et al., 2002; Dai et al., 2005; Novikov et al., 2006). All these studies confirmed existing of the increased plutonium concentrations in aquifer at the different distances downstream from the sources (up to several km). Some data suggest that colloidal transport of Pu is not a substantial process (Dai et al., 2002; Dai et al., 2005). Other results confirm that colloids are responsible for a long-distance migration of plutonium in aquifer (Penrose et al., 1990; Kaplan et al., 1994; Kersting et al., 1999; Novicov et al., 2006). Differences between the studies results can be attributed both to the experimental protocol and site specific conditions.

Despite these researches, our knowledge and ability for the correct prediction of Pu behavior in aquifer are still limited.

The aim of the study was to improve our understanding of plutonium transport in the shallow aquifer.

2. Methods

2.1. SOURCE

The study site is located in the territory of the shallow waste burial called "Red Forest", which was created in 1987 during the ChNPP cleanup operations (Fig. 1). The site includes trench #22 and adjacent territories. The 70 m long and 2–3 m deep trench contains radioactive waste (humus topsoil and rests of the contaminated trees) and thus represents the main source of radionuclides and stable chemical elements in the sand aquifer near its location. According to our estimation, in 2000 the trench contained ^{137}Cs ≈ 600 GBq, ^{238}Pu ≈ 2.4 GBq, ^{239,240}Pu ≈ 4.3 GBq (Bugai et al., 2005). Microscopic analyses of waste particles and leaching experiments had shown that 10–30% of the radioactive inventory was associated with chemically extra-stable Zr-U-O particles. The largest fraction of activity (associated with the fuel component radionuclides of Chernobyl fallout) in the trench (30–60%) is currently associated with relatively slowly dissolving non-oxidized UO_2 fuel particles (Dewier et al., 2004).

The trench bottom is periodically flooded by a shallow (3 ± 1 m) groundwater table. The real groundwater velocity in the aquifer is estimated at 10 m/year.

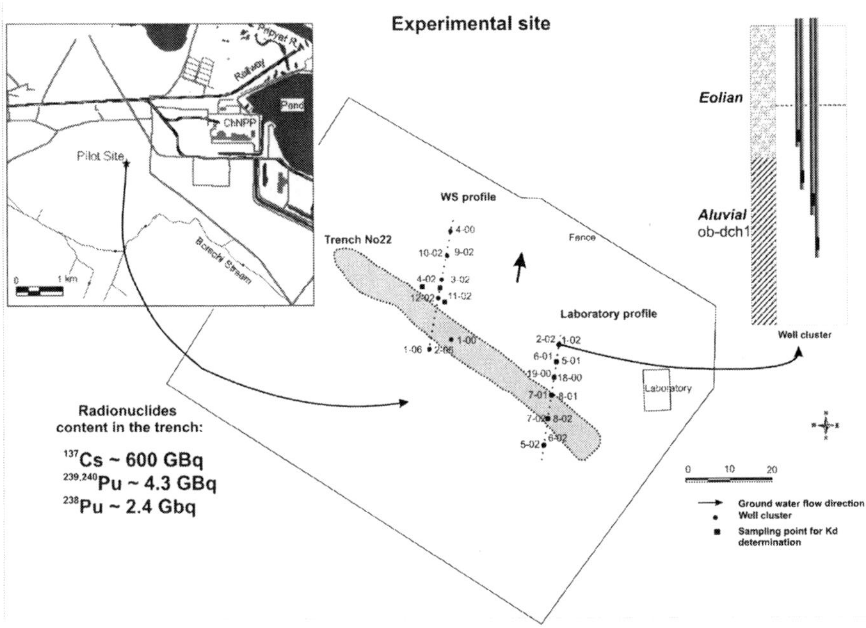

Figure 1. Location of the Pilot Site and monitoring wells sampled in this study.

2.2. SAMPLING

Water sampling for determination of total plutonium activity was carried out with the system that consisted of a vacuum pump, connecting hoses, an online filter and a water volume meter. Cotton wound cartridge filters (0.5 μm pore rate, Westdale Filters Limited, UK) were used for online water filtration. Volume of samples varied in the range of 10–30 L for total Pu determination and in the range of 2–4 L for ultrafiltration, depending on the expected radionuclides activity in the water samples. Water flow rates during sampling varied from 0.15 to 0.3 L/min.

2.3. ULTRAFILTRATION

The procedure consisted of sequential filtrations through 0.2 μm, 30 kDa, 5 kDa and 1 kDa (just for the last water sample) membranes using a Millipore ultrafiltration cell (8,400). Argon was applied to the ultrafiltration cell to provide the necessary pressure of 0.4 MPa. The specific activity of colloid-associated fractions of Pu (30 kDa–0.2 μm; 5–30 kDa and 1–5 kDa) were calculated from Pu activity measured in the filtrate and retained by each membrane during filtration.

2.4. IN-SITU K_d ESTIMATION

In this investigation in-situ values of Pu distribution coefficients between solid aqueous phases in the waste burial (unsaturated zone) and in an upper part of aquifer downstream the trench along the "WS" well profile were estimated.

Distribution coefficient of a radionuclide in the trench body was estimated as the ratio between specific activities of the radionuclide in a burial soil sample and in the soil solution. Ceramic soil water samplers (SWS), which had been installed into the trench body at the different depths, were used for soil solutions collecting. Burial soil samples were collected from the same depth that SWS had been installed.

Obtained K_d values in aquifer were based on measured Pu activity in eolian sand cores and ground water samples that were collected at the same points. The sampling was performed by the sampling system "Cobra" equipped with the specific finals.

Both soil water and ground water samples were filtrated through 0.45 μm filter before radionuclide determination. Radionuclides specific activities of solid phases were based on dry weight.

2.5. PLUTONIUM DETERMINATION

Standard radiochemical procedures based on precipitation of the high-valence metals hydroxides with following ion-exchange chromatography were used for the plutonium isotopes extraction from the collected water samples. ^{242}Pu isotope was used as a spike. A part of the procedures was carried out at the site. Due to the big volumes of water samples Pu isotopes were co-precipitated with hydroxides in field conditions. Supernatant solution was discharged and the residue was used for next procedures.

After the stabilization in 4-valence form, plutonium was absorbed selectively from the iron hydroxide (anion-exchanging resin AB–17-8, Ukraine) which had been diluted by nitric acid. The very thin and homogeneous spectrometric sources for measurements of ^{238}Pu, ^{239}Pu and ^{240}Pu are prepared by electrolytic deposition of the isotopes on the plates of stainless steel.

All alpha-spectrometer measurements were done using Soloist α-spectrometer with Soloist-U0300 detector (EG&G ORTEC).

2.6. GROUND WATER CHEMISTRY

Ground water chemistry in the aquifer close to the trench (Table 1) is determined by the disintegration of buried waste material (litter and trees) and following migration of the disintegration products into the saturated zone. In general, concentration of the main dissolved ions, dissolved organic carbon and values of E_h in the eolian aquifer were maximal in the wells located close to the source, and decreased downstream of the trench.

TABLE 1. Basic ground water chemistry in the shallow aquifer near the trench (2006)

	Unit	Range
pH		4.8–6.3
E_h	mV	75–320
DOC	mg/L	0.1–6.5
HCO_3^-	mg/L	0.2–28.5
SO_4^-	mmol/L	0.06–0.31
Na^+	mmol/L	0.01–0.06
K^+	mmol/L	0.07–0.70
Mg^{2+}	mmol/L	0.02–0.03
Ca^{2+}	mmol/L	0.05–0.30

3. Experimental results

3.1. SPATIAL PLUTONIUM DISTRIBUTION IN AQUIFER

In 2005–2006 sampling of ground water for determination of the total plutonium spatial distribution in the aquifer was carried out. Samples were collected from two wells profiles, which were located along the ground water flux direction (Fig. 1). Plutonium isotopes were reliable measured in all collected samples. Obtained plutonium isotopes specific activities in the filtered water samples (<0.5 μm) are presented on Table 3. Specific activity of Pu decreased downstream of the source. Maximum values of Pu specific activity were obtained in the upper part of the aquifer in several meters from the trench border (well 12-02-1), and minimum Pu activities were measured in the wells upstream of the trench. The character of Pu spatial distribution indicates that the trench was the main source of radionuclides in the aquifer (Fig. 2).

Plutonium activity levels in ground water depended on location and depth of the wells and varied from 0.3 to 300 mBq/L. Radionuclides specific activity distribution in the waste inside the trench is not homogeneous, and this heterogeneity of the source leads to different radioactive contamination of the aquifer along "WS" and "Laboratory" well profiles.

In several wells located along "Laboratory profile" (wells 8-02, 7-02), ^{241}Am specific activities in ground water were obtained. Maximum activity was found in the upper part of the aquifer (4.1 mBq/L). In general, activity ratios between americium and plutonium isotopes in these wells were higher than the average value for the Chernobyl fuel component. These data confirm the higher mobility of americium in the Environment.

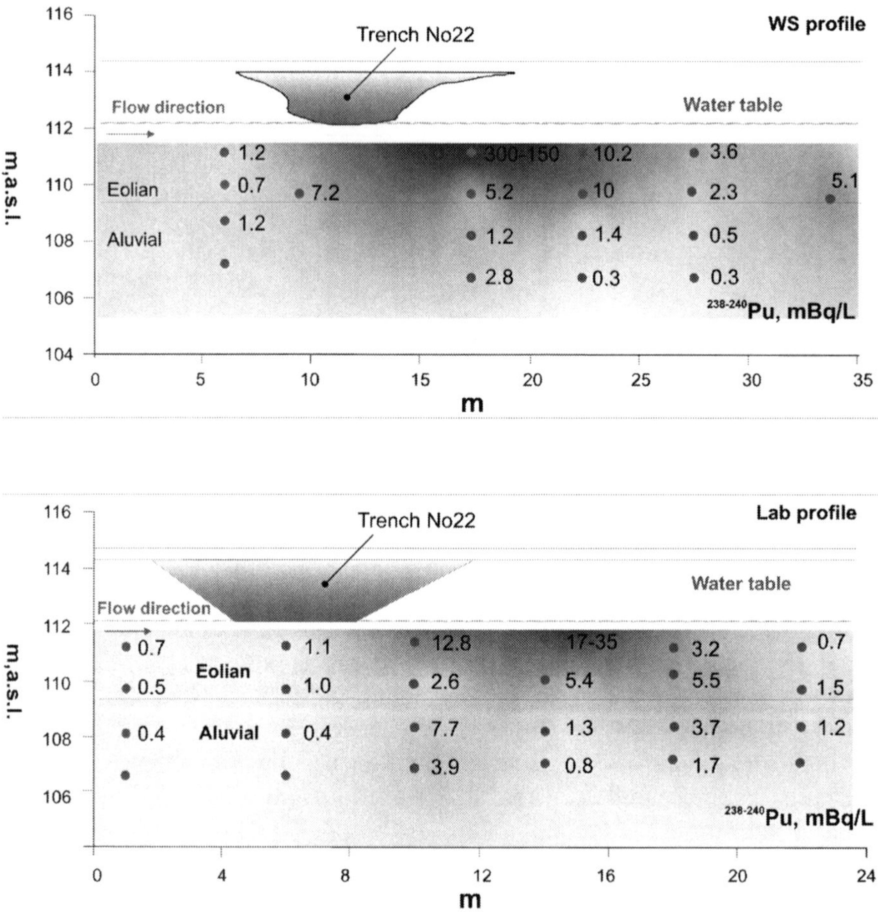

Figure 2. Plutonium spatial distribution in the shallow aquifer at the waste burial.

3.2. PLUTONIUM ISOTOPIC RATIO

Radioactive waste material in the trench is presented mainly by the fuel component of Chernobyl fallout with fixed ratios between radionuclides activities. In our case we used $^{239,240}Pu/^{238}Pu$ isotopic ratio to confirm the trench origin of plutonium in the shallow aquifer at the site. Mean of the ratios in all measured ground water samples was 2.2 (STD = 0.35) while the average value of activities of these isotopes in the Chernobyl fuel at the moment of measurement was 1.9 (Fig. 3)

Chernobyl fuel also contains radionuclides, which are progenitors of plutonium isotopes (^{244}Cm, ^{242}Cm and ^{243}Am). The predecessors can migrate downstream of the source and in the process decay to progeny Pu isotopes. Ratio between plutonium isotopes in this case would be differed from that in the fuel fallout. ^{239,240}Pu/^{238}Pu activity ratio of 0.03 was obtained for plutonium that originated from the decay curium and americium. This estimation was based on the average specific activities of transplutonium elements in the fuel.

Obtained experimental results on ^{239,240}Pu/^{238}Pu activity ratio in the ground water samples confirmed that in the site conditions Pu in the aquifer was mainly originated from the waste burial.

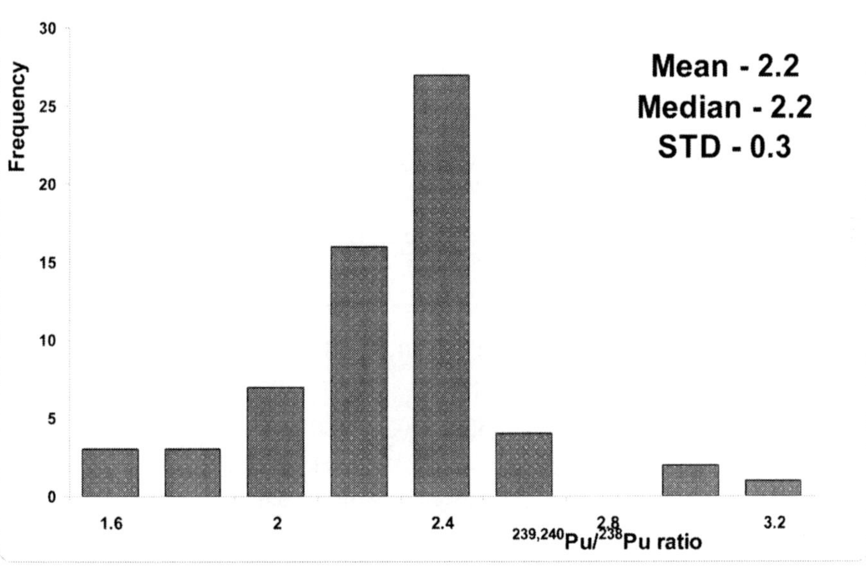

Figure 3. ^{239,240}Pu/^{238}Pu activity ratios in the ground water samples.

3.3. DISTRIBUTION COEFFICIENTS

Values of distribution coefficients (K_d) of plutonium in waste material (Table 2) and in eolian sand (Table 3) were estimated in field conditions. The obtained plutonium K_d value in the aquifer (7–47 L/kg) are two orders of magnitude less than those in the source material ((1–2)·10^3 L/kg). The data confirmed that Pu fractions in the aquifer were much more mobile than those in the source. Direct comparison of their values would be incorrect because of different ratios between solid and aqueous phases in the source (unsaturated zone) and in the aquifer. Usual humidity of the waste material in the trench body is 6–7 Wt% (liquid:solid about 0.08). In the aquifer liquid:solid ratio is about 0.4.

The waste burial does not have any artificial barriers preventing spreading of the radionuclides and other chemical compounds. Thus the radionuclides migration is fully determined by their chemical forms and characteristics of the waste material of the trench and underlying rock which can be considered as the natural barriers reducing vertical and lateral migration of radionuclides. Some Pu fractions spread relatively fast from the source and, in the case of their chemical stability in the aquifer, Pu may reach the Pripyat River in the far future.

TABLE 2. In-situ radionuclides distribution coefficients in the trench body (unsaturated zone)

Depth, m	^{137}Cs, L/kg	^{239,240}Pu, L/kg	^{238}Pu, L/kg
1.75	7,600 ± 1,800	1,650 ± 500	1,600 ± 500
1.25	5,200 ± 920	1,900 ± 520	2,200 ± 630
0.75	8,600 ± 1,600	1,400 ± 430	1,800 ± 600

TABLE 3. Plutonium specific activities in solid and aqueous phases from the upper part of aquifer and K_d values derived from these data

Point #	Sampling date	Phase	A, mBq/L/kg			K_d, L/kg	
			^{239,240}Pu	^{238}Pu		^{239,240}Pu	^{238}Pu
1	13.06.2006	Ground water	112 ± 16	50 ± 7	–	6.8 ± 1.4	6.6 ± 1.4
		Eolian sand	805 ± 130	350 ± 60			
2	17.04.2007	Ground water	18 ± 3	8 ± 1	–	42 ± 12	47 ± 14
		Eolian sand	780 ± 190	380 ± 90			
3	17.04.2007	Ground water	94 ± 20	41 ± 9	–	20 ± 6	24 ± 8
		Eolian sand	1,900 ± 420	980 ± 240			

3.4. COLLOID-ASSOCIATED PLUTONIUM

Several groundwater samples, which had been collected in the upper part of the aquifer during 2006–2007, were sequentially filtrated in-situ through membranes with different pore sizes ranged from 0.22 μm to 1 kDa. The work aimed to obtain size fractionation of mobile Pu species, which had been observed in the aquifer. Obtained results are present in Fig. 4. The size fractionation data on the plutonium activities in the collected colloids suggest that significant part of Pu (50–80%) passes through the 5 kDa filter.

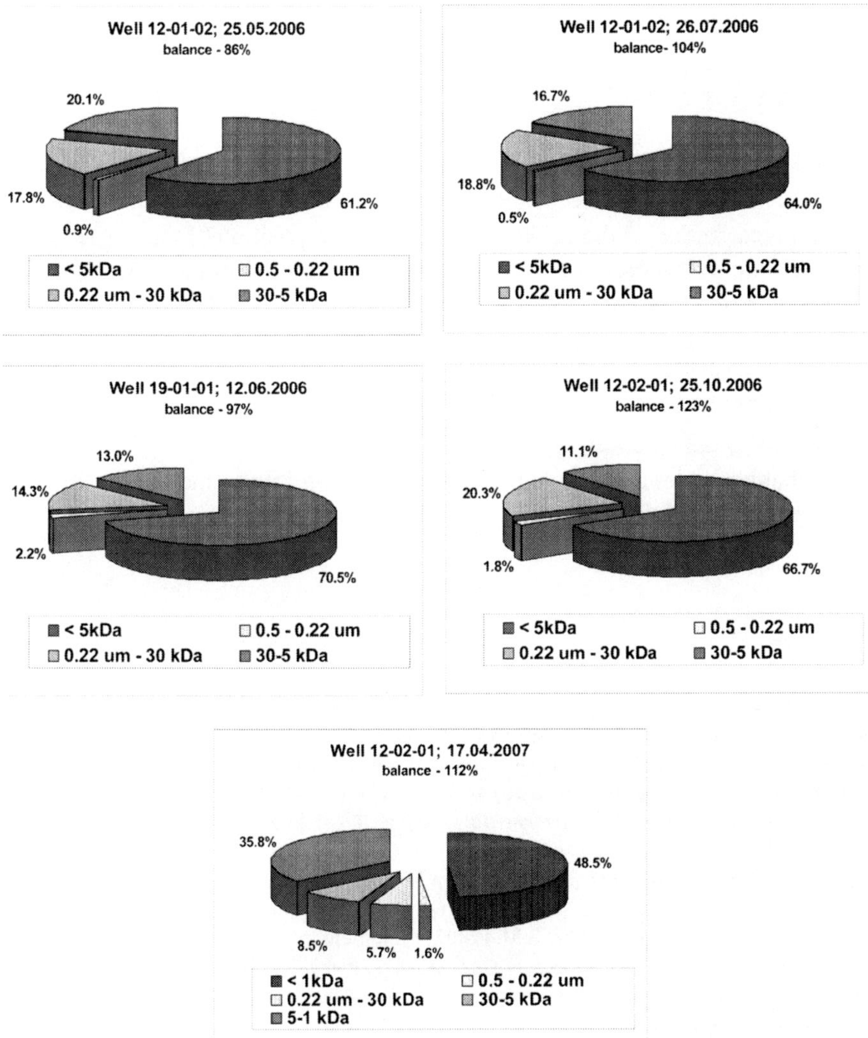

Figure 4. Plutonium size fractionation in the shallow ground water samples.

Distribution of the plutonium activity between fractions in the samples processed in the different periods was quite similar and did not depend on the sampling point location.

Degradation of the waste leads to formation of the natural organic matter, which affects the mobility of Pu in the subsurface. It was shown that low molecular weight, hydrophilic components of natural organic matter in the shallow aquifer can migrate rapidly (McCarthy et al., 1993, 1996). Probably, in our case Pu complexates with organic ligands that enhances its transport through the environment. Obtained data on total organic carbon concentration in the aquifer showed a weak correlation with Pu specific activity.

4. Conclusions

- Obtained spatial distribution of plutonium in the aquifer appears to be highly directional, reflecting the orientation of the hydraulic gradient.
- Characterization of the initial radionuclide composition of the waste showed that main part plutonium in the aquifer is originated from the trench.
- The obtained in-situ plutonium K_d values (so-called "total" K_d) in the aquifer (7–42 L/kg) are two orders of magnitude less than those in the source material ((1–2)·10^3 L/kg);
- About 30% of Pu in the upper part of aquifer is associated with 0.2 μm–5 kDa colloidal fraction.
- Probably, colloid-facilitated transport plays an important role in the "fast" migration of Pu in the condition of the experiment.

References

Buddemeier, R.W. and Hunt, J.R. (1988) Transport of colloidal contaminants in groundwater: radionuclide migration at the Nevada Test Site, *Applied Geochemistry*, **3**: 535–548.

Bugai, D., Kashparov, V., Dewiere, L., Khomutinin, Yu., Levchuk, S., and Yoschenko, V. (2005) Characterization of subsurface geometry and radioactivity distribution in the trench containing Chernobyl clean-up wastes, *Environmental Geology*, **47**(6): 869–881.

Dai, M.H., Kelley, J.M., and Buesseler, K.O. (2002) Sources and migration of plutonium in groundwater at the Savannah River site, *Environmental Science & Technology*, **36**: 3690–3699.

Dai, M.H., Buesseler, K.O., and Pike, S.M. (2002) Plutonium in groundwater at the 100K-Area of the U.S. DOE Hanford Site, *Journal of Contaminant Hydrology*, **76**: 167–189.

Dewiere, L., Bugai, D., Grenier, C., Kashparov, V., and Ahamdach, N. (2004) ^{90}Sr migration to the geo-sphere from a waste burial in the Chernobyl exclusion zone, *Journal of Environmental Radioactivity*, **74**(1–3): 139–150.

Kersting, A.B., Efurd, D.W., Finnegan, D.L., Rokop, D.J., Smith, D.K., and Thompson, J.L. (1999) Migration of Plutonium in Groundwater at the Nevada Test Site, *Nature*, **397**: 56–59.

Kaplan, D.I., Bertsch, P.M., Adriano, D.C., and Orlandini, K.A. (1994) Actinide association with groundwater colloids in a coastal plain aquifer, *Radiochim. Acta*, **66/67**: 181–187.

McCarthy, J.F., Williams, T.M., Liang, L., Jardine, P.M., Palumbo, A.V., Jolley, L.W., Cooper, L.W., and Taylor, D.L. (1993) Mobility of natural organic matter in a sandy aquifer, *Environmental Science & Technology*, **27**: 667–676.

McCarthy, J.F., Gu, B., Liang, L., Mas-Pla, J., Williams, T.M., and Yeh, T.-C.J. (1996) Field tracers tests on the mobility of natural organic matter in a sandy aquifer, *Water Resources Research*, **32(5)**: 1223–1228.

Novikov, A.P., Kalmykov, S.N., Utsunomiya, S., Ewing, R.C., Horréard, F., Merkulov, A., Clark, S.B., Tkachev, V.V., and Myasoedov, B.F. (2006) Colloid transport of Plutonium in the far-field of the Mayak production association, Russia, *Science*, **314**: 638–641.

Penrose, W.R., Polzer, W.L., Essington, E.H., Nelson, D.M., and Orlandini, K.A. (1990) Mobility of plutonium and americium through a shallow aquifer in a semiarid region, *Environmental Science & Technology*, **24**: 228–234.

RESTORATION OF RADIATION EVENTS OF THE PAST BY ANTEDATED LAKES BOTTOM SEDIMENT LAYERS

ELENA KVASNIKOVA*[1], SERGEY GORDEEV[2]
[1]*Institute of Global Climate and Ecology, Moscow, Russia*
[2]*Scientific & Production Enterprise, "Radon", Moscow Department, Russia*
**phone: ?????, e-mail: ranexa.kv@g23.relcom.ru; rem@radon.ru*

Abstract: The principles and survey procedure of the identification of past radiation events on the basis of antedated layers of pond bottom sediments are described. Results of experimental investigations (2005–2006) of sediment samples taken in the Timiryazev Garden pond in Moscow and in the pond of a village Podkhozheye in Moscow district are presented. Dating of bottom layers have been carried out using ^{210}Pb. This radionuclide is used as a tracer to estimate the sedimentation rate in ponds (0.3–0.4 cm/year.). It is shown that the content of ^{137}Cs, the main anthropogenic radionuclide in all fallout of the second part of the 20th century, is relatively higher at the depth 7–8 cm below the bottom surface. This can be explained by the Chernobyl fallout. A second maximum of 11–13 and 17–22 cm (for each of the ponds) can be explained by a global origin. The origin of both of these maxima is confirmed by radionuclide ratios, which are typical for global and Chernobyl depositions at the Moscow region.

Keywords:

1. Introduction

Reconstruction of past radiation events is a scientific task. It can improve the estimation of doses. An analysis of the vertical distribution profiles of some radionuclides in undisturbed soils and/or sediments can give a background for such calculations. Lake or pond sediments can be used for this purpose. The task of this article is to demonstrate successful demonstrations a method proposed by us to reconstruct radiation events.

D.H. Oughton and V. Kashparov (eds.), *Radioactive Particles in the Environment*,

2. Background of the study

^{137}Cs is an artificial radionuclide found in most of the radioactive contaminated areas formed in the second part of the 20th century. ^{137}Cs resulting from global fallout from the end of 50s to the beginning of 60s can be estimated by an activity of 4 kBq/m^2 with a deviation of 30% for midlatitudes of the Northern hemisphere (Boltneva et al., 1997; Kogan et al., 1991). Following the the 1963 moratorium on atmospheric testing of nuclear weapons, the decrease of activity due to the decay was pertibated by fallout from the Chinese and French explosions Thus global levels have been relatively stable until the beginning of the 80s (Izrael, 2002). Radionuclide ratios in global fallout are 0.63 for ^{90}Sr/^{137}Cs (Kogan et al., 1991) and 0.021 for ^{137}Cs/$^{239+240}$Pu (Myasoedov and Goryachenkova, 1991). In 1980–1986 global levels reduced due to decay (with a $T_{½}$ = 30 years for ^{137}Cs and $T_{½}$ = 28 years for ^{90}Sr). ^{137}Cs contamination of the Russian Plain at the latitude of Moscow has been estimated at 3 kBk/m^2 just before Chernobyl accident. This area was contaminated by Chernobyl depositions of 1986 (Izrael, 2002).

Since ^{90}Sr and transuranium radionuclides in the Chernobyl release were refractory (non-volatile), ratios ^{90}Sr/^{137}Cs for distances over 200 km from the source are 10 times smaller than in global fallout, and this ratio diminishes with distance (^{90}Sr/^{137}Cs ~ 0.01–0.02) (Stukin, 1991). Ratios of $^{239+240}$Pu/^{137}Cs in Chernobyl release over 200 km from the source are 100 times smaller than in global fallout ($^{239+240}$Pu/^{137}Cs < 0.00025).

Ecological half-lifes of ^{137}Cs from global and Chernobyl events in soils and water are connected with landscape. ^{137}Cs is a weakly soluble element and is strongly fixed by soil particles. ^{137}Cs distribution in the top layer of different soils can be described by quasi-diffusion or convection-diffusion models (Prokhorov, 1981). A typical ^{137}Cs profile was studied in soil on the grassy catchment slope of a Timiryazev Garden pond in Moscow. Results showed that the specific activity maximum was found at the depth of 3–4 cm (16.9 Bq/kg). The observation of the maximum deeper than the surface can be explained by (1) vertical migration, (2) surface soil transport due to erosion, (3) wind transport. Erosion and wind transport are negligible in our case.

^{137}Cs is easily measurable in the total depth of studied profile (with an observational accuracy better than 12%). At the present time it is not possible to separate the Chernobyl deposition from the total by ^{134}Cs/^{137}Cs ratio (^{134}Cs $T_{½}$ = 2.06 years.). The soil profiles studied reflects migration conditions of radionuclide behaviour in soils, but we can not identify the contribution of each radiation events in the summary in each layer.

Sediment sections in lakes and ponds give better data for dating events. Aerosols contained radionuclides fell on the water surface, and the input is well preserved by the year of sedimentation. On the condition that the studied lake has not been influenced by human activity during the last half of century, it is possible to date sediment layers.

^{137}Cs activity levels must be comparable with levels at the catchment top chosen as a model site. It is necessary to take in consideration that distortions of vertical ^{137}Cs profile can be caused by sediment roiling (Brunskill and Wilkinson, 1987), bioturbation and dissipation of activity by the depth (Davis et al., 1984). From our experience, the main uncertainties are connected with irregularity of sediments in the upper 2–3 cm. In bottom sediments, the zero point of sediment surface is a very relative level.

3. Using ^{210}Pb to antedate sediment layers

^{210}Pb is widely used as a marker in studying erosion-accumulative processes. It is noted in (He and Walling, 1997; Walling and He, 1996) that ^{210}Pb fallout from the atmosphere is a quasi-permanent process. So, it is possible to estimate soil erosion rates for a period of decades.

^{210}Pb is a natural radionuclide in the chain of ^{238}U with a $T_{½}$ =22.26 years. ^{226}Ra from this chain decays with production of the rare gas ^{222}Rn, this spreads up into the atmosphere where it decays to $^{210}Pb_{atm}$. This radionuclide falls out from the atmosphere with natural aerosols. Thus in ^{210}Pb soil profile two components are visible: $^{210}Pb_{atm}$ derived from the atmosphere and $^{210}Pb_{ter}$ formed from ^{226}Ra present terrestrial ground. Furthermore, in ordert to antedate sediments, it is necessary that ^{210}Pb is a very feeble migrant (Lima et al., 2005).

Features of ^{210}Pb measurement and possibilities of its using as a marker are discussed in (Markelov et al., 2005; Robbins, 1978). $^{210}Pb_{atm}$ can be measured of the total ^{210}Pb and of ^{226}Ra for to subtract than a part of $^{210}Pb_{ter}$ connected with ^{226}Ra. X-ray spectrometry is used to detect $^{210}Pb_{total}$. ^{226}Ra is measured on gamma-spectrometer by intermediate short-lived ^{214}Pb (or ^{214}Bi). To form a radiation equilibrium of ^{214}Pb (or ^{214}Bi) with ^{226}Ra, it is necessary to keep hermetically packed samples for about 3 weeks.

To date a sediment layer we used a model of permanent rate of deriving $^{210}Pb_{atm}$. Concentration of $^{210}Pb_{atm}$ changes with time in each layer according to the following equation:

$$A = A(0)e^{-kt}, \tag{1}$$

Where A(0) – is a specify acrivity of $^{210}Pb_{atm}$ in the top sediment layer, t – time, $k = 0.693/T_{1/2}$.

Dating of a sediment layer at depth x is calculated using the equation:

$$t = \frac{1}{k}\ln\frac{A(0)}{A}. \tag{2}$$

The $^{210}Pb_{atm}$ fraction easily visible on the graph of ^{210}Pb vertical distribution in soils or sediments.

4. Lakes suitable for dating sediment layers

Our task is to develop a method for retrospective reconstruction of past radioactive contamination at the areas having combined contamination sources. It is very important to choose a suitable site for this purpose (Schertz et al., 2006). It is clear that it is necessary to choose a lake with a negligible run-off from a catchment. Bottom sediments show large variability by their sorbtion feature. Silts have a better absorbing capacity.

A sediment accumulation in the lake can be accompanied by run-off from the catchment. This fact is not suitable incase when we would like study only sediments deriving from the atmosphere (Schertz et al., 2006). So, a lake suitable for our task is a lake with a silt bottom, a small catchment with a matted soil on the slopes and with a negligible anthropogenic influence.

We have found two ponds with gentle matted slopes (Fig. 1). Their upper bottom sediment layers 0–5 cm are presented by silt particles of <1 ηм (>50% of the weight); maximal ^{137}Cs concentrations were fixed at the depth 7–20 cm where silt ground has particles 1–5 ηм (>75% of the weight).

(1) Pond in Timiryazev Garden in Moscow on the river Schutchya

(2) Pond in the village Podkhozheye, Serebryanoprudsky region of Moscow district on the river Polosnya.

Figure 1. Studied lakes.

History of radioactive contamination in Moscow and Moscow region includes two main events: global contamination with a maximum of deposition in 1963 and weak Chernobyl deposition of 1986. The two lakes chosen for study have a different input of global and Chernobyl deposition in the total contamination.

5. Description of lakes

Lake 1. The pond in the Timiryazev Garden in Moscow was created in the 19th century and after that it was never cleaned up (55°40′11″; 37°30′02″). A catchment of 1.2 km^2 includes the following areas: 7% – town buildings; 5% – matted slopes of Schuchya river valley; 60% – gentle slopes under protected forest. Soils: sod-podzols. History of landuse: building construction works in the 1960s at the river valley above the pond; forest area of the catchment is protected and belongs to Timiryazev academy since the beginning of the 20th Century. Mainly run-off water flows from here.

Twelve integral samples on the depth 0–30 cm (by tube sampler d = 8 cm) and 1 colomn layer by layer (Fig. 1) have been taken in the forest area of the catchment. ^{137}Cs contamination is 2.3 ± 0.41 kBq/m^2; $V = \sigma/a_{cp} = 19\%$.

Three colomns have been taken layer by layer in the pond bottom. ^{137}Cs contamination is 2.0 ± 0.2 kBq/m^2. Values of the total ^{137}Cs at the catchment and in the pond bottom are similars. This fact confirms a very feeble influence of run-off from catchment to the pond.

Pond 2. The pond in the village Podkhozheye, Serebryanoprudsky region of Moscow district (at the South part of Moscow district) was created in the end of the 1950s, and after that it was never cleaned up (54°07′30″ 38°31′10″). A catchment of 2 km^2 includes the following areas: 7% – village houses with gardens; 13% – matted slopes of Polosnya river valley; 80% – gentle slopes under arable land. Soils: chernozems and soddy soils on the valley slopes under grass. Run-off water flows mainly from arable land on gentle slopes of catchment, but soil transported from here can not reach the pond, being stopped by village. The village can be also source of material transported to the pond, but it also cannot reach the pond to be stopped by the dense grass of the pond slopes. So, the most part of transported soil re-accumulates inside the catchment and does not reach the pond. A pond catch the more thick soil particles derived from the catchment.

Twenty-four integral soil samples on the depth 0–30 cm (by tube sampler d = 8 cm) have been taken at the two arable slopes of catchment. ^{137}Cs contamination is 9.6 ± 1.35 kBq/m^2; $V = \sigma/a_{cp} = 21\%$ on the left site of catchment, and 8.1 ± 0.68 kBq/m^2; $V = \sigma/a_{cp} = 12\%$ on the right site of catchment.

Two colomns have been taken layer by layer in the pond bottom. ^{137}Cs total contamination is 8.9 kBq/m^2. Values of the total ^{137}Cs at the catchment and in the pond bottom are similars. This fact confirms a very feeble influence of run-off from catchment to the pond.

6. Bottom sediment sampling

A main requirement for bottom sediment sampling to date layers is an absence of deformation of a sediment core during sampling procedure. There are some ways for this. One of the best way is a core freezing before his taking out (Lima et al., 2005). This method invokes a very complicated sampler device using liquid nitrogen to freeze core.

An other way was carried out by us. We used a sampler-tube (1 m of a length) with a fang closing endlong. It is necessary to rotate a tube to close it and take up at the surface.

The best result we received in winter by sampling from the ice. So, we used feature of Russian climate for conventional sampling. The sampler is opening near the site of sampling, and the naturally freezing core is cutting by a knife on layers with thickness of 0.5–1 cm.

7. Experimental study of ^{137}Cs and ^{210}Pb contain in sediments. Discussion

Sediment sampling was carried out in 2005–2006 in the two ponds, choosing the main calm areas without flow and run-off influence. Sampling has been carried out in winters from ice. The γ-spectrometry device type Canberra with a high pure Germanium detector and X-ray spectrometer have been used.

^{137}Cs vertical distribution are given on the Figs. 2 and 3 for the more representative cores in two ponds.

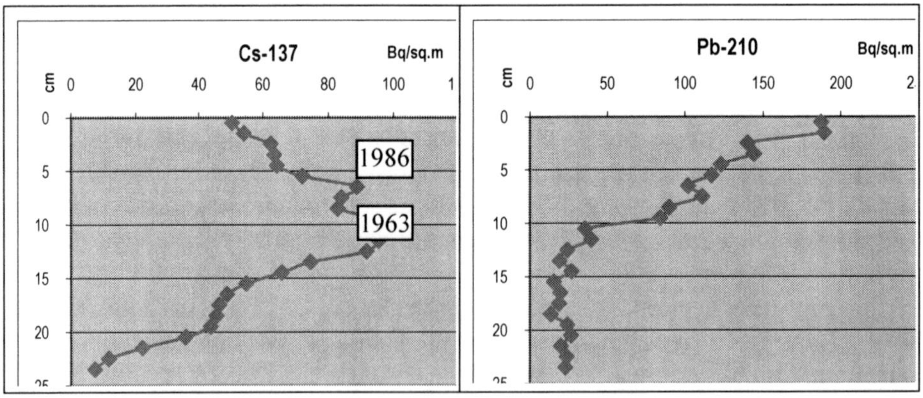

Figure 2. Timiryazyev Garden pond in Moscow: vertical distribution in bottom sediments of studied radionuclides, 2005.

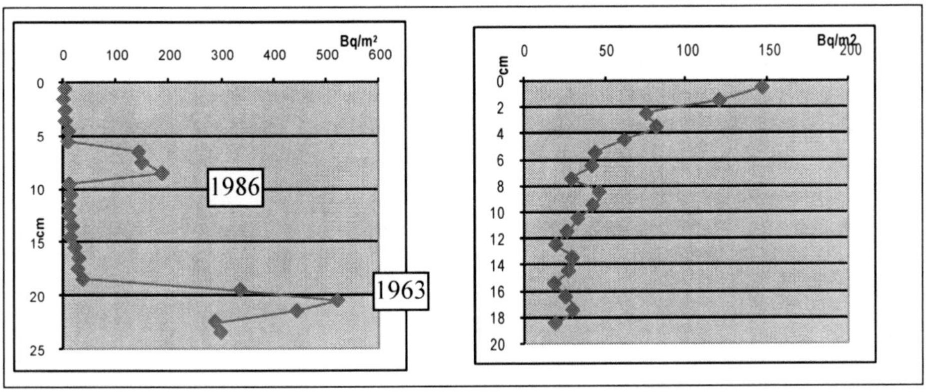

Figure 3. Podkhozheye village pond in Southern part of Moscow district: vertical distribution in bottom sediments of studied radionuclides, 2006.

Sedimentation rate has been estimated using ^{210}Pb vertical distribution for both of ponds as 0.3–0.4 cm/year. A specify activity increased with a depth in both of cases. The upper maximum is visible at the depth of 6 cm in Timiryazev Garden pond and at the depth of 8 cm in Podkhozheye village pond. It corresponds to 1986 – a year of Chernobyl deposition.

A period of 42 years passed after 1963 is forthcoming at the depth of 15 cm in the case of estimated sedimentation rate. Really we detect a second maximum near this calculative value: at the depth of 11–13 cm in Timiryazev Garden pond and at the depth of 18–20 cm in Podkhozheye village pond. Uncertainty in global maximum identification is connected probably with (a) long-term outlook of the event, (b) with a unevenness of sedimentation rate in ponds in the second part of the 20th century due to the instability of economy in Russia and the resulting considerable changes in landuse.

8. Conclusion

Samples from layers antedated as Chernobyl and global depositions were analysed for the content of ^{90}Sr and $^{239+240}$Pu.

Literature data on radionuclide ratios in global deposition are 0.63 for ^{90}Sr/^{137}Cs and 0.021 for $^{239+240}$Pu/^{137}Cs. The ratio of ^{90}Sr/^{137}Cs is stable with time due to similar $T_{½}$ for both of isotopes. In view of considerably different $T_{½}$ for $^{239+240}$Pu and ^{137}Cs, their ratio in 2006 can be estimated as $^{239+240}$Pu/^{137}Cs = 0.055. In this study, we obtained ratios in lake sediment samples from 17–20 cm of 0.27–0.31 for ^{90}Sr/^{137}Cs and 0.03 for $^{239+240}$Pu/^{137}Cs. This can be considered to be similar to those for global fallout, and confirm the origin of depositions.

In contrast, radionuclide ratios in Chernobyl deposition are 0.01–0.02 for $^{90}Sr/^{137}Cs$ and 0.00025 for $^{239+240}Pu/^{137}Cs$. This reflects the fact that the ^{90}Sr and $^{239+240}Pu$ released by Chernobyl accident was as refractory species, compared to the greater release of the volatile ^{137}Cs. The ratios measured in the present study sediment samples were 0.0065 for $^{90}Sr/^{137}Cs$ and 0.00041 for $^{239+240}Pu/^{137}Cs$. These are are also similar to anticipated values.

The method demonstrated in this example study from the Moscow region can be used to identify radiation events of the past in other locations.

References

Boltneva L.I., Izrael Yu.A., Ionov V.A., and Nazarov I.M. (1977) ^{137}Cs and ^{90}Sr global contamination and doses of external irradiation in the USSR. *Atomic Energy*, **42**(5), 355–360. (in Russian).

Brunskill J.G. and Wilkinson P. (1987). Annual supply of U-238, U-234, Th-230, Ra-226, Pb-210, Po-210 and Th-232 to lake Ontario from terrestrial and atmospheric sources. *Canadian Journal of Fishery & Aquatic Sciences*, **44**, 215–230.

Davis R.B., Hess C.T., et al. (1984). ^{137}Cs and ^{210}Pb dating of sediments from soft-water lakes in New England (USA) and Scandinavia, a failure of ^{137}Cs dating. *Chemistry & Geology*, **44**, 151–185.

He Q. and Walling D.E. (1997) The distribution of fallout ^{137}Cs and ^{210}Pb in undisturbed and cultivated soils. *Applied Radiation Isotopes*, **48**, 677–690.

Izrael Yu.A. (2002) *Radioactive Fallout after Nuclear Explosions and Accidents. Elsevier*, 281 p.

Kogan R.M., Nazarov I.M., and Fridman Sh.D. (1991) Theory of environmental γ-spectrometry. Moscow: *Energoatomizdat* 233 (In Russian).

Lima A.L., Hubeny J.B., Reddy C.M., Ring J.W., Hughen K.A., and Eglinton T.I. (2005) High-resolution historical records from Pettaquamscutt River basin sediments: ^{210}Pb and varve chropnologies validate record of ^{137}Cs released by the Chernobyl accident. *Geochimica et Cosmochimica Acta*, **69**(7), 1803–1812.

Markelov M.V., Golosov V.N., and Stukin E.D. (2005) Possibilities of using of ^{210}Pb and ^{137}Cs as markers to estimate erosion rates. *Meteorology and Hydrology*, 4, 66–84.

Myasoedov B.F. and Goryachenkova T.A. (1991) Forms of occurrence of plutonium in soils. *Journal of Radioanalytical and Nuclear Chemistry*, **203**(1),153–157.

Pennington W., Cambray R.S., and Fisher E.M. (1973) Observation on lake sediments using fallout ^{137}Cs as a tracer. *Nature*, **242**, 324–326.

Prokhorov V.M. (1981) Migration of radioactive contaminants in soils. Physics and chemical features and modeling. Moscow. *Energoizdat* 98.

Robbins R.A. (1978) Geochemical and geophysical application of radioactive lead. *The Biogeochemistry of Lead in the Environment.* Ed. J.O. Nriagu. Elsevier, Amsterdam, pp. 286–383.

Schertz M., Michel H., Barci-Funel G., and Barci V. (2006) Transuranic and fission product contamination in lake sediments from an alpine wetland, Boreon (France). *Journal of Environmental radioactivity*, **85**, 380–388.

Stukin E.D. (1991) Characteristics of primary and secondary caesium-radionuclide contamination of the countryside following the Chernobyl NPP accident. In: *Proceedings of Seminar on Comparative Assessment of the Environmental Impact of Radionuclides Released during Three Major Nuclear Accidents: Kishtym*, Windscale, Chernobyl. Luxembourg, 1–5 October 1990. Commission of the European Communities, *Radiation Protection-53*, EUR 13574, **1**, 255–302.

Walling, D.E. and He. Q. (1996) Using fallout lead-210 measurements to estimate soil erosion on cultivated land *Soil Science Society of American Journal*, **63**(5), 1404–1412.

^{241}Ams AND ^{137}Cs IN SOILS OF THE BELARUS PART OF THE CHERNOBYL ZONE

VIACHASLAU ZABROTSKI*, ALIAKSANDR DUTAU, YURI BONDAR
Polessje State Radiation-Ecological Reservation, Tereshkova str. 7. Hoiniki, Gomel region, 247600, Belarus
**phone: 375 2346 92190, e-mail: vzabrotski@tut.by*

Abstract: Some peculiarities of radioactive contamination of forestlands of the Polessje State Radiation-Ecological Reservation by ^{241}Am, ^{137}Cs and other radionuclides are presented.

Keywords: americium-241, cesium-137, strontium-90, plutonium, forestland, fuel particles, Chernobyl.

1. Introduction

It is generally accepted that radioactive contamination of the near Chernobyl zone is mainly due to the dispersed particles of nuclear fuel. ^{241}Am and ^{241}Pu (its precursor) were ejected from the Chernobyl reactor together with particles of UO_2. For that reason, by measuring the gamma-emission of ^{241}Am, we can conclude about the presence (today or in the past) of fuel particles in a sample.

The data presented were obtained from analysis of samples taken during radiation surveys of forest areas of the Polessje State Radiation-Ecological Reservation. The Reservation was established in September 1988 on the Belarus part of the territory, after contamination by the Chernobyl accident. It extends 70 km from east to west and about 50 km from north to south. The area of the Reservation equals 2162 km^2. More than 20 thousand people in 118 settlements lived on that territory before Chernobyl.

Today the main function of the Reservation is the management of the exclusion zone and scientific research. About 50 scientists and technicians work in the scientific part of the Reservation. One of the divisions is spectrometry and radiochemistry laboratories. The main task of the laboratory from the practical point of view is the examination of the forest to determine the degree of contamination of the wood by ^{137}Cs. If contamination of the wood is less than permissible levels (for example 740 Bq/kg for firewood) the trees could be cut and used in economy. All territory of the Reservation is divided between 16 forest areas. In some of them economical activity is prohibited.

In the initial steps of examination, the test site (its square varied from 0.1 to tenths of hectares) was chosen by the forester. Initially we checked the homogeneity of the site in relation to the dose rate. Standard samplers with diameter 4 cm and length 20 cm was used to take four soil samples at the corners of the rectangle with sides from 30 to 50 m. These samples were combined to make a bulk samples. A global positioning system was used to register the exact place of sampling. The forest areas which were visited by workers of the laboratory are marked on the map (Fig. 1).

The gamma-spectrometer with high purity germanium detector with beryllium window and Genie-2000 software (Canberra) were used to determine the content of ^{241}Am and ^{137}Cs in the soil. We also used the β-spectrometer with plastic scintillator produced by "Atomtex" (Belarus) to measure the ^{90}Sr concentration instrumentally. We began to make the instrumental analysis of soil samples since 2006 when we received the last-named gamma-spectrometer in the framework of IAEA technical cooperation project.

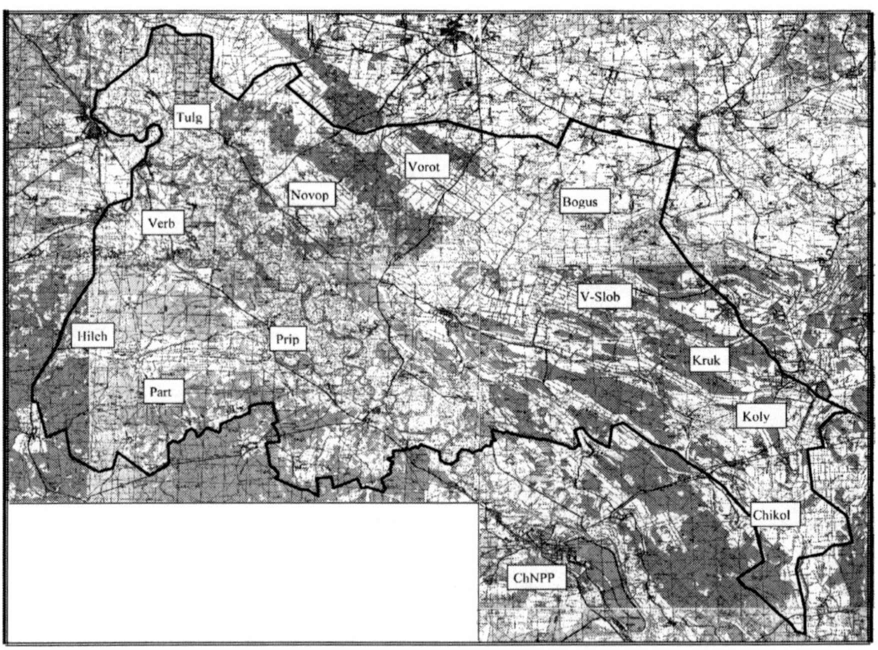

Figure 1. Arrangement of forest areas on the territory of Polessje State Radiation-Ecological Reservation. Forestries: Koly – Kolybanskoe, Kruk – Krukovskoe, Chiko – Chikolovichskoe, V – Slob, V. – Slobodskoe, Prip – Pripaytskoe, Bogush – Bogushevskoe, Part – Partizanskoe, Novop – Novopokrovskoe, Vorot – Vorotetskoe, Hilch – Hilchanskoe, Verb – Verbovichskoe, Tulg – Tulgovichskoe.

Some of the results received in the work are presented in Table 1 (the forest areas are situated in the order of increasing the distance to ChNPP). It can be shown that there is certain correlation (with coefficient 0.57) between distance to ChNPP and quantity of the samples with activity concentration of ^{241}Am ([^{241}Am]) less than minimal detectable activity (MDA). We can also claim that the more the distance to ChNPP the less the activity concentration of ^{241}Am in soil samples.

TABLE 1. Activity concentration of ^{241}Am, ^{137}Cs, ^{90}Sr in soil. Data were received in radiation examination of forest areas of Polessje State Radiation-Ecological Reservation in 2006–2007

Forestry	No of samples	Part of samples with [^{241}Am] less than MDA	$^{241}Am_{mean}$ (Bk/kg)	$^{241}Am_{max}$ (Bk/kg)	$^{137}Cs_{mean}$ (Bk/kg)	$^{137}Cs_{max}$ (Bk/kg)	$^{90}Sr_{mean}$ (Bk/kg)
Kolybanskoe	10	0	8.8	15.6	560	1,110	275
Krukovskoe	16	0	10.5	31.8	2,020	9,630	595
Chikolovichskoe	6	0	14.8	41.9	708	1,654	615
V.-Slobodskoe	18	0	8.1	13.1	1,710	3,130	820
Pripaytskoe	1		2.2	2.2	1,159	1,159	340
Bogushevskoe	47	2	8.9	18.3	2,088	6,300	865
Partizanskoe	7	29	2.6	4.3	716	1,055	172
Novopokrovskoe	6	0	12.2	20.1	2,010	2,750	1,066
Vorotetskoe	31	6.5	7.3	14.2	2,970	6,250	1,675
Hilchanskoe	41	65	2.5	3.95	2,080	8,210	535
Verbovichskoe	5	0	2.7	5.8	1,084	1,840	98
Tulgovichskoe	24	33	3.7	5.6	1,412	2,367	658

The experimental data received let us to evaluate the upper limit of density of contamination of the area by ^{241}Am. For the Tulgovichskoe forestry it varies in the range 0.002–0.006 Ku/km^2, whereas for the Chikolovichskoe forestry – in the range 0.04–0.1 Ku/km^2. These magnitudes are comparable with legislative norms which establish the possibility of living on territory contaminated by ^{238}Pu, ^{239}Pu, and ^{240}Pu.

It is well known that content of ^{241}Am in environment is increasing at the moment and finally will exceeds that of the ^{239,240}Pu in 2.5 times. According to the data (Kudrjashov et al., 2007) ratio ^{241}Am/^{239,240}Pu equals 1.1. In our work we received more high value of the ratio which equals 1.40 ± 0.15 (0.15 is one standard deviation). These data were received in analysis of 17 soil samples taken from the upper soil level. Gamma-spectrometry was used to determine the concentration of ^{241}Am. Following procedure (Zabrotski and Bondar, 2006) was used to measure the content of ^{239,240}Pu in soil:

- Dissolution of the sample in microwave oven
- Isolation of Pu on the anion exchange column
- Precipitation of the Pu with NdF3
- Alpha-spectrometry

We carried out a special experiment to find evidence of existence of hot particles in soil. The joint soil sample (about 1.7 kg) was thoroughly mixed and divided into 10 subsamples. Each of them was measured to determine ^{241}Am and ^{137}Cs concentrations. Relative standard deviations for activity concentration of ^{241}Am and ^{137}Cs were equal approximately 10 and 2 % correspondingly. We propose two options to interpret these results:

1. The hot particles are destroyed.
2. The hot particles are very fine and numerous.

It should be noted that we registered the ^{241}Am in the majority of the samples taken at the territory of Reservation in 2006–2007. The activity concentration of ^{241}Am changed from 2 Bq/kg (typical MDA for accumulation time 10–20 h) to the 2000 Bq/kg for upper centimeters layers of soil profiles.

We analyzed the results obtained during examination and found that samples taken in Kolybanskoe and Chikolovichskoe forest areas have characteristics of contamination which are different from the samples taken in other areas. To make that difference more clear we divided all results in two groups and compare them. First group includes samples taken from Kolybanskoe and Chikolovichskoe forest areas and second group includes the samples taken from all other forest areas. We constructed the distribution of the samples on the value of ^{137}Cs

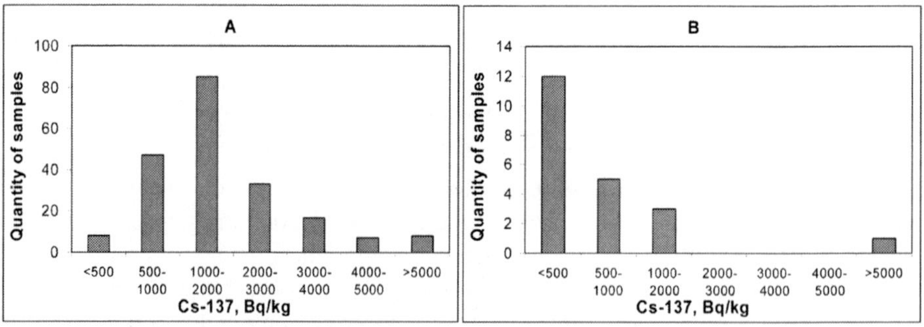

Figure 2. The distribution of the samples on the value of ^{137}Cs activity concentration: A – in the majority of the forest areas, B – in Kolybanskoe and Chikolovichskoe forest areas.

concentration (Fig. 2), ^{241}Am concentration (Fig. 3) and on the magnitude of ^{241}Am/^{137}Cs ratio (Fig. 4).

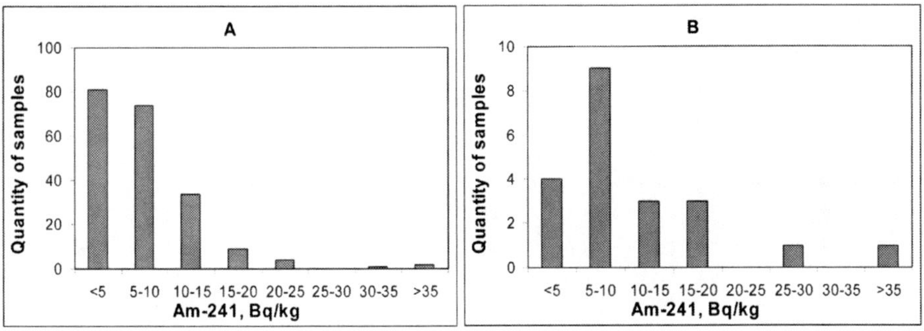

Figure 3. The distribution of the samples on the value of ^{241}Am activity concentration: A – in the majority of the forest areas, B – in Kolybanskoe and Chikolovichskoe forest areas.

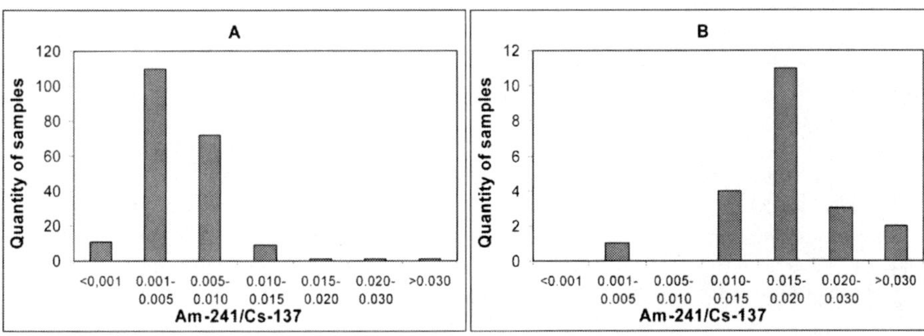

Figure 4. The distribution of the samples on the value of ^{241}Am/^{137}Cs ratio: A – in the majority of the forest areas, B – in Kolybanskoe and Chikolovichskoe forest areas.

According to the data of Fig. 2 the samples taken from two selected forest areas are characterized by relatively low activity concentrations of ^{137}Cs. We suppose that this is the main reason why these samples have magnitudes of ^{241}Am/^{137}Cs ratio which are different from the majority of the samples (Fig. 4). Really the distribution of the samples on the value of ^{241}Am concentration does not differ substantially from the corresponding distribution of the main part of samples (Fig. 3).

When we constructed the distribution of the samples according to the value of the ^{137}Cs/^{90}Sr ratio (Fig. 5) we found that it is the same as seen in earlier studies (Kashparov et al., 2000). This can be interpreted as proof of the reliability of the instrumental method used for determination of ^{90}Sr. The above-stated assumptions concerning the low content of ^{137}Cs in the samples from Kolybanskoe and Chikolovichskoe forest areas are also supported by Fig. 5

because the maximum of the distribution of ratio $^{137}Cs/^{90}Sr$ for these two forest areas is decreasing.

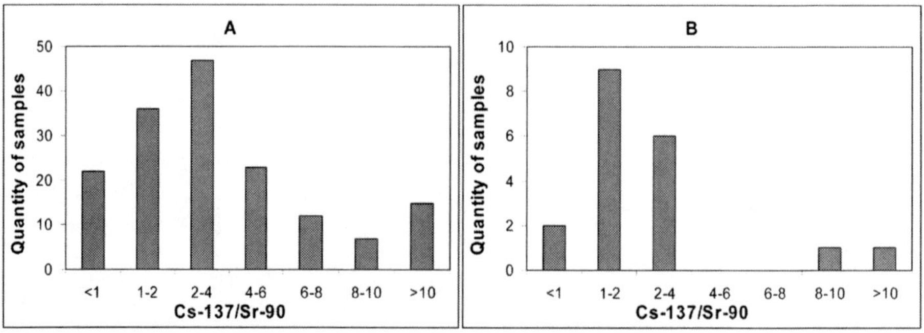

Figure 5. The distribution of the samples on the value of $^{137}Cs/^{90}Sr$ ratio: A – in the majority of the forest areas, B – in Kolybanskoe and Chikolovichskoe forest areas.

References

Kashparov, V. A., Lundin, S. M., Khomutinin, Yu. V., et. al. (2000) Soil contamination with ^{90}Sr in the near zone of the Chernobyl accident, *Radiochemistry* **42**(6), 550–559. (In Russian).

Kudrjashov, V., Mironov, V., and Konoplya, E. (2007) Radiation and Chernobyl: Transuranium elements on the territory of Belarus. (Gomel. Institute of radiology), 128 p. (In Russian).

Zabrotski, V. and Bondar, Yu. (2006) Combined procedure of determination of ^{238}Pu, $^{239,240}Pu$, ^{241}Pu, and ^{90}Sr in the soil, *Radiochemistry* **48**(1), 87–91 (In Russian).

PART 3: DATABASES, PLATFORMS AND MEASUREMENT TECHNIQUES

THE "HOT PARTICLES" DATA BASE

MARINA ZHURBA*[1], VALERY KASHPAROV[1], NOUREDDINE AHAMDACH[2], BRIT SALBU[3], VASYL YOSCHENKO[1], SVYATOLSLAV LEVCHUK[1]
[1]*Ukrainian Institute of Agricultural Radiology NAU, Mashinostroitelej str., 7, Chabany, Kiev' Region, 08162, Ukraine*
[2]*Institute for Radioprotection and Nuclear Safety (IRSN), BP6, 92266 Fontenay aux Roses cedex, France*
[3]*Norwegian University of Life Sciences, P.O. Box 5003, 1432 Aas, Norway*
**phone: +38044 5267531, e-mail: marina@uiar.kiev.ua*

Abstract: During 1987–1996, more than 1,800 Chernobyl hot particles were selected by scanning thin soil layers with a dosimeter. Size and radionuclides activity in hot particles were measurement. The "Hot particles" database (in Microsoft Access) was created by integrating this information and complemented with data on hot particles sampled in 1986 in Poland and inside of the "Shelter". This database is widely used in ex-USSR and other countries, and it is available for interested professionals. This database contains the supplemental information related to Chernobyl nuclear fuel: specific activity of radionuclides, activity and radionuclide ratios in the different stage after accident and for the different depth of depletion of nuclear fuel. As well as a photo-gallery of FP were generated with an electron microscope is accessible. The results of works in the near zone of ChNPP accident are compiled on the CD "Radioactive contamination of the 30-km zone" and the Database "Hot Particles" released by UIAR. Obtained regularities of the fuel particles behavior in the Environment create the basis for prognosis of the radiological situation in the ChNPP Exclusion zone.

Keywords: radionuclides, hot particles, database, nuclear fuel, Chernobyl accident, radioactive contamination.

1. Introduction

The radioactive fallout at Chernobyl is composed of a fuel component, i.e., particles of finely-dispersed nuclear fuel (fuel particles); and a condensed component, formed as a result of the condensation of volatile fission products (radioisotopes of I, Te, Cs and, to a small extent, Sr and Ru, which were released during the high temperature annealing of nuclear fuel) on the surfaces of different carriers (Kashparov, 2003; Kashparov et al., 2001, 2003). Fuel particles (FP) within the radioactive fallout are considered to be a specific peculiarity of the ChNPP accident. In and around the 30-km zone of the accident, about 2,000 km^2 were contaminated mainly by FP that were contained in the main part of the release from the reactor of non-volatile radionuclides including biologically significant ^{90}Sr and radioisotopes of plutonium and americium. Fuel particles were also found at long distances from ChNPP in many European countries (Pollanen et al., 1997). Particles of condensed cesium were formed in one remote zone as a result of ^{137}Cs deposited with precipitates.

During the last half century, a number of fundamental studies have been carried out, as a result of nuclear weapon tests and large radiation accidents, on the behavior of artificial radionuclides in the environment after their global fallout. There have also been model experiments with water soluble radionuclide species and various radioactive particles. However, the first results that were carried out on fuel traces at Chernobyl have revealed how inapplicable this earlier work is. For example, radionuclides in fuel traces of radioactive fallouts were characterized as having a significantly lower mobility and biological availability when compared to the condensed form of fallouts at Chernobyl. Therefore, in the first years after the Chernobyl accident, predicted data on radionuclide migration in the near zone were overestimated when compared to previously obtained irregularities (mainly concerning ^{90}Sr). Fuel particles have been dissolving over time and are still dissolving now, resulting in an increase in plant contamination with ^{90}Sr and its migration to surface and ground waters. At the present time, the basic radioactive flow-out from the territory of the exclusive zone is caused by ^{90}Sr leaching from fuel particles.

Lack of knowledge on the behavior of radionuclides released to the environment and the composition of the matrix of irradiated nuclear fuel particles did not one allow one to correctly estimate the radiological situation in the near zone during the ChNPP accident and to predict its change in the future, as well as to optimize the application of countermeasures.

From the point of view of the formation of the radiological situation (influence of external irradiation to humans and animals, inhalation of internal irradiation, and peroral intake of radionuclides into organisms), the physical and chemical characteristics of FP-associated radionuclides are mostly expressed in terms of:

- The formation of radioactive contamination of the territory
- The inhalation intake of radionuclides during the passage of radioactive clouds and the resuspension by wind, in the radionuclide metabolism in organisms and the formation of doses
- The migration of radionuclides in soil that determine the dynamics of the external irradiation dose rate and the contamination of surface and ground waters, as well as the changes in the biological availability of radionuclides in time that determine the levels of contamination in plants and agricultural products and the intake of radionuclides into human organisms
- The peroral intake of radionuclides into the organisms of agricultural animals, radionuclide metabolism and radioactive contamination of agricultural products

The urgency for investigating the environmental behavior of radionuclides, especially those contained in Chernobyl fuel particles, is indicated by all the factors mentioned above. Such information will be useful not only for eliminating the consequences of the Chernobyl accident, but also for its potential use in other accidents where irradiated nuclear fuel particles are released and for the needs of radioactive waste storage. Release of fuel particles, similar to those of Chernobyl, can appear during accidents: at nuclear fuel cycle enterprises, in transport activities, in the dismantling of nuclear reactors, at spent nuclear fuel facilities, etc. A special significance for studies of fuel particle formation and their behavior in the environment is also evident from the unprecedented terrorist acts in 2001 in the USA and by the threats to use "dirty" bombs in terrorist attacks.

2. Methods

During 1987–1989, more than 1,200 relatively large (size >10 μm, activity >100 Bq) Chernobyl hot particles were selected by scanning thin soil layers with a dosimeter, and about 500 additional hot particles were selected in the period 1989–1996. In the near zone at Chernobyl, such particles are mainly (up to 97%) composed of particles of finely-dispersed nuclear fuel – fuel particles (FP). The specific activity of these particles corresponds to the activity of the fuel at the moment of the accident, excluding the volatile and highly mobile radionuclides of $^{134,137}Cs$. The presence of fissionable material in the particles was demonstrated using neutron-activation analysis, electron probe micro-analysis, and laser mass-spectrometry (particle enrichment with ^{235}U was found to be about 1–2%). Size and activity of radionuclides in hot particles were measurement by optical microscopy and by α-, β-, γ-spectrometry. The burn-up distribution of fuel particles larger than 10 μm indicates that, at the moment of accident, particles were released from a limited part of the reactor where burn-up values were lower in comparison with the average fuel burn-up in the 4th Unit of ChNPP (Kashparov et al., 1996). Ratios between ^{90}Sr, ^{95}Zr, ^{106}Ru, ^{125}Sb, ^{134}Cs, ^{137}Cs, ^{144}Ce, ^{154}Eu, ^{155}Eu and transuranium elements (TUE) in fuel particles were estimated using experimental data that allow the calculation of the territory contaminated by the main, radiologically important radionuclides included in the fuel component of the fall-out. The "Hot particles" database (in Microsoft Access) was created by integrating this information and complemented with data on hot particles sampled in 1986 in Poland and inside of the "Shelter". This database is widely used in ex-USSR and other countries, and it is available for interested professionals.

3. Database description

The "Hot particles" database (Fig. 1) contains the next information for each hot particle: number and shifr, coordinate of sampling (angle and distance from ChNPP) and size (min and max), type (fuel or condensed) and burn-up, description (view, color, shape, and structure), measurement date and radionuclides activity (Fig. 2). Also there is an indications of the institutions presented the results.

Estimation of nuclear fuel burn-up in HP is determined as a ratio of the different radionuclides activity (Kuriny et al., 1993). Menu of database can be used for the edition and input of data (Fig. 2) or the initial tables can be opened directly in in Microsoft Access.

Microsoft Access allows to create the needed queries for the formation of a data access and for the calculation of nuclear derivatives which characterize the HP, such as a fractionation of radionuclides (Fig. 3).

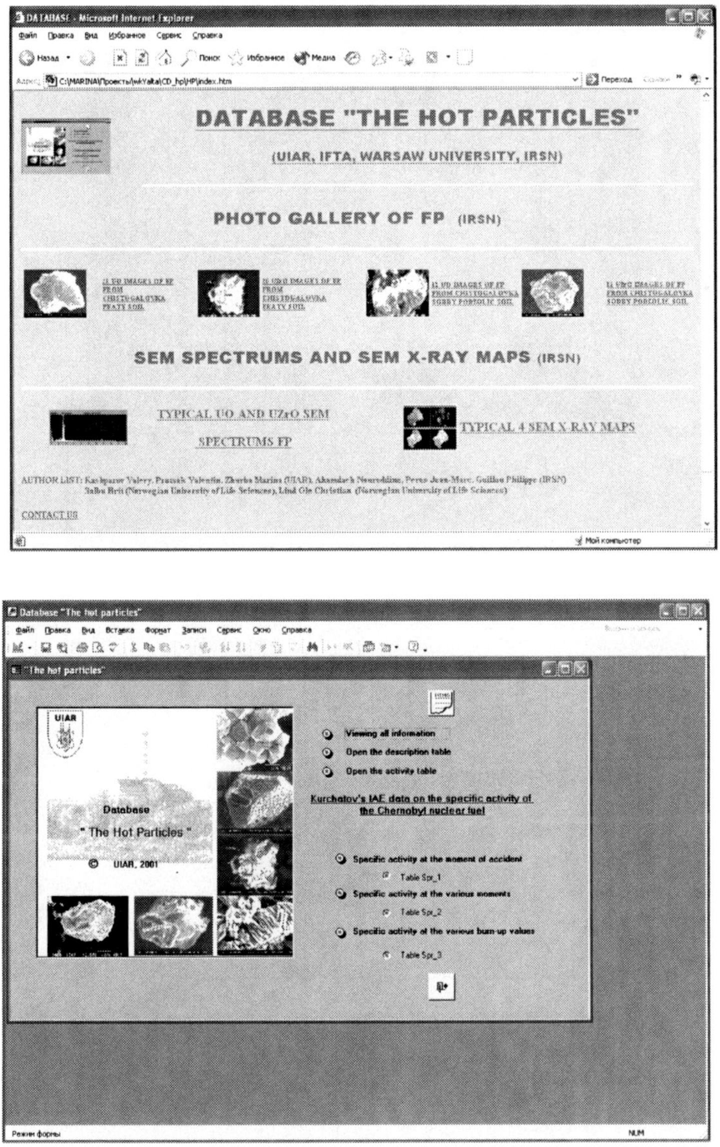

Figure 1. The "Hot particles" database menu.

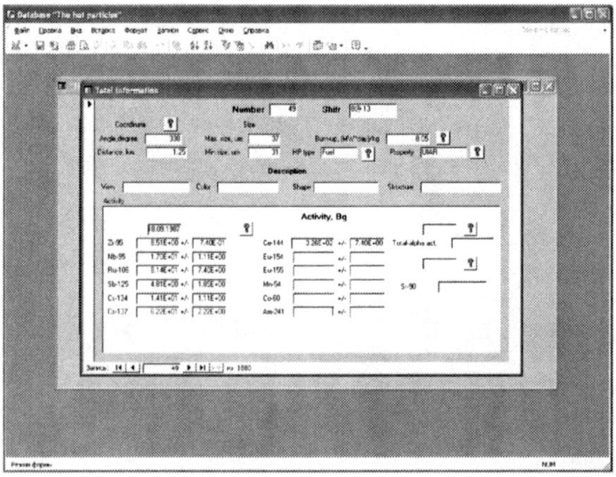

Figure 2. The "Hot particles" database input.

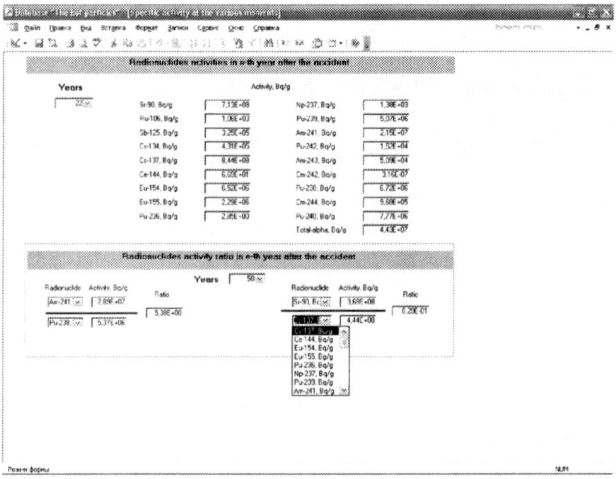

Figure 3. The supplemental information related to Chernobyl nuclear fuel in the "Hot particles" database.

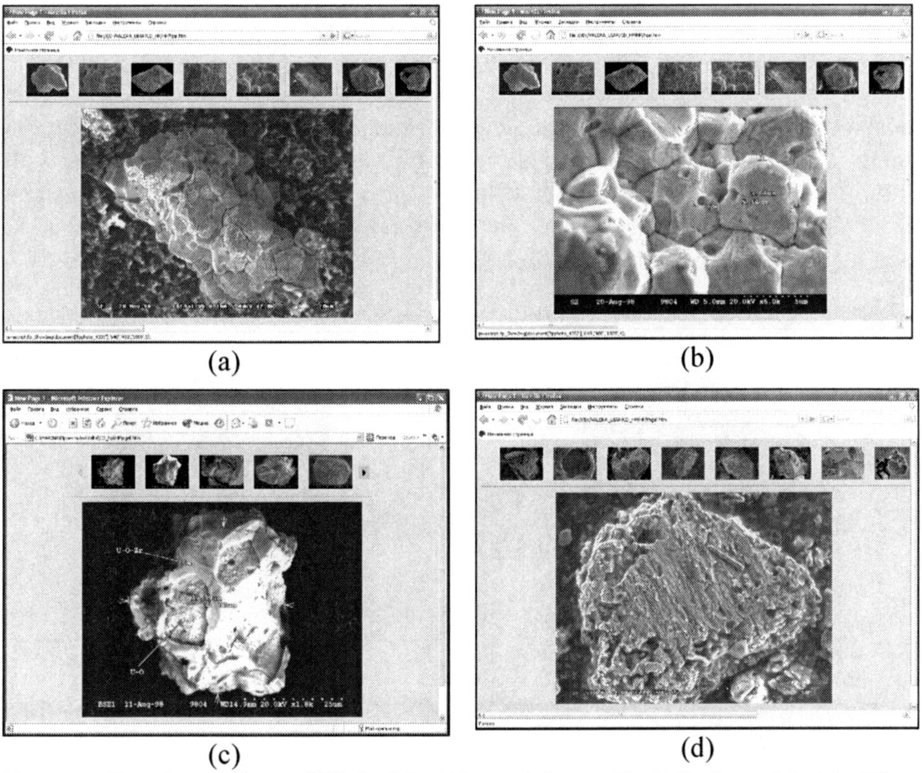

Figure 4. The photo-gallery of FP in DB: UO_2 particles (a, b); U-Zr-O particle (c); UO_{2+x} particles (d).

Fuel particles with only a low degree of transformation were formed during the first stage of the Chernobyl accident as a result of the mechanical destruction of the fuel during the explosion (04/26/86). They were released mainly in the western direction (Salbu et al., 2001). These are pieces of nuclear fuel with a well-formed "grain"-structure, deposited as separate grains (crystallites) of UO_2, and their nature is determined by the technology of the manufacturing process (Fig. 4). The external view of these low-transformed Chernobyl fuel particles, after a relatively long contact with soil (up to 20 years), is practically the same as that of particles of irradiated nuclear fuel just after removal from the reactor at the completion of their exploitation. This means that the dissolution rate of UO_2 particles in the environment is very low (Kashparov et al., 2004).

At later stages of the accident, fuel particles with a highly transformed matrix were formed. Oxidization in air led to destruction of the nuclear fuel along the grain edges. Oxygen diffusion into the grains caused the formation of various oxides of uranium, and due to the differences in density and crystal grate of these oxides, the UO_2 surface was subjected to cracking. The fuel could be partially melted by the high temperatures attained during the accident

(Kashparov, 2003). The melted fuel was probably contacted by the various construction and fire suppression materials. As a result, complex chemical compositions appeared, especially on the surfaces of fuel particles. As a rule, coarse melted fuel particles contain zirconium, which can be explained by the melting of uranium oxides at higher temperatures, as compared to particles with uranium-zirconium compositions (Figs. 4 and Fig. 5). Chemically extra-stable U-Zr-O particles were formed in the first moment of the accident on 26.04.86 and then were deposited within the narrow western trace.

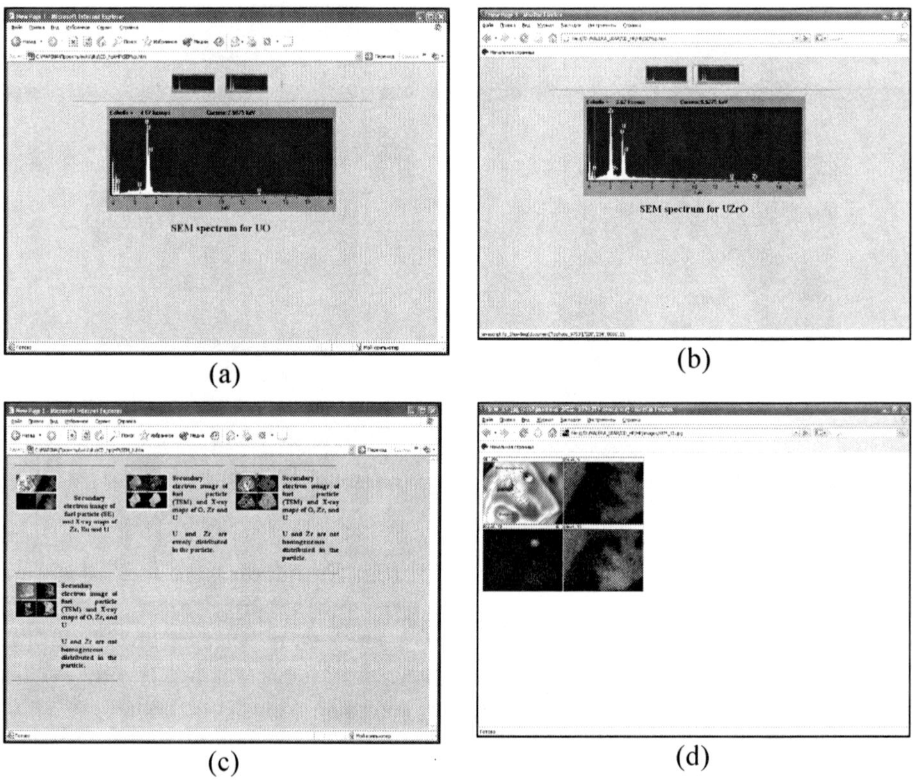

Figure 5. SEM spectrums U-O particle (a) and U-Zr-O particle (b); SEM X-ray maps of particles (c, d).

4. Conclusion

The results of works in the near zone of ChNPP accident are compiled on the CD "Radioactive contamination of the 30-km zone" and the Database "Hot Particles" released by UIAR. Obtained regularities of the fuel particles behavior in the Environment create the basis for prognosis of the radiological situation in the ChNPP Exclusion zone.

The data obtained were used to specify activity correlations of the fuel component radionuclides of Chernobyl fallout and to create the maps of the 30-km ChNPP zone terrestrial density of contamination with ^{90}Sr, ^{154}Eu, ^{238}Pu, $^{239+240}Pu$ and ^{241}Am.

I present time cooperation with Byelorussian scientists is developing in the frame of NATO project, where the methodology elaborated in Ukraine will be applied for mapping of contaminated with Chernobyl fuel particles territory of Belarus. It is planning a use of the data base for modeling and elaborating a long-term prognosis on the further development of radiological situation inside the "Shelter" due to respirable fractions of radioactive aerosols.

The database "Hot Particles" can be used as a base or a component of integrated database at the properties of radioactive particles from the different sources (nuclear tests, radiation accident, etc.), for the newly-discovered hot particles and for the prevention of illegal production and distribution of the radioactive materials.

References

Kashparov, V.A. (2003) Hot particles at Chernobyl. *Environmental Science and Pollution Research* **10/1**, 21–30.

Kashparov, V.A., Ahamdach, N., Zvarich, S.I., Yoschenko, V.I., Maloshtan, I.N., and Dewiere, L. (2004) Kinetics of dissolution of Chernobyl fuel particles in soil in natural conditions. *Journal of Environmental Radioactivity* **72/3**, 335–353.

Kashparov, V.A., Ivanov, Yu.A., Zvarich, S.I., Protsak, V.P., Khomutinin, Yu.V., Kurepin, A.D., and Pazukhin, E.M. (1996) Formation of hot particles during the Chernobyl nuclear power plant accident. *Nuclear Technology* **114/1**, 246–253.

Kashparov, V.A., Lundin, S.M., Khomutinin, Yu.V., Kaminsky, S.P., Levtchuk, S.E., Protsak, V.P., Kadygrib, A.M., Zvarich, S.I., Yoschenko, V.I., and Tschiersch, J. (2001) Soil contamination with ^{90}Sr in the near zone of the Chernobyl accident. *Journal of Environment Radioactivity* **56/3**, 285-298.

Kashparov, V.A., Lundin, S.M., Zvarich, S.I., Yoschenko, V.I., Levtchuk, S.E., Khomutinin, Yu.V., Maloshtan, I.N., and Protsak, V.P. (2003) Territory contamination with the radionuclides representing the fuel component of Chernobyl fallout. *The Science of the Total Environment* **317/1–3**, 105–119.

Kuriny, V.D., Ivanov, Yu.A., Kashparov, V.A., Loschilov, N.A., Protsak, V.P., Yudin, E.B., Zhurba, M.A., and Parshakov, A.E. (1993) Particle associated chernobyl fall-out in the local and intermediate zones. *Annals of Nuclear Energy* **20/6**, 415–420.

Pollanen, R., Valkama, I., and Toivonen, H. (1997) Transport of radioactive particles from the Chernobyl accident. *Atmospheric Environment* **31/21**, 3575–3590.

Salbu, B., Krekling, T., Lind, O.C., Oughton, D.H., Drakopoulas, M., Simionovichi, A., Snigireva, I., Snigirev, A., Weitkamp, T., Adams, F., Janssens, K., and Kashparov, V. (2001) High energy X-ray microscopy for characterization of fuel particles. *Nuclear Instruments and Methods in Physics Research A* **467–468**, 1249–1252.

THE EXPERIMENTAL PLATFORM IN CHERNOBYL: AN INTERNATIONAL RESEARCH POLYGON IN THE EXCLUSION ZONE FOR SOIL AND GROUNDWATER CONTAMINATION

NATHALIE VAN MEIR[1], DIMITRY BUGAÏ[2], VALERY KASHPAROV[3]
[1]*Institute for Radioprotection and Nuclear Safety, Laboratoire d'Etude des Transferts dans les Sols et le sous-sol, IRSN/DEI/SARG, Paris, France*
[2]*Institute of Geological Sciences, Kiev, Ukraine*
[3]*Ukrainian Institute of Agricultural Radiology, Kiev, Ukraine*

Abstract: This review article introduces an experimental site situated within the Chernobyl exclusion zone, that is equipped to study (1) release of radionuclides from dispersed hot fuel particles, (2) transport of these radionuclides through the unsaturated zone and finally (3) their transport in the phreatic aquifer. This paper presents how the site is equipped and what methodology was followed in order to understand and reproduce the observed ^{90}Sr contaminant plume downstream from a shallow waste trench in an area about 2.5 km west from the Chernobyl Nuclear Power Plant (ChNPP), called the "Red Forest". Main results include an inventory and description of the physical and chemical properties of the fuel particles encountered in the waste trench, a valid reproduction of flow through the unsaturated zone, a first model to describe flow and transport of ^{90}Sr in the aquifer. These results also indicate the main factors that need better understanding.

Keywords: Chernobyl, groundwater, fuel particles, Sr-90, aquifiers.

1. Introduction

In 1999 an experimental site was set up at about 2.5 km west of the Chernobyl Nuclear Power Plant (ChNPP) in an area with hundreds of shallow waste trenches (Fig. 1). These trenches were dug in 1987 to clean up the rests of dead irradiated "Red Forest" trees and to get rid of the surrounding upper soil layer on which small hot fuel particles (HFP) were deposited during the explosion and the subsequent fire. The site described within this paper centres around one such a trench, trench n°22, and covers an area of roughly 100 × 100 m (Fig. 1). The trench, that has a length of about 70 m, a width of 6–8 m and a depth of 2–3 m, is positioned perpendicular to the water flow direction and is flooded at its base from time to time. Measurements of ^{90}Sr in the aquifer indicated a contaminant plume downstream from the trench. The main objective behind setting up the site was, therefore, to study the release from radionuclides (RN) from the trench, presumably from hot fuel particles, to follow RN-transport in the unsaturated zone and the subsequent contaminant spreading in the aquifer. Three main zones of interest were thus defined; (1) the trench, (2) the unsaturated zone outside the trench, and (3) the aquifer. Each zone was equipped with monitoring equipment and each entity investigated allowing the determination of the main parameters for flow and transport. In return for this investment effort we now possess observations over time of changes in contaminant concentration, changes in *in situ* flow regime, and a thorough understanding of the composition and physical and chemical behaviour of the HFP.

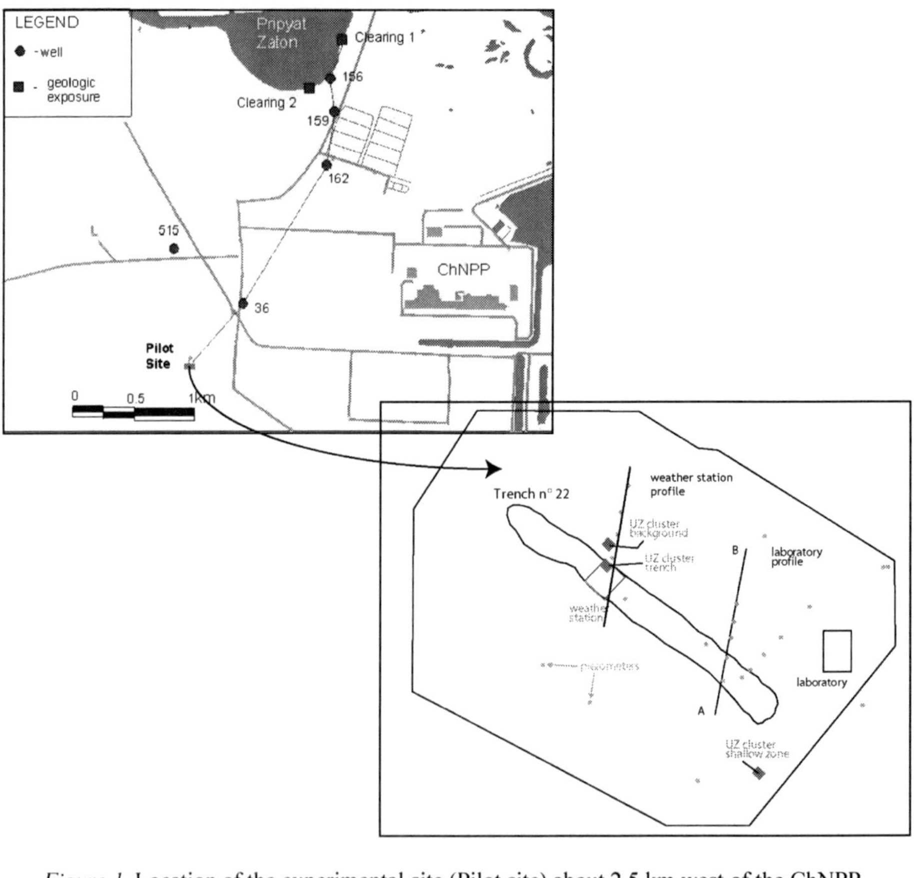

Figure 1. Location of the experimental site (Pilot site) about 2.5 km west of the ChNPP.

This paper first describes the site composition and set-up in more detail, followed by a description of the HFP and a short overview of the main results. It does not foresee to describe every step taken, but wishes to give an overview of the methodology followed and of the current understanding on how radionuclides were released and transported to the aquifer. The site is comparable to other sites with surface burial of radioactive waste, e.g. The Hanford Site (Pace et al., 2007; Steefel et al., 2003; Zhuang et al., 2007) or White Oak Creek (Garten Jr, 1999; McCarthy et al., 1998), but differs much in the origin of the waste buried, the surrounding geology (highly permeable sand), and in trench design. The fact, that the waste trench is situated in a relatively simple geological situation, contributes to a more thorough separation of hydraulic from transport phenomena.

2. Site layout and set-up

2.1. GEOLOGY

The site elevation is relatively flat with variations between 113.5 and 115 m amsl, the highest parts of which correspond with the trench area. From top to bottom we first encounter a thin anthropogenic layer of imported clean sand (0–20 cm), followed by Quaternary deposits up to a depth of about 30 m from the surface. At 30 m depth we find marls of Eocene age (Kiev Marls) that form the basis of the phreatic aquifer and can be considered as impermeable in our context. The Quaternary deposits can be further subdivided into an upper eolian sandy layer, that is partly unsaturated and partly saturated, and a lower alluvial unit (Fig. 2), that is always saturated. The eolian sand has a higher hydraulic conductivity than the alluvial sand and transport of the contaminant plume is, therefore, more pronounced in the upper part of the aquifer (Fig. 2).

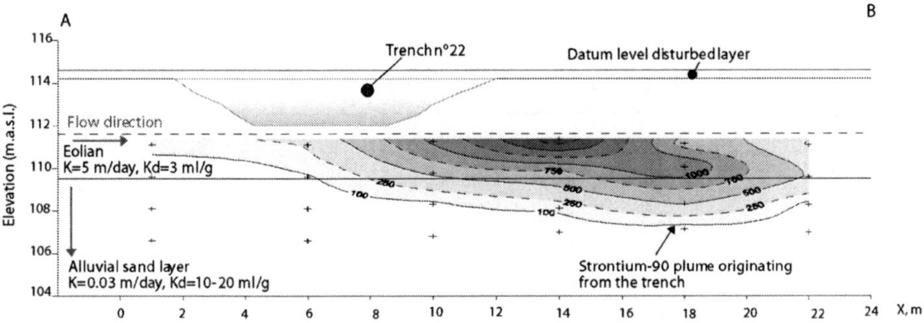

Figure 2. Laboratory cross-section from south (A) to north (B) showing the observed ^{90}Sr levels in 2003.

2.2. MONITORING SET-UP

The site has been set up to monitor the three zones defined in the introduction; the trench, the unsaturated zone outside the trench and the aquifer. In all three zones we wish to determine water flow direction and velocity and transport velocity of radionculides.

To this end, the unsaturated zone, with an average depth of 2.5–3 m, has been equipped with sensors at different depths (Tables 1 and 2) for water content (time domain reflectometry) and temperature, suction pressure (tensiometers) and soil water samplers allowing to derive radionuclide activity and major chemistry in the unsaturated zone (Fig. 1). There are three different sensor clusters on site (Fig. 1); (1) the trench (called TWS), (2) the background unsaturated zone (undisturbed eolian sand, called BGA), and (3) a topographically lower UZ area where the UZ is thinner (called TLA). Tensiometers are only inserted from a depth of 0.75 m onward, as they are frost-sensitive. This information together with data derived from an on-site weather station allow calculating flow and getting an idea about transport in the unsaturated zone. The weather station registers precipitation, air humidity, hours of sunshine, wind speed and wind direction, as well as temperature resulting in a calculation of the potential evapotranspiration (PET) on site.

TABLE 1. Inserted instruments in the TWS and BGA sites

Depth (mbgl)	Water content	Suction pressure
0.25	Delta T ML2x	No
0.50	Delta T ML2x	No
0.75	Delta T ML2x	No
1.25	Delta T ML2x	UMS T6
1.75	Delta T ML2x	UMS T6
2.25	Delta T ML2x	UMS T6

TABLE 2. Inserted instruments at the TLA site

Depth (mbgl)	Water content	Suction pressure
0.25	Delta T ML2x	No
0.50	Delta T ML2x	No
0.75	Delta T ML2x	UMS T6
1.00	Delta T ML2x	UMS T6

To monitor water table elevation changes, general aquifer geochemistry and radionuclide activity changes, piezometers have been installed in many different places and with filter screens at different depths. Larger piezometers (2″ diameter) have been equipped with pressure sensors, and several smaller diameter piezometers (1″) were installed along the principal flow direction (north) in two main cross-sections; the Laboratory and Weather Station (WS) cross-section (Fig. 1). Every dot along these cross-sections on Fig. 1 implies in fact four different screen depths, two in the eolian part and two in the alluvial part of the aquifer. These cross-sections are sampled regularly for groundwater geochemistry changes and radionuclide activity changes (mainly for ^{90}Sr and recently also for plutonium isotopes). Hydraulic parameters have been derived from on-site slug tests and a pumping test. The main flow direction was inferred from several tracer tests with non-reactive tracers. These also provided first estimates for on-site dispersivity values.

2.3. TRENCH DELINEATION

The trench is the main source for the ^{90}Sr plume observed downstream from this trench. It is, therefore, necessary to delineate the trench well and to estimate the total inventory of radioactivity present at one time. An extensive shallow drilling campaign was carried out on a regular grid over the suspected trench contour, in each borehole gamma radiation was measured. These were then calibrated for ^{137}Cs. A total radioactivity of 600 GBq of ^{137}Cs was back-calculated for the year 1987. The measurements showed that the trench bottom is quite irregular and that the north-western part (close to the WS cross-section) is deeper than the south-eastern part (close to the Laboratory cross-section). They also highlighted the heterogeneity of the trench radioactivity distribution. The difference in ^{90}Sr plume activity and shape observed in the two aquifer cross-sections is therefore not surprising (Figs. 2 and 3).

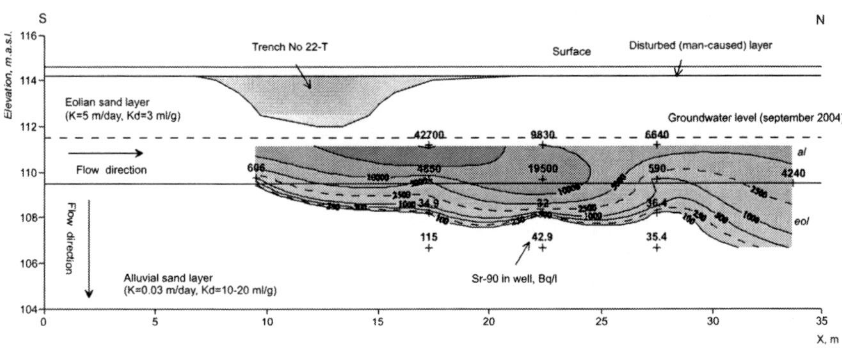

Figure 3. Weather Station (WS) cross-section showing the ^{90}Sr activities in 2003.

3. The hot fuel particles (HFP) in trench n°22

Even though some ^{90}Sr in the trench is not associated with hot fuel particles, the majority is. HFPs were separated from trench soil samples and subsequently investigated with a scanning electron microscope (SEM). They were also subjected to different stability and dissolution tests. In the Chernobyl exclusion zone three types of fuel particles are distinguished (Kashparov et al., 2004):

Chemically extra stable particles, supposedly ZrUO that formed as a result of high temperature annealing of UO_2 in the presence of zirconium in construction material. These particles were formed in the first moments of the accident (26.04.1986) and were then deposited in the narrow western trace that passes through the experimental site.

Non-oxidised chemically stable fuel particles, UO_2, of the first release (26.04.1986) that were formed as a result of a mechanical destruction of the nuclear fuel (Salbu et al., 2001). These particles created the narrow western trace that passes through the experimental site.

Chemically less stable particles, UO_{2+x}, that formed as a result of nuclear fuel oxidation in the period from 26.04.1986 to 05.05.1986. These particles mainly formed the northern and southern fuel fallout traces.

All three types of particles have been identified by SEM in the samples from the trench on site. Their initial relative distribution in the trench is estimated at 0.19 ± 0.15 for the ZrUO group, 0.57 ± 0.15 for the UO_2 class and 0.21 ± 0.15 for the oxidised particle group, UO_{2+x}. Their dissolution rates were determined for different values of pH (Fig. 4), which clearly shows a pH-dependency as well as different dissolution rates for each group. In fact, the ZrUO particles are considered as insoluble. For a pH of 5, which is the pH encountered in the trench solution, the dissolution constants for the release of ^{90}Sr are 0.018 (±0.001)/year for the UO_2 class and 0.28 (±0.05)/year for the more soluble UO_{2+x} class. This means that today (21 years after the accident) all of the UO_{2+x} particles have dissolved and have released radionuclides to the soil solution.

4. Flow and transport in the unsaturated zone

In a first attempt to describe flow and transport of ^{90}Sr in the unsaturated zone, a simplified concept combining one-dimensional flow with a Kd (distribution coefficient: the ratio of the sorbed phase concentration to the solution phase concentration in equilibrium) value to reproduce transport was used. Flow is described by Richard's equation and the nonlinear relation (characteristic curves) between suction pressure and water content is described by the van Genuchten equation (van Genuchten, 1980). These characteristic curves were set up based on field observations of suction pressure and water content and laboratory experiments conducted in parallel (Fig. 5). These curves show the difference in flow characteristics between the trench material (TWS) and the undisturbed eolian UZ sand (BGA).

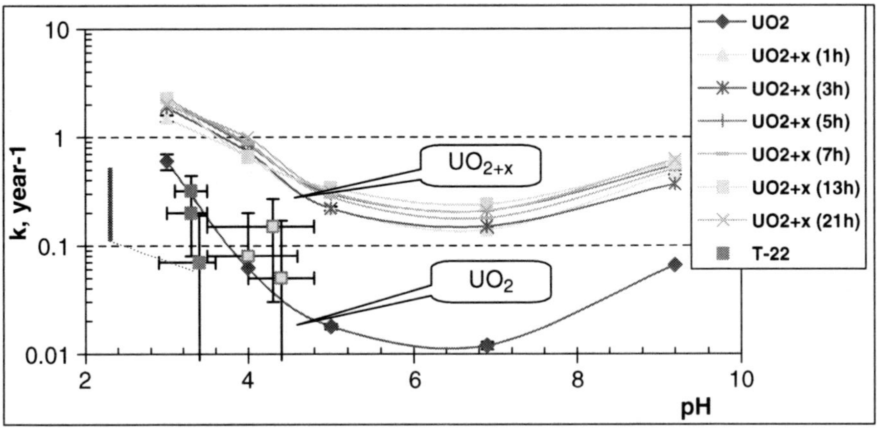

Figure 4. Dissolution kinetics as a function of pH of two different HFP types.

A transport model is developed that combines the fractionated release of ^{90}Sr from the different fuel particles with an average water content and two different porous media; the trench material and the eolian sand. This means that for flow, the two media have different characteristic curves and for transport they have different values for the distribution coefficient (Kd; trench, eolian sand 2–3 ml/g). The output of this model forms the input of ^{90}Sr in the aquifer. In fact, the observations in some piezometers were used to calibrate the values for Kd (Fig. 6).

Complementary to the observations on site, laboratory experiments are set up to study the transport aspect in more detail. Distribution coefficients and the isotherm for ^{90}Sr have been determined in batch experiments, flow through reactors and column experiments using Chernobyl eolian sand (Szenknect, 2003; Szenknect et al., 2005). In all experiments and in all modelling a constant geochemistry is assumed.

5. Flow and transport in the aquifer

The hydraulic horizontal gradient of the eolian aquifer is derived from automatic water pressure measurements in different piezometers. A value between 0.001 and 0.003 is determined. This value along with the horizontal hydraulic conductivity inferred from slug tests and a pumping test at 3–5 m/day, gives an average effective velocity between 5 and 15 m/year in the eolian sand. In the alluvial part of the aquifer, the main hydraulic gradient is in the vertical direction with a value of 0.03, but hydraulic conductivities are much lower; 0.5 m/day in the horizontal direction and about one order of magnitude smaller for the vertical hydraulic conductivity. The flow direction is north, but may vary during periods of very heavy rainfall, because of a non-equilibrium distribution

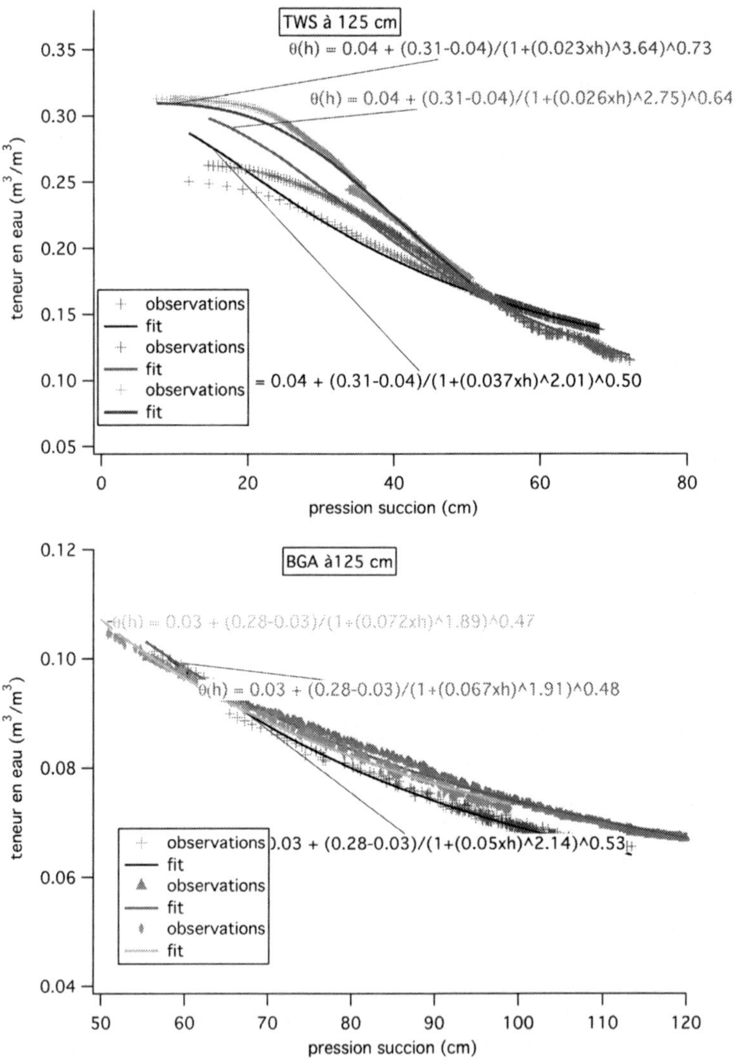

Figure 5. The derived characteristic curves h(θ) for two observation sites; the trench soil (TWS) and the undisturbed eolian sand (BGA).

of hydraulic heads. This in formation allows a first hydraulic modelling of flow in the aquifer along a cross-section under steady-state conditions, when an average recharge of 300 mm/year is considered (annual rainfall is between 600 and 700 mm/year).

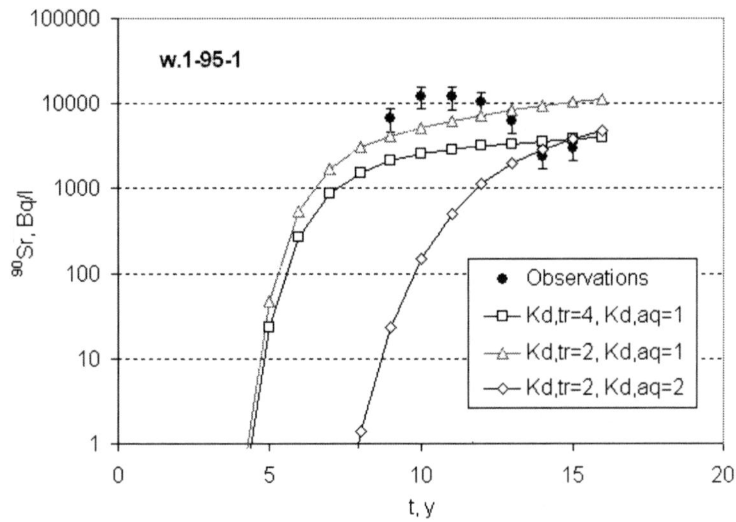

Figure 6. Calibration of the 1D unsaturated flow model for Kd with observations of ^{90}Sr activity in a piezometer (1-95-1).

In a first approximation, transport is considered to be ruled solely by a constant retardation factor. This distribution coefficient determining the retardation factor is adjusted by trial and error fitting of the contaminant plume. Values for the ^{90}Sr distribution coefficient are 0.5 ml/g for the eolian aquifer and between 1 and 4 ml/g in the alluvial part of the aquifer (Fig. 2). These values are much lower than those derived from laboratory experiments (between 25 and 43 ml/g). However, most laboratory experiments were carried out using synthetic water typical of chemical conditions upstream from the trench. Longitudinal dispersivity values are derived from tracer tests with non-reactive tracers (e.g., chloride) to be in the order of 2–3 cm. For modelling purposes, this value was increased to 10 cm as the model area considered is bigger, and dispersivity is a scale-dependent parameter (see e.g. Gelhar and Axness, 1983; Molz et al., 1983; Pickens and Grisak, 1981; Rajaram and Gelhar, 1995). A ratio of 10 is considered between longitudinal and transverse dispersivity.

The modelling result (Bugai et al., 2005) presented in Fig. 7 follows from using the concentration output of the unsaturated one-dimensional model as concentration input for the saturated modelling. The steady state simulation is carried out with Modflow-MT3D (Harbough and MacDonald, 1996; Zheng and

Wang, 1999). The general tendency of the spreading ^{90}Sr contaminant plume is well characterised. Discrepancies are addressed in the discussion below.

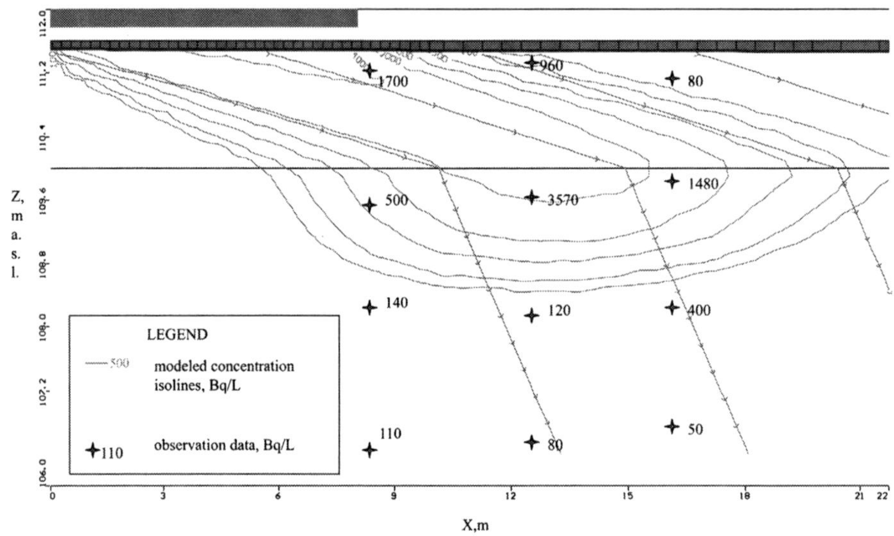

Figure 7. Best result obtained for 2D modelling of the release of ^{90}Sr from the trench to the aquifer.

The hydraulic regime can indeed be considered as a steady state regime for long simulation periods, as hydraulic boundary conditions have not changed substantially. Steady state flow is also considered for the unsaturated zone, which implicitly implies that the water table has not moved substantially during the simulation time (15 years from 1986 to 2001). Observations of water table elevation in some older piezometers on site and in the surrounding area, indicate however, that the water table has risen and fallen over time and has at some points flooded the lower part of the trench. This would change the hydraulics of the system, as the two entities, trench and eolian sand, have different flow curves.

Rephrasing the second assumption, the model considers a constant distribution coefficient over time. The question remains, however, how valid this assumption is under field conditions. Looking at major ion concentration variations in the aquifer upstream and downstream from the aquifer, a clear trench signal is present in ions such as calcium, sulphate, magnesium and stable strontium. The concentration of these elements is higher, up to an order of magnitude, downstream from the trench compared to values upstream. It can be safely concluded that they have a trench origin. If the solution originating from the trench would be constant, the assumption of a constant Kd is correct. But detailed monitoring of concentrations in the piezometers clearly shows chemical changes over time, forcing the eolian sand to re-equilibrate and implying changing

Kd-values over time. The calibrated Kd-value derived from modelling should, therefore, be considered as a mean value over the simulated period. Because the downstream water has higher concentrations of calcium and stable strontium than those considered in laboratory experiments, it is normal that the modelled Kd is lower than the experimentally derived value.

Finally, the question remains if the less soluble and less oxidised particles of the UO_2 class undergo further oxidation over time or not. This could also change the input solution to the aquifer over time.

6. Conclusion and future perspectives

The paper gives an overview of the set-up and the main results of an experimental site in the Chernobyl exclusion zone around a shallow waste trench. In order to understand the release of radionuclides from buried hot fuel particles, an extensive monitoring system was installed and several sampling campaigns carried out. The HFPs have been isolated and described by SEM and dissolution testing. Three categories are discerned, one extra stable class considered insoluble, a more soluble class, present for about 57% with a dissolution constant of 0.018/year, and a third class, comprising about 21% of the buried particles in the trench with a dissolution constant of 0.28/year for a constant pH of 5. The unsaturated zone is equipped with tensiometers, water content sensors and soil water samplers, allowing the derivation of the main parameters needed to calculate flow. The aquifer is monitored for water level changes and geochemistry. Several hydraulic tests and tracer tests have allowed deriving the main parameters needed for calculating flow in the aquifer. The combined information has led to a first satisfactory steady state flow and transport model starting from release of radionuclides from the HFPs to propagation of the ^{90}Sr contaminant plume in the aquifer. At the same time, the modelling results have revealed new questions that are being addressed in the current phase of the project. Namely, how do transient changes in water level influence the release of RN from the trench, how to explain the geochemical changes in the aquifer, are these changes related to maturing organic matter in the trench, and how has the observed plutonium in the aquifer been transported? Today, longer timeseries of observations exist and more emphasis is set on the chemical aspects of the transport. This should lead to more satisfactory answers to the questions and hypotheses raised in the discussion paragraph.

References

Bugai, D., Dewiere, L., Kashparov, V., Ardois, C., and Stammose, D. (2005) ^{90}Sr migration from the low-level waste burial sand aquifer at the Chernobyl NPP site: Hydrogeological, geostatistical and geochemical model integration and calibration, Migration 2005, Avignon, France, pp. 203–204.

Garten Jr, C.T. (1999) Modeling the potential role of a forest ecosystem in phytostabilization and phytoextraction of ^{90}Sr at a contaminated watershed. *Journal of Environmental Radioactivity*, **43**(3): 305–323.

Gelhar, L.W. and Axness, C.L. (1983) Three-dimensional stochastic analysis of macrodispersion in aquifers. *Water Resources Research*, **19**(1): 161–180.

Harbough, A.W. and MacDonald, M.G. (1996) User's documentation for MODFLOW-96, an update to the U.S. Geological Survey modular finite difference ground-water flow model. Open-File Report 96-485, U.S. Geological Survey, Reston, Virginia.

Kashparov, V.A. et al. (2004) Kinetics of dissolution of Chernobyl fuel particles in soil in natural conditions. *Journal of Environmental Radioactivity*, **72**(3): 335.

McCarthy, J.F., Czerwinski, K.R., Sanford, W.E., Jardine, P.M., and Marsh, J.D. (1998) Mobilization of transuranic radionuclides from disposal trenches by natural organic matter. *Journal of Contaminant Hydrology*, **30**(1–2): 49–77.

Molz, F.J., Güven, O., and Melville, J.G. (1983) An examination of scale-dependent dispersion coefficients. *Ground Water*, **21**(6): 715–725.

Pace, M.N., et al. (2007) Transport of Sr^{2+} and $SrEDTA^{2-}$ in partially-saturated and heterogeneous sediments. *Journal of Contaminant Hydrology*, **91**(3–4): 267–287.

Pickens, J.F. and Grisak, G.E. (1981) Scale-dependent dispersion in a stratified granular aquifer. *Water Resources Research*, **17**(4): 1191–1211.

Rajaram, H. and Gelhar, L.W. (1995) Plume-scale dependent dispersion in aquifers with a wide range of scales of heterogeneity. *Water Resources Research*, **31**(10): 2469–2482.

Salbu, B., et al. (2001) High energy X-ray microscopy for characterization of fuel particles. *Nuclear Instruments and Methods in Physics Research*, **467–468**: 1249–1252.

Steefel, C.I., Carroll, S., Zhao, P., and Roberts, S. (2003) Cesium migration in Hanford sediment: A multisite cation exchange model based on laboratory transport experiments. *Journal of Contaminant Hydrology*, **67**(1–4): 219.

Szenknect, S. (2003) Transferts de radioéléments en zone non saturée. Etude expérimentale et modélisation appliquées au Site Pilote de Tchernobyl. PhD Thesis, Université Joseph Fourier-Grenoble I, Grenoble/Fontenay-aux-Roses, 286 pp.

Szenknect, S., Ardois, C., Gaudet, J.-P., and Barthès, V. (2005) Reactive transport of ^{85}Sr in a Chernobyl sand column: Static and dynamic experiments and modeling. *Journal of Contaminant Hydrology*, **76**(2005): 139–165.

van Genuchten, M.T. (1980) Closed-form equation for predicting the hydraulic conductivity of unsaturated soils. *Soil Science Society of America Journal*, **44**(5): 892.

Zheng, C. and Wang, P.P. (1999) Chapter 3: Overview of solution techniques, MT3DMS: A modula three-dimensional multispecies transport model for simulation of advection, dispersion and chemical reactions of contaminants in groundwater systems, documentation and users' guide.

Zhuang, J., McCarthy, J.F., Tyner, J.S., Perfect, E., and Flury, M. (2007) *In situ* colloid mobilization in Hanford sediments under unsaturated transient flow conditions: Effect of irrigation pattern. *Environmental Science and Technology*, **41**(9): 3199–3204.

DECONVOLUTION OF ALPHA SPECTRA FROM HOT PARTICLES

ROY PÖLLÄNEN*[1], TERO KARHUNEN[1], TEEMU SIISKONEN[1], HARRI TOIVONEN[1], ANDREAS PELIKAN[2]
[1]*STUK – Radiation and Nuclear Safety Authority, P.O. Box 14, FI-00881 Helsinki, Finland*
[2]*Dienstleitungen in der automatischen Datenverarbeitung und Informationstechnik, Austria*
**phone: ?????????, e-mail: roy.pollanen@stuk.fi*

Abstract: A computer code known as AASIFIT is developed to unfold complex alpha spectra. Peak shapes used in the fitting are obtained from the simulations. In addition to activities of the nuclides present in the sample, the code can provide source characterization. AASIFIT is applied for a nuclear bomb particle collected in Thule, Greenland. It is shown that direct alpha spectrometry provides more accurate activity estimate for $^{239+240}$Pu than gamma-ray spectrometry.

Keywords: direct alpha spectrometry, deconvolution, hot particle.

1. Introduction

In alpha spectrometry, the properties of the detector and the characteristics of the source have a notable influence on the detected alpha particle energy spectrum. In addition, materials between the source and the detector, the measurement geometry and coincidences between alpha particles and other particles emitted by the source may considerably affect the alpha peak shapes. During recent years intensive work has been done to study the peak shapes and to develop numerical analysis of the spectra and fitting codes. "Analytical models to represent the shape of a monoenergetic alpha peak are necessary for spectral analysis, but can also help to understand the physical phenomena involved in the detection process, from emission inside the source to interaction with the detector" (García-Toraño, 2006).

Computer codes developed for unfolding the alpha particle energy spectra can usually cope with a thin sample processed radiochemically for counting. In that case the shape of the peaks is determined mainly by the properties of the detector, which makes spectrum analysis independent on the source characteristics. A number of computer codes and peak shape models with various levels of complexity have been developed during recent years (see García-Toraño, 2006, and the references therein). In these codes the shape is often treated as convolution of a Gaussian distribution with low-energy side (double/triple) exponential tail, although other shapes are used, too.

In direct alpha spectrometry, when radiochemical source processing is not performed, the peak shape may be strongly influenced by the source properties. The shape may then deviate drastically from that of an ideal Gaussian peak, which evokes the question of correct peak shape to be used in the spectrum analysis. Spectrum unfolding cannot be performed without this information. Additional problem is that several nuclides with a block of multiple peaks may be present in the spectrum and the shape of the peaks of different nuclides may also vary. In general, there are no analytical models for peak shapes and, thus, a totally new approach is needed to unfold the spectra obtained with direct alpha spectrometry.

A novel computer code to analyze the alpha spectra is introduced here. Although the code is designed to unfold complex spectra and especially those from direct alpha spectrometry, it can be used for conventional spectra obtained from radiochemically processed samples as well. A prominent feature is that, in addition to calculation of the nuclide activities, the code can be used to characterize source properties. The peak shapes are modelled using the Monte Carlo simulation software known as AASI (Siiskonen and Pöllänen, 2005) in which a fitting module is linked to perform the spectrum unfolding. The software package is demonstrated here using a nuclear bomb particle from the Thule accident (Eriksson, 2002) as an example.

2. Thule particle characteristics

The nuclear bomb particle used in the present demonstration was isolated from a sea sediment sample collected in Thule, Greenland, in 1997. Pöllänen et al. (2007) used gamma-ray spectrometry, scanning electron microscopy (SEM) with X-ray microanalysis and direct alpha spectrometry together with the Monte Carlo simulations to characterize the properties of this particle. A brief summary of the analysis results are presented in the following.

The presence of ^{241}Am (activity 0.43 ± 0.04 Bq), ^{235}U (0.003 ± 0.001 Bq) and ^{239}Pu (4 ± 2 Bq) was detected in the gamma-ray spectrum. On the basis of the SEM pictures, the particle was irregular in shape with approximate dimensions given in Fig. 1. The X-ray microanalysis revealed that the particle was a mixture of U and Pu.

Direct high-resolution alpha spectrometry (450 mm^2 PIPS detector) also revealed the presence of $^{239+240}$Pu and ^{241}Am with approximate activities of 3.3 Bq and 0.5 Bq, respectively. The sum activity of alpha-particle emitting radionuclides was estimated to be 3.8 ± 0.3 Bq. The density of the particle was estimated with the aid of the Monte Carlo simulations and approximate value of 8.8 g cm^{-3} was obtained. However, these numbers are based only on simulations and no fitting tools were available in the spectrum analysis. Here, the sophisticated spectrum analysis tool is applied to characterize the particle.

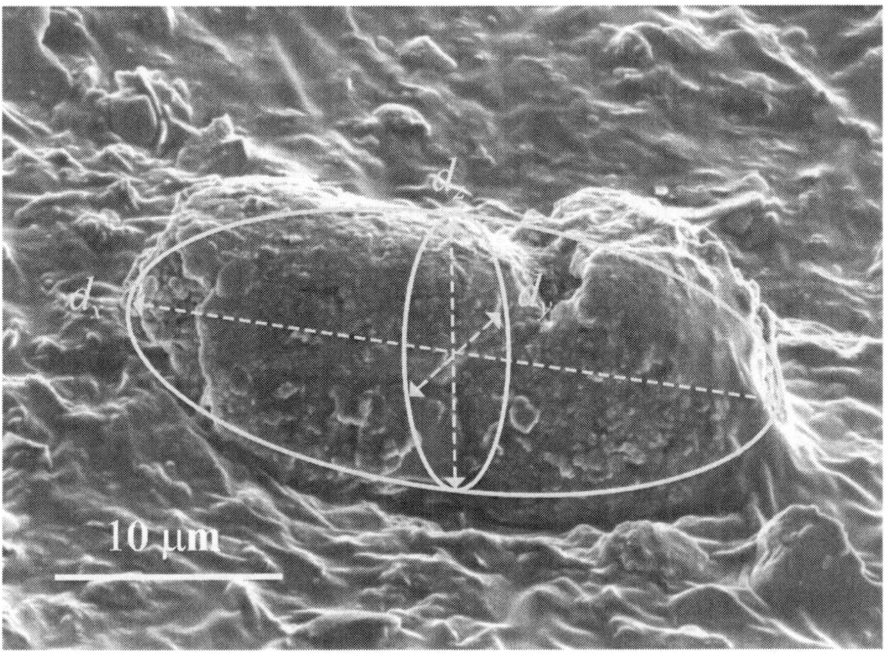

Figure 1. Secondary electron image of the Thule particle considered here. The particle surrounded by inactive sediment particles is attached on the glue of a tape Approximate dimensions d_x = 27 μm, d_y = 16 μm and d_z = 9 μm are estimated from the SEM pictures (this one and other pictures not shown here).

3. Spectrum unfolding

3.1. GUIDELINES TO DESIGN THE NEW SPECTRUM UNFOLDING CODE

A spectrum from a three-dimensional source may be totally different compared with that from a radiochemically processed sample. At least the following differences may exist:

1. Several alpha-particle emitting radionuclides from different elements may be present in the sample and, thus, multiple overlapping peak groups may be present in the spectrum.
2. The shape of the alpha peaks may be totally different compared with that obtained from a thin sample. This is due to the alpha particle energy absorption in the source material. In the case of a thin sample the peak width (in terms

of FWHM) may be of the order of 15 keV (depending of the detector) but for a thick sample it may be even a hundred times larger or more.

3. The peak shape may also depend on the radioelement in question. The reason for this is that radioelements may be distributed differently in the sample. For example, some elements may be present on the surface of the sample whereas others are homogeneously distributed in the three-dimensional sample.
4. Although energy vs. channel calibration is the same when measuring a thin or a thick source, the concepts like peak "location", "centroid" or "energy" are not optimal for characterizing the peaks from a thick source.

It is obvious that direct alpha spectrometry cannot be applied for a source whose effective thickness is large (e.g., tens of micrometers or more) but it may be a feasible method provided that the sampling is designed properly. For example, it can be used for air filter samples (Siiskonen and Pöllänen, 2004), low-thickness samples (Pöllänen et al., 2005) and hot particles (Pöllänen et al., 2007).

A prerequisite for a viable method for spectrum unfolding is that the phenomena mentioned above (items 1–4) are properly taken into account for designing a computer code. Correct peak shape in the fitting process is one of the most challenging problems for unfolding the spectra. Unfortunately, there are no generic analytical models that can be applied for a three-dimensional source. The best option would be to get the shapes from measurements but this is seldom possible. Another possibility is to generate them using simulations as shown here.

The following guidelines are followed for the new computer code (c.f., the list above):

1. Possible presence of multiple peaks in the spectrum is taken into account by constructing a data library which contains all relevant radionuclides. Peak overlapping is taken into account by fitting the family of the peaks of a radionuclide rather than fitting individual peaks. The yields of individual peaks in the groups are taken from the data library.
2. Peak shapes are generated through simulations. However, in this case accurate information about the measurement conditions and knowledge of the source characteristics are necessary. AASI is able to consider a variety of measurement geometries and source properties in the simulations.
3. Each radionuclide has its own specific shape. In the peak shape simulation each nuclide is considered separately and every transition mentioned in the data library is taken into account in the simulation. Thus, histograms rather than predetermined analytical peak shapes are considered in the fitting.
4. Energy calibration is taken into account in the fitting by letting the group of peaks of a single radionuclide walk along the energy axis. The user must

have some background information about the origin of the source and its properties for selecting the nuclides to be used in the spectrum analysis. Otherwise false identifications are possible.

3.2. SIMULATION OF THE ALPHA PARTICLE ENERGY SPECTRA

AASI computer code was originally designed to simulate alpha particle energy spectra in alpha spectrometry but the domain of the software is now enlarged to spectrum analysis. In the simulations the code takes into account the alpha particle energy loss in the source material, the material between the source and the detector and in the detector dead layer (Fig. 2). In the following simulations/measurements there are no absorbing material layer between the source and the detector. The PIPS detector response to monoenergetic alpha particles is assumed to have a Gaussian part with a two-component exponential tailing on the low-energy side of the full-energy peak.

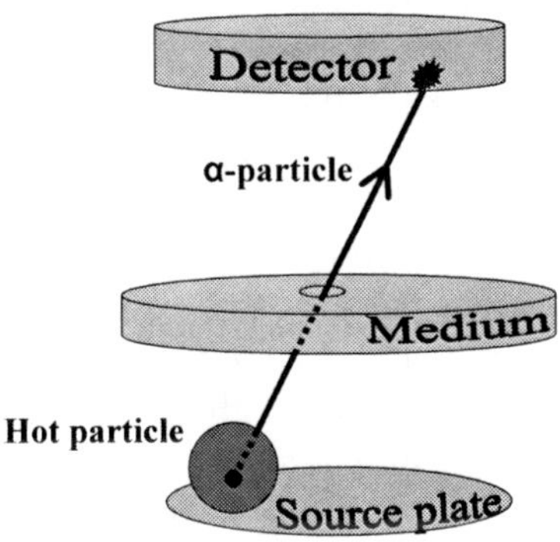

Figure 2. Simplified operation principle of AASI in the simulations. A spherical hot particle (not to scale) is on a source plate. An alpha particle generated in the source is travelling through the source material (here the hot particle) and the medium and finally enters the detector. Energy loss in the material of known properties (dimensions, composition, density) is calculated using known stopping power dE/dx. In the final stage, the detector response is taken into account.

AASI is able to take into account varying source characteristics. The source material and the objects emitting the alpha particles are not necessarily the same. In general, the source matrix is assumed to be homogenous with known dimensions and elemental composition whereas the distribution of radionuclides incorporated in the source matrix may vary. For example, hot particles with known elemental composition and size distribution act as a source matrix and radionuclides within the particles may be homogeneously distributed in the particles or on their surfaces. In addition, the particles may be embedded in another matrix, such as an air filter, with uniform or exponential depth distribution. In AASI, hot particle shapes can be either spherical or elliptical with user-given dimensions. Other shapes must be modelled as a combination of these shapes. Even though the shape of the particle presented in Fig. 1 is somewhat irregular, it is modelled in the following as an ellipsoid.

3.3. AASIFIT - A NOVEL DECONVOLUTION CODE FOR ALPHA SPECTROMETRY

AASIFIT computer code to deconvolute the alpha particle energy spectra can be used in various ways. We demonstrate here how the code can be used for particle characterization using the properties of the Thule particle mentioned above as a starting point. Since dimensions of the particle are well known, the focus here is to have a justified estimation for the density.

When investigating source particle characteristics through fitting, the procedure may proceed as follows (all phases are performed in AASIFIT).

- Open the measured spectrum for the analysis.
- Pick out nuclide(s) from the library to be used in the fitting and define the number of alpha particle emissions from the source to be used for generating simulated peaks in the fitting.
- Choose correct measurement geometry used in the simulations.
- Give pre-estimated source particle characteristics, i.e., dimensions, density and elemental composition, to be used in the simulations.
- Select appropriate alpha particle energy region (region-of-interest) from the measured spectrum that is used in the fitting.
- Choose minimum and maximum values for the quantity (particle density in this example) to be estimated through fitting. Step size between the minimum and maximum value may also be defined if simplified stepping algorithm is preferred.
- Start combined simulation and fitting.

AASIFIT generates simulated peak group shapes for the nuclides in question using the input parameter combination given by the user. The simplified operation scheme is as follows: At first, user-defined minimum value for the density is used in the fitting (Levenberg-Marguardt algorithm is used). The code calculates χ^2-value for the fit and stores it. Secondly, new nuclide-specific peak group shapes are generated using the next value for density (minimum value + step size), and fitting and χ^2-calculation is performed again. This process is repeated until maximum value for the density is reached. Note that in addition to the simple linear stepping, bisection stepping process may also be used. In the final stage, that value of the density is selected which gave the smallest χ^2-value in the fitting, and surface areas of the peaks (or nuclide-specific activities) are reported.

The reduced chi-square, a quality-fitting index, is computed as follows (CTBTO, 2007):

$$\chi^2 = \frac{1}{n}\sum_{i=1}^{h}\frac{1}{y_i}(f_i - y_i)^2, y_i > 0, \tag{1}$$

where
y_i = counts in channel i,
f_i = value of the fitted shape in channel i,
l = first channel of analysis range,
h = last channel of analysis range,
n = number of channels with nonzero counts.

The result of the abovementioned process is illustrated in Fig. 3. The fitting area is limited to 1–5.6 MeV because below 1 MeV no useful information can be obtained from the spectrum. Here, only ^{239}Pu and ^{241}Am are assumed to be present in the particle, since in terms of activity the contribution of other nuclides is negligible.

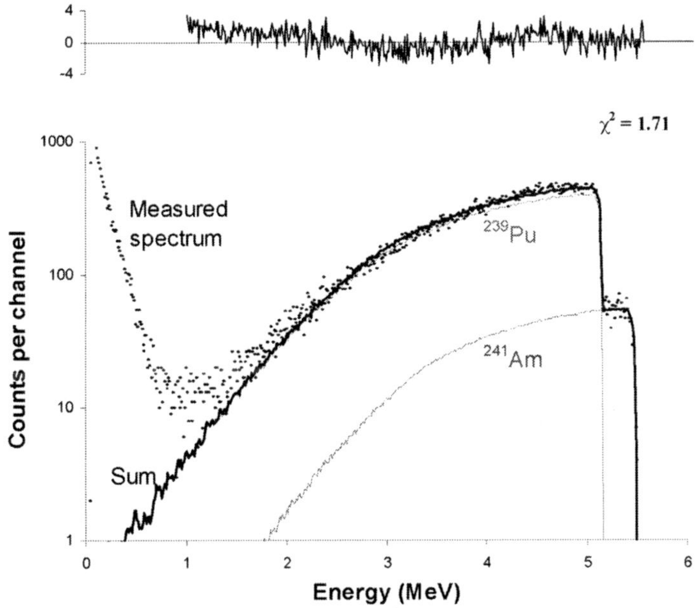

Figure 3. The peaks of ^{241}Am and ^{239}Pu (thin lines) and their sum (thick line) fitted to the measured spectrum (dots, see Pöllänen et al. (2007) for detailed description of the measurement). The reduced residual, i.e. the difference between the measured and simulated values, is shown at the top (in terms of standard deviation of the number of counts).

Simulated peaks are not smooth in contrast to those from analytical models. In fact, a simulated peak is a histogram whose area is fitted to the measured data. The larger the number of alpha particle emissions in the simulations, the smaller the roughnesses in the histograms. In general, the small number of alpha particle emissions in the simulations ensures the rapid execution of the code and the large value of χ^2, whereas the high number of emissions means slower simulations and smaller values of χ^2 in the fitting.

4. Density estimation of the Thule particle

As mentioned above, Pöllänen et al. (2007) evaluated values for the density of the Thule particle using the Monte Carlo simulations. However, the values were only tentative because no analysis tools were used for spectrum unfolding. The development of the spectrum analysis tool enables now more reliable estimations.

The estimation of the density is performed from the measured spectrum presented in Fig. 3. In the simulations the number of alpha particle emissions from ^{241}Am and ^{239}Pu is 5×10^6 which means that the total number of counts in the simulated peaks is 1.4×10^6 (the geometrical detection efficiency $\varepsilon = 0.276$ when source-detector distance is 6 mm). The density of the elliptical particle is estimated as a function of the *z*-dimension (Fig. 1) of the particle because this dimension cannot be evaluated accurately from the SEM pictures. The upper limit of the thickness of the irregularly-shaped Thule particle is evaluated from a measurement performed at the source-detector distance of 41 mm. In this measurement geometry the maximum energy loss of the alpha particles was about 4 MeV which gives for the upper limit $d_{z,max} = 13$ µm.

According to the SEM pictures, the z-dimension of the elliptical particle is 9 ± 2 µm (Pöllänen et al., 2007) and these limits (11–7 µm) give for the density $\rho = 7.4 - 10.6$ g cm^{-3} (Fig. 4). In the simulations the elemental composition of the particle is assumed to be UO_2. The calculated peak areas for ^{241}Am and ^{239}Pu do not sensitively depend on the selected combination of ρ and d_z (Table 1).

TABLE 1. The estimated density, χ^2 value of the fit and peak areas for $^{239+240}$Pu and ^{241}Am as a function of the Thule particle d

d_z (µm)	ρ (g cm^{-3})	χ^2	$^{239+240}$Pu peak area (counts)	^{241}Am peak area (counts)
4	17.5	1.25	67,750 ± 380	8,750 ± 260
5	14.3	1.24	67,550 ± 380	8,920 ± 260
6	12.2	1.29	67,310 ± 390	9,100 ± 270
7	10.6	1.39	67,210 ± 400	9,080 ± 280
9	8.5	1.71	67,050 ± 450	8,940 ± 310
11	7.4	2.19	66,400 ± 500	9,450 ± 360
13	6.5	2.71	66,170 ± 560	9,330 ± 400
15	5.9	3.05	65,960 ± 600	9,350 ± 420

The prerequisites for successfull density estimation are the ability to model the particle in the simulations and capability of performing fast and reliable fitting. The value of χ^2 characterizes goodness of the fit (and nothing more), but it cannot be solely used as a measure for a satisfactory outcome. The models used in the simulations are equally important as evidenced in the following. As shown in Table 1, χ^2 decreases with decreasing d_z which means that the simulated peak shapes are close to the measured shapes for small d_z. However, small d_z values (<7 μm) are not justified on the basis of the SEM pictures. The discepancy arises from the fact that the model of elliptical source in the case of irregularly-shaped particles is not necessarily the best option for the particle shape. Unfortunately, it is not possible to incorporate all possible shapes in the simulations. Thus, small χ^2 values alone cannot be used to justify the selection of the best combination of ρ and d_z. Information obtained from other analysis methods (here SEM) and the models used in the simulations are important as well.

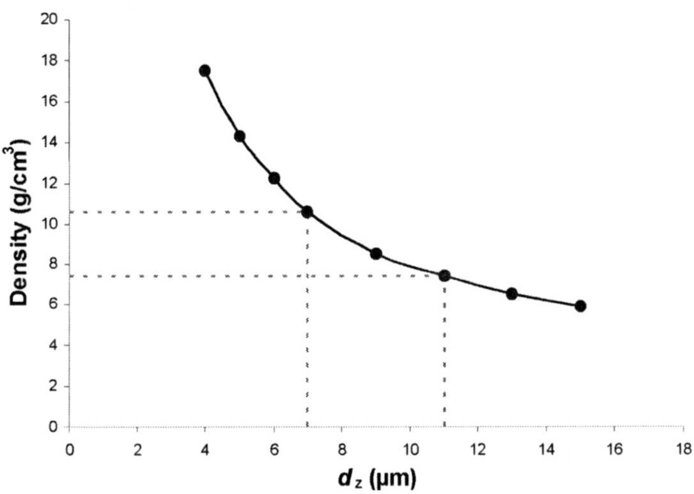

Figure 4. The estimated Thule particle densities as a function of thickness, d_z (see Fig. 1), other dimensions remain unaltered. The estimation is based on simulated peak shapes of ^{239}Pu and ^{241}Am, which are fitted to the measured spectrum.

5. Activity estimation

In contrast to the density estimation, goodness of the fit can be solely used as an appropriate indicator for the activity estimation. An excellent agreement was found between the measured data and the fit (Fig. 5).

For each fit presented in Table 1, the activities can be calculated dividing the peak areas by data acquisition time (73,100 s) and geometrical detection efficiency (ε = 0.276). To estimate the activities presented in Table 2, only those fits are taken into account for which χ^2 is less than 1.3.

A systematic error for the geometrical detection efficiency is possible because the Thule particle cannot be accurately placed in the vacuum chamber of the alpha spectrometer. Uncertainty on the source to detector distance is of the order of ±0.5 mm and that of the lateral shift ±2 mm (off the symmetry axis of the detector). These variations give an additional uncertainty of 6%.

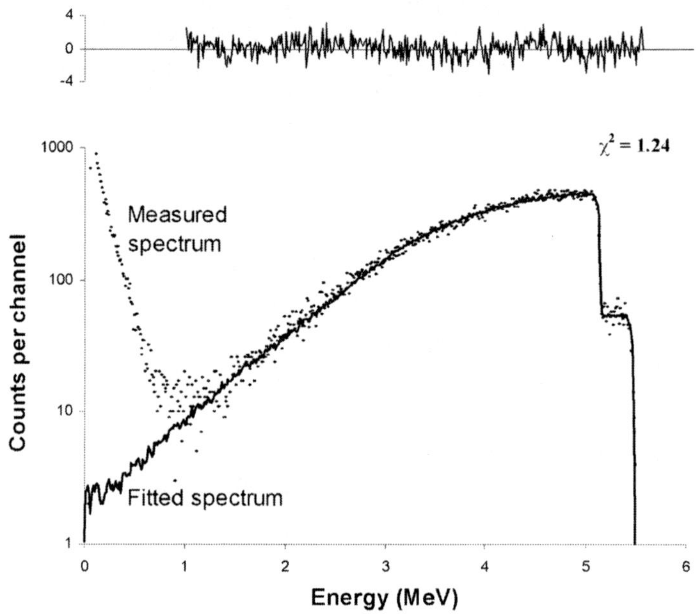

Figure 5. The simulated sum spectrum of ^{241}Am and ^{239}Pu (thick line) fitted to the measured spectrum (dots) and the reduced residual.

TABLE 2. The activities obtained with gamma-ray spectrometry (Pöllänen et al., 2007) and direct alpha spectrometry (the uncertainty includes systematic error). The activity of ^{235}U cannot be obtained using direct alpha spectrometry

Nuclide	γ-Ray spectrometry	Direct α-spectrometry
$^{239+240}$Pu	4 ± 2 Bq	3.36 ± 0.20 Bq
^{241}Am	0.43 ± 0.04 Bq	0.44 ± 0.03 Bq
^{235}U	0.003 ± 0.001 Bq	–

6. Conclusions

AASIFIT computer code is developed to unfold complex alpha particle energy spectra. In addition to evaluate the activities of the nuclides present in the sample, the code can be used to investigate the source properties. Instead of analytical peak shapes, the code uses simulated peaks which are fitted to the measured spectrum.

AASIFIT is demonstrated here for the Thule hot particle. Owing to the notable alpha particle energy absorption in the radioactive particle, there are no clear peaks in the measured spectrum. However, AASIFIT was able to unfold this spectrum which is far too complicated for other codes.

In the simulations the particle is assumed to be elliptical in shape with the z-dimension 7–11 μm and these limits give for the density ρ = 10.6–7.4 g cm^{-3}, respectively. The upper limit of the z-dimension is estimated from the measured maximum alpha particle energy loss and it is of the order of 13 μm. This number is in accordance with the SEM pictures.

AASIFIT can calculate the number of alpha particles totally absorbed in the source material. This is a unique feature of the code which makes estimation of nuclide-specific activities possible for thick sources. The results obtained with direct alpha spectrometry coincide well with those from gamma-ray spectrometry. Both methods give comparable results for ^{241}Am but direct alpha spectrometry is superior for $^{239+240}$Pu.

Goodness of the fit alone cannot be necessarily considered the only measure to obtain reliable results from the AASIFIT code. The models and the source properties used in the simulations must also be justified. In the simulations it is often possible to use such an input parameter combination which may give simulated peaks that are coincidentally identical to those obtained from the measurements. Thus, the relevance of the selected parameter combination must be proven case by case.

Acknowledgements

The participants of the Thule expedition (1997) are kindly acknowledged for the samples. The auhors are grateful to Mr. Mikael Moring and Mr. Jyrki Juhanoja for taking the SEM picture.

The AASI code is available free of charge from the following link:
http://www.stuk.fi/tutkimus/programs/aasi/en_GB/aasi/

References

CTBTO (preparatory commission for the comprehensive nuclear-test-ban treaty organization). (2007) Tutorial of Radionuclide Analysis and Evaluation Software Aatami – 4.10. Office of the Executive Secretary, Evaluation Section, 15.3.2007, 1–195.

Eriksson, M. (2002) On weapons plutonium in the Arctic Greenland (Thule, Greenland). PhD thesis. Risø National Laboratory, 1–146.

García-Toraño, E. (2006) Current status of alpha-particle spectrometry. *Appl. Radiat. Isot.* **64**, 1273–1280.

Pöllänen, R., Siiskonen, T., Moring, M., and Juhanoja, J. (2007) Direct alpha spectrometry for characterising hot particle properties. *Radiat Measurem.* **42**, 1666–1673.

Pöllänen, R., Siiskonen, T., and Vesterbacka, P. (2005) High-resolution alpha spectrometry from thick sources. *Radiat. Measurem.* **39**, 565–568.

Siiskonen, T., and Pöllänen, R. (2004) Simulation of alpha particle spectra from aerosol samples. *Appl. Radiat. Isot.* **60**, 947–953.

Siiskonen, T. and Pöllänen, R. (2005) Advanced simulation code for alpha spectrometry. *Nucl. Instr. Meth. Phys. Res.* **A550**, 425–434.

THE USE OF ^{236}U AS A TRACER OF IRRADIATED URANIUM

VLADISLAV MIRONOV*[1], SERGEY PRIBYLEV[1], VLADISLAV ZHURAVKOV[1], JANNA MATUSEVICH[1], MICHAEL HOTCHKIS[2], DAVID CHILD[2]
[1]*International Sakharov Environmental University, 23 Dolgobrodskaya str., Minsk, 220009, Belarus*
[2]*Australian Nuclear Science and Technology Organization, Lucas Heights, NSW 2234, Australia*
**phone: ???????, e-mail: mvladislav@bk.ru*

Abstract: A brief review of the use of ^{236}U as a tracer of irradiated uranium is given. Data on vertical migration of irradiated uranium in various types of soils, the destruction degree of fuel particles determined by carbonate leaching of ^{236}U (VI), the nuclear fuel burn-up fraction in Chernobyl originated fallout and the total contamination of territory of Belarus by irradiated uranium are presented.

Keywords: fuel particle, uranium, neptunium, mass-spectrometry, carbonate leaching, isotopic ratio, nuclear fuel burnup.

1. Introduction

Among all radionuclides introduced into the biosphere as a result of the accident at the Chernobyl nuclear power plant (ChNPP) in April 1986, the biggest part in terms of mass is clearly uranium.

Considerable attention has been focused on irradiated uranium contamination in the region near the Chernobyl reactor (the exclusion zone) (Boulyga et al., 2002; Mironov et al., 2002). Information on contamination by Chernobyl originated uranium in the more remote regions of Belarus (far-field locations more than 150 km from the reactor) is absent. One of the aims of our work is to define the limits of the distribution of Chernobyl originated uranium.

The majority of actinides in the primary fallout was in the form of inert fuel particles (FP). These small fragments of irradiated fuel characteristically range in size from fractions of micrometers up to hundreds of micrometers. They consist of a uranium oxide matrix (UO_2) with various other radionuclides embedded in this uranium matrix. The radionuclide composition of fuel particles is close to that of the irradiated nuclear fuel of the 4-th power unit, but with fractionation of the volatile products.

Under the influence of different active components of the soil complex and oxidation of four-valency uranium, destruction of FPs occurs, which causes changes in their dispersion and the release of radionuclides from the fuel matrix. As a result of these processes the fraction of radionuclides in mobile forms increases, making them biologically available to plants and consequently more dangerous for humans. The study of fuel particle migration in the soil as well as rate of radionuclide leaching under different conditions seems to be a central problem for forecasting the behaviour of the radioactive fallout in the ChNPP estranging zone.

The technique that we use for an estimation of degree of FP destruction is based on application of soft carbonate leaching of uranium (VI) without dissolution of UO_2 matrix. For the purposes of this work we need to differentiate uranium originating from ChNPP and the uranium existing in the soil before radioactive fallout.

According to a previous study the level of surface soil contamination in evacuated areas in Belarus with Chernobyl originated uranium corresponds approximately to the concentration range of natural uranium which is 0.1–2.0 grams per tonne (Shagalova, 1986). This fact makes accurate determination of irradiated "Chernobyl" uranium in contaminated soil samples by measuring the $^{235}U/^{238}U$ isotope ratio difficult.

The presence of ^{236}U in environmental samples represents a good indication that nuclear material had been introduced into the vicinity. In naturally occurring uranium, a ^{235}U (n,γ) ^{236}U reaction is produced by neutrons generated by cosmic irradiation, spontaneous fission or (α, n) reaction on light nuclei, and the $^{236}U/^{238}U$ isotope ratio is expected to be about 10^{-14} at/at in the environment (Mironov et al., 2002). In spent nuclear fuel the $^{236}U/^{238}U$ ratio can exceed 10^{-3} at/at depending on the reactor type and the history of fuel burnup. Due to the difference of ^{236}U concentration in natural uranium and spent fuel uranium (>7 orders of magnitude) ^{236}U is a useful "fingerprint" of uranium originating from a nuclear reactor or accelerator driven system (Boulyga et al., 2002). Recently, the determination of ^{236}U was recognized by International Atomic Energy Agency (IAEA) as an important signature to differentiate sources of contamination.

We have used the advantages of ^{236}U determination firstly for studying the vertical migration of Chernobyl originated uranium in soils which are typical for a 30-km zone around ChNPP, secondly for an estimation of burnup of nuclear fuel that was released as a result of accident at ChNPP and finally for an estimation of fuel particle destruction rate.

2. Methods

Soil samples were collected from reference sites located in radioactive spots of the 30-km zone around ChNPP in 1998–2000. Samples were taken from soddy-podzolic sabulous, sandy and peaty soils as the territory of Belarus belongs to soddy-podzolic zone, and in the southern part of the country large areas are occupied with peaty and sandy soils. Soil samples were selected using the trench method at a depth of up to 60 cm.

TABLE 1. Geographical coordinates and description of sampling locations

	Chistogalovka	Chistogalovka	Pripyat	Malinovka
Coordinates	510 22′39″N 300 02′26″E	510 22′34″N 300 02′00″E	510 25′08″N 300 04′23″E	530 30′46″N 310 37′35″E
Soil type	Soddy-podzolic sabulous	Turfy	Sandy	Soddy-podzolic
Density of soil layer (0–5 cm), g/cm^3	1.30	0.26	1.41	1.40
pH (KCI)	3.65	4.87	4.00	3.80
C_{org}	1.55	29.4	0.11	1.63

Some soil samples were also collected from a location in SE Belarus, 255 km NE of Chernobyl (Malinovka village, Mogilev region). This far field location lies within the Bryansk-Belarus Spot centered 200 km NNE of the reactor. Soil samples were collected using the same trench method at a depth of up to 40 cm.

Geographical coordinates and some of the soil characteristics of sampling locations selected are summarized in Table 1.

The low levels of ^{236}U in the environment and the similarity of the ^{236}U alpha-particle energies to the more abundant ^{235}U samples means traditional alpha-counting is not practicable. The alpha emission of ^{235}U includes energies 4.502 MeV (relative intensity 1.7%) and 4.435 MeV (relative intensity 0.7%) which interfere with ^{236}U emissions (main energies of 4.494 MeV and 4.445 MeV). Additional limitation can be caused by low energy tailing of alpha emissions from ^{234}U and ^{235}U. Application of special spectra unfolding software for resolving interferences in alpha spectrometry can allow determination of $^{236}U/^{238}U$ isotopic ratios down to 2.6×10^{-5} (relative uncertainty of about 10%).

Mass-spectrometric methods (MS) are considered to be the most sensitive for ^{236}U measurements in environmental samples. The type of mass-spectrometer selected for an analysis will depend on the required sensitivity, expected $^{236}U/^{238}U$ isotope ratio, and the concentration of ^{236}U in the samples.

$^{236}U/^{238}U$ isotope ratios >10^{-6} at/at can be determined by inductively coupled plasma mass-spectrometry (ICP-MS). Uranium isotopic ratio measurements in soil samples from the 30-km zone were carried out by means of this method. Three kinds of different ICP-MS were used: a double-focusing sector field ICP-MS ELEMENT (Finnigan MAT, Bremen, Germany), a quadrupole-based ICP-MS Elan-6000 (Perkin Elmer, Sciex Corporation, Ontario, Canada) and a quadrupole-based ICP-CC-MS (Platform ICP, Micromass Ltd., Manchester, UK) with a hexapole collision cell. Helium was introduced into the collision cell of the ICP-CC-MS as a buffer gas. The measured uranium isotopic ratios were corrected taking into account the mass discrimination factor (assuming a linear correlation), determined experimentally by measuring CCLU-500 standard solution ($^{234}U/^{238}U$ = 0.011122, $^{235}U/^{238}U$ = 0.99991, $^{236}U/^{238}U$ = 0.002789) (Nuclear Research Center, Prague, Czech Republic), hydride rate UH^+/U^+ ratios and the dead time of the detector. Values of two last parameters were determined by use of NBS U-020 standard solution (National Institute of Standards and Technology (NIST, USA).

ICP-MS provides good sensitivity and accuracy of isotopic measurements, therefore this technique is one of the most suitable methods for routine $^{236}U/^{238}U$ isotope ratio measurements in environmental samples (Mironov et al., 2002). However, this method is not applicable for measurements of $^{236}U/^{238}U$ isotope ratios in environmental samples at levels less than 10^{-6} at/at. High abundance sensitivity for $^{236}U/^{238}U$ ratios, down to 10^{-12} is feasible by accelerator mass spectrometry (AMS) which is a well-established technique for uranium isotopic measurement (Hotchkis et al., 2000).

AMS was used for measurements of isotopic ratio $^{236}U/^{238}U$ in soil samples from far field areas of contamination, where $^{236}U/^{238}U$ isotope ratio was expected to be <10^{-6} at/at. Samples were prepared by high temperature Lithium Metaborate/Tetraborate fusion, followed by dissolution of the fusion bead and extraction of actinides using an Eichrom TRU resin (Eichrom Industries). The purified uranium fractions were then co-precipitated with iron oxide and mixed with niobium powder which improves the electrical and thermal properties of the sample surface. The $^{236}U/^{238}U$ isotope ratios were then measured using AMS.

These measurements were performed by the ANTARES FN Tandem accelerator at ANSTO, Australia using the actinides beam line described previously (Hotchkis et al., 2000). Measurements were carried out with a terminal potential of 4MV on the 5+ charge state species. ^{236}U was counted in an ionization chamber, while ^{238}U was measured off axis in a secondary electron multiplier.

The method of determination of the degree of fuel particle destruction is based on application of soft carbonate leaching of uranium (VI) from air dried soil samples with a solution of 10% Na_2CO_3–5% $NaHCO_3$ for 90 minutes, avoiding dissolution of matrix UO_2 (Patent C1 BY, 2002). Using ^{236}U analysis, the content of Chernobyl originated uranium in the whole sample is determined, and then in the part of sample which has passed into carbonate solution, i.e., the fraction present as uranium (VI). Knowing the ratio of these two values it is possible to determine degree of destruction of the fuel particles.

Concentrations of natural and technogenic uranium were calculated from experimental data on the content of ^{236}U and ^{238}U in soil profiles using a technique described elsewhere (Mironov et al., 2005).

In this work for the purpose of mapping of radioactive contamination GIS technology, version 3.1 ARC/VIEW, was used.

3. Results and discussion

Some of the results on vertical migration research of natural and Chernobyl originated uranium in soil profiles for sampling locations 1–3 are shown in Fig. 1. For natural uranium various tendencies of redistribution in soils (ascending, descending and combined) are observed.

Non-uniformity of distribution of uranium in various kinds of soil is determined not only by the level of uranium in the underlying bedrock, but also by peculiarities of processes of soil formation in various bioclimatic and soil zones. From the data in Fig. 1 it is obvious that distribution of uranium of Chernobyl origin in various soil types is characterized by the descending tendency. In all types of soil the basic part (more than 70%) of Chernobyl originated uranium is concentrated in the top 10 cm soil layer.

Results of measurements of uranium in soil samples selected from far-field contaminated areas are presented in Table 2.

Table 2 shows our results for measurements of $^{236}U/^{238}U$ ratios in a soil profile from our Malinovka experimental site. These results have been combined with other published data including results of analyses of soil samples collected in Russia and Belarus (near-field). The $^{236}U/^{238}U$ isotope ratios in natural uranium ores are given as well.

As evident from the data obtained, the $^{236}U/^{238}U$ isotope ratio in far-field locations is found to be in the range 6.19×10^{-8} at/at to 1.64×10^{-7} at/at. It was mentioned above that the $^{236}U/^{238}U$ natural isotope ratio is expected to be about 10^{-14} at/at in the environment (Mironov et al., 2002). Consequently these samples from the Malinovka sampling location show an unambiguous presence of reactor fuel indicated by the enhanced ratio of $^{236}U/^{238}U$. It should be noted that irradiated uranium contamination of the soil surface in the region near the Chernobyl reactor (Masany, Kulazhin) is also characterized by high non-uniformity.

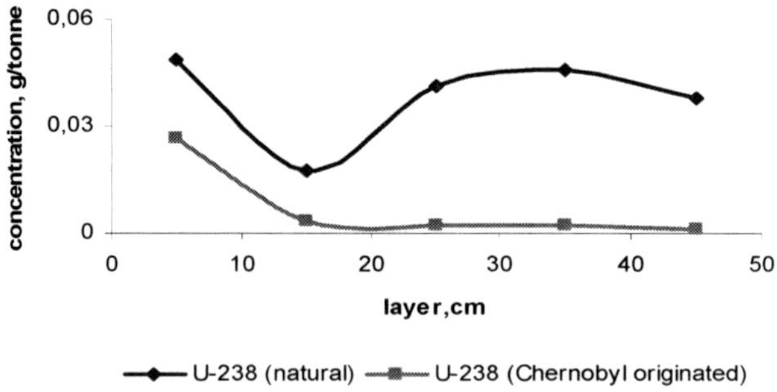

(a)

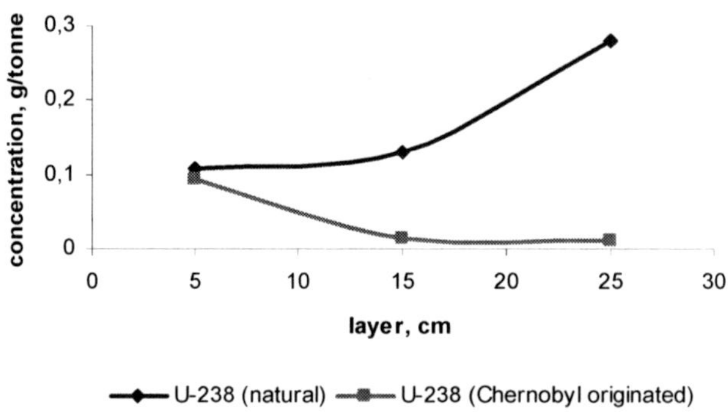

(b)

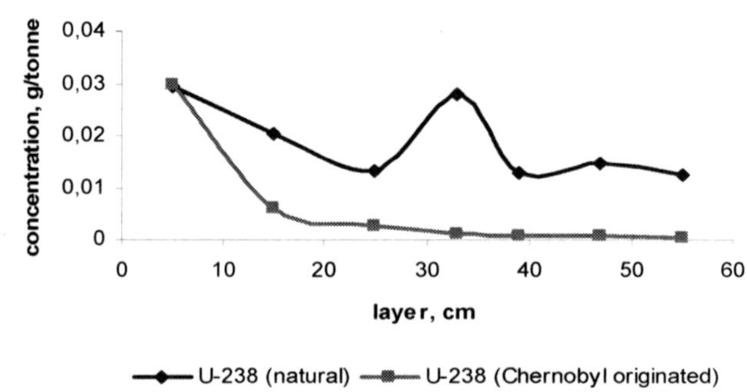

(c)

Figure 1. Distribution of uranium in soil profiles: (a) Chistogalovka (soddy-podzolic sabulous soil); (b) Chistogalovka (turfy soil); (c) Pripyat (sandy soil).

TABLE 2. Results of determination of $^{236}U/^{238}U$ ratios in samples from the Malinovka site

Samples	Depth (cm)	$^{236}U/^{238}U$ (1E-09)	^{236}U (at/g)
	0–5	70.8 ± 3.5	1.62×10^7
Belarus, Malinovka	5–10	164.1 ± 7.2	3.45×10^7
Soil profile	10–15	74.5 ± 7.1	
~255 km from	15–20	78.4 ± 6.4	
Chernobyl	20–30	62.6 ± 11.7	
	30–40	61.9 ± 6.9	
Natural uranium ores (Mironov et al., 2002)		0.012–0.56	
Russia, soil, ~165 km from Chernobyl (Pribylev et al., 2005)			1.2×10^{10}
Belarus, Kulazhin, 18 km from Chernobyl (Mironov et al., 2002)	0–10		From 1.9×10^7 to 2.6×10^{10}
Belarus, Masany, 12 km from Chernobyl (Mironov et al., 2002)	0–10		From 1.0×10^{11} to 5×10^{13}

Our results (Malinovka experimental site) have been converted to ^{236}U concentrations, taking into account the concentration of ^{238}U measured by alpha-spectrometry (1.14 Bq/kg for the 0–5 cm layer and 1.04 for the 5–10 cm layer).

By considering the $^{237}Np/^{236}U$ isotope ratio in the fuel particles(Loshchilov et al., 1992; Mironov et al., 2000) and the ^{236}U concentration (Table 2), it is possible to estimate a total amount of ^{237}Np atoms per gram of soil. ^{237}Np concentration in soil for the 0–5 cm and 5–10 cm layers is expected to be 0.76×10^6 at/g and 1.62×10^7 at/g respectively. Thus, ^{237}Np concentration in far-field areas of contamination is comparable with ^{236}U. It should be noted that mobility of neptunium in soil-to-plant system is 100 times higher than mobility of plutonium. This makes neptunium of as similar radiological concern in terms of bioavailability as plutonium and americium. The problems previously posed by determination of ^{237}Np by mass-spectrometry (Kenna, 2002; Lindahl et al., 2004) may be overcome by using accelerator mass spectrometry. Currently there is no published data on the ^{237}Np content in the environmental media of Belarus.

Destruction of FPs can be described as a sequence of two processes: oxidation of U(IV) to U(VI) and subsequent transition of U(VI) into a solution as stable carbonate complexes. Oxidation is the heterogeneous process taking place on FP surface. Ions of trivalent iron, soil oxygen etc. can act as oxidizers. U(VI) exists as uranyl ions in a soil solution in neutral and acid medium. The uranyl ion is characterized by a high propensity to form highly soluble and stable complex compounds like $[UO_2(CO_3)_3]^{4-}$ (stability constant is 10^{22}) (Moriyama et al., 2002).

The solubility of UO_2 is insignificant and does not depend on pH of the medium in a broad interval from 2 up to 12 when oxidizing agents are absent. It is possible to calculate the degree of destruction of FPs from the portion of uranium dissolved into carbonate solution, i.e., fraction of U(VI).

Concentration of oxidizers in soil is much higher than concentration of uranium of Chernobyl origin; therefore it is possible to use a chemical kinetics formula for first order reactions, considering that reaction of oxidation of 4-valency uranium is pseudomolecular:

$$\kappa = \tau^{-1} \times \ln \frac{J}{J_0} \tag{1}$$

where: J/J_0 is the fraction of radionuclide remaining in the FP or parts of non-destructed FP for the time τ after the accident.

The results of carbonate leaching of ^{236}U have shown that until now more than 60% of Chernobyl originated uranium is in the structure of non-oxidized fuel matrix (Table 3). No noticeable influence of pH of soil solutions on the destruction rate of FP was observed. It is quite possible that the influence of this soil characteristic is not seen however because of insignificant fluctuations of values of pH (4–6) in the soil samples analyzed.

TABLE 3. Carbonate leaching of $^{236}U(VI)$ from different soils in near-field areas

№	Type of soil	% of radionuclide in non-oxidized form
1	Soddy-podzolic sabulous	64 ± 4
2	Peaty	75 ± 2
3	Sandy	70 ± 6

On the basis of the experimental data, constants of FP destruction rate (k, year^{-1}) in various types of soil were calculated. The values of constants of FP destruction rate, determined by carbonate leaching of Chernobyl originated ^{236}U (year^{-1}), are within the limits of 0.019–0.030/year.

Previously the method of determination of FP destruction rate was determined by examination of ^{90}Sr content (Kashparov et al., 1999). The essence of the method consists of the following: ^{90}Sr is extracted from the soil samples by 2M solution of ammonium acetate over 24 h; the remaining part of ^{90}Sr in the residual FPs is determined and the rate constant of FP destruction is calculated.

The defects of the acetate leaching method are as follows:

a. There are many physicochemical forms of ^{90}Sr in soil with low solubility which is difficult to register quantitatively.
b. Strontium has a high migration rate in soil; it is necessary to do soil sampling to a depth of more than 50 cm.
c. The uneven ^{90}Sr distribution in throughout UO_2 (in FPs).

d. It is a time consuming analysis (>24 h).

It is relevant to note that acetate leaching of strontium from soil allows us to estimate an output fraction of this radionuclide from a fuel particle. It is important in terms of radiation safety because strontium forms species which are very mobile and accessible to plants. However destruction of FPs is more preferably estimated on the uranium itself.

According to the official data, as a result of the accident at the ChNPP, approximately 3% of the nuclear fuel in the 4th power unit was released into the atmosphere in the form of finely-dispersed particles which were distributed to significant distances from the reactor accident. Considering the great heterogeneity of nuclear fuel burnup in the active zone of the high-power channel-type (HPCT) reactor at the moment of the accident, the small part of the fuel thrown out into the atmosphere, and specificities of time course of the accident, significant differences in the nuclear fuel burn-up fractions are possible for different sites exposed to radioactive fallout. Therefore determination of real burn-up fraction for specific sites is necessary to allow correct determination of correlation coefficients of radionuclides in fallout examined.

A technique has been developed that allows determination of burn-up fraction and amount of uranium fuel in fallout after the accident at ChNPP using ^{236}U. The basis of the technique is as follows: calculated correlations between isotopes of uranium in an active zone of HPCT-1000 reactor depending on burn-up fraction; experimental data on natural and technogenic ^{236}U, ^{235}U, ^{238}U isotopic ratios in soil samples. Dependences between concentrations of isotopes of uranium in soil samples are reflected with the following formulae:

$$\frac{C_2\left({}^{236}U\right)}{C_1\left({}^{238}U\right)+C_2\left({}^{238}U\right)}=\alpha \tag{2}$$

$$\frac{C_1\left({}^{235}U\right)+C_2\left({}^{235}U\right)}{C_1\left({}^{238}U\right)+C_2\left({}^{238}U\right)}=\beta \tag{3}$$

$$\frac{C_1\left({}^{235}U\right)}{C_1\left({}^{238}U\right)}=\delta \text{ (natural isotopic ratio)} \tag{4}$$

$$\frac{C_2\left({}^{235}U\right)}{C_2\left({}^{236}U\right)}=154\times B^{-1,346} \tag{5}$$

$$\frac{C_2\left({}^{238}U\right)}{C_2\left({}^{236}U\right)}=4180\times B^{-0,817} \tag{6}$$

where C_1 – natural uranium concentration in soil, g/kg; C_2 – concentration of Chernobyl originated uranium in soil, g/kg; B – burn-up fraction.

Measurements of uranium isotopic ratios α, β and concentrations C_2 (^{236}U) are carried out with the help of mass-spectrometry techniques. The burn-up fraction B in soil samples examined is determined from the Eq. (7):

$$154 \times B^{-1,346} - 4180 \times \delta \times B^{-0,817} = \frac{(\beta - \delta)}{\alpha} \quad (7)$$

Thus, the total amount of irradiated uranium in soil can be determined from the concentration of ^{236}U using the ratios mentioned above. This was used for studying vertical migration of uranium and determination of nuclear fuel burn-up fraction.

It has been shown, that in the 30-km zone around ChNPP the average burn-up fraction value is (9.4 ± 0.5) MW.day/kgU. This is 17% lower than average calculated burn-up fraction value in 4-th block of ChNPP at the moment of the accident. In the area of Khoiniki (70 km from the ChNPP) decrease of burn-up fraction up to (7.7 ± 0.4) MW·day/kgU is observed. It means that correlation factors for determination of contamination of territory with transuranic elements can be used only for 30-km zone around ChNPP. Outside this zone it is necessary to continue research on nuclear fuel burn-up fractions in each area subject to radioactive fallout.

A map of contamination of the territory of Belarus with irradiated uranium has been created. For this purpose we have used the data on surface contamination of soil in the southern areas of Gomel region by plutonium (1992) (Kudrjashov, 1998).

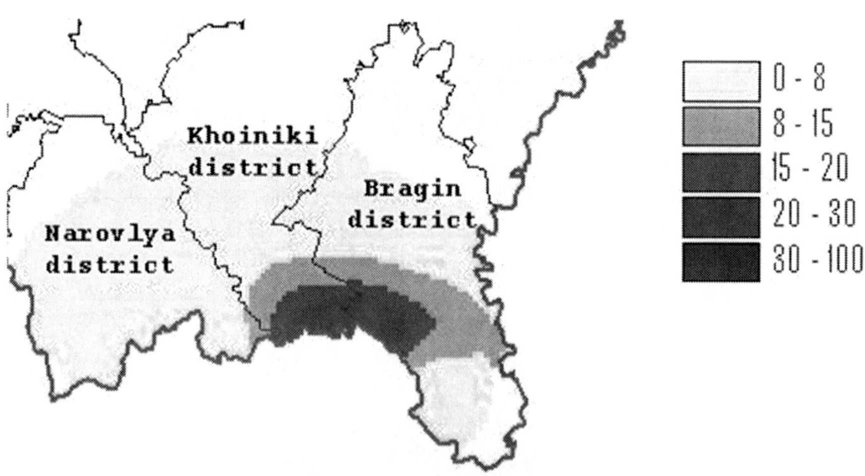

Figure 2. Uranium contamination of southern districts of Gomel region, Bq/m^2.

^{238}U content was calculated from ^{239,240}Pu and ^{238}Pu content using isotopic ratios ^{238}U/^{239,240}Pu and ^{238}U/^{238}Pu for FP (numerically equal 0.00095 and 0.00206 accordingly) (Kudrjashov, 1998). The contamination density (Bq/m^2) of ^{238}U is shown in Fig. 2.

According to different independent estimations, Belarus territory was contaminated with irradiated uranium of between 1.5 and 4.1% of the total content of fuel in the active zone at the moment of the accident (Pribylev et al., 2005).

4. Conclusion

Use of ^{236}U as a tracer of irradiated uranium allows us to address a number of actual problems: to directly identify the irradiated uranium and also to study its vertical migration; to determine the destruction degree of fuel particles, as well as to make calculations in order to determine nuclear fuel burn-up fractions and the amount of uranium fuel in radioactive fallout.

Also, use of the AMS technique for measurements of ^{236}U allows us to identify unambiguously the irradiated uranium in the regions of Belarus that are remote from the Chernobyl nuclear power plant.

References

Boulyga, S.F., Matusevich, J.L., Mironov, V.P., Kudrjashov, V.P., Halicz, L., Segal, I., McLean, J.A., Montaser, A., and Becker, J.S. (2002) Determination of ^{236}U/^{238}U isotope ratio in contaminated environmental samples using different ICP-MS instruments, *J. Anal. At. Spectrometry*, **17**, 958–964.

Hotchkis, M., Child, D., Fink, D., Jacobsen, G.E., Lee, P.J., Mino, N., Smith, A.M., and Tuniz, C. (2000) Measurement of ^{236}U in environmental media, *Nucl. Instrum. Methods Phys. Res*, B **172**, 659.

Kashparov, V.A., Oughton, D.H., Zvarich, S.I., Protsak, V.P., and Levchuk, S.E. (1999) Kinetics of fuel particle weathering and ^{90}Sr mobility in the Chernobyl 30-km exclusion zone, *Health Phys*, **76** (3), 251–259.

Kenna, T.C. (2002) Determination of plutonium isotopes and Np-237 in environmental samples by inductively coupled plasma mass spectrometry with total sample dissolution, *JAAS*, **17**, 1471–1479.

Kudrjashov, V.P. (1998) Contamination of territory of Belarus with transuranic elements, PhD thesis, Minsk.

Lindahl, P., Roos, P., Eriksson, M., and Holm, E. (2004) Distribution of Np and Pu in Swedish lichen samples (*Cladonia stellaris*) contaminated by atmospheric fallout, *J. Env. Rad*, **73**, 73–85.

Loshchilov, N.A., Kashparov, V.A., Yudin, Ye.B., and Protsak, V.P. (1992) Fractionation of radionuclides in chernobyl originated fuel hot particles, *Radiochemistry*, **5**, 125–134.

Mironov, V., Kudrjashov, V., Ananich, P., and Zhuravkov, V. (2000) *Proceedings of the International Conference on Radioactivity after Nuclear Explosions and Accidents*, Moscow, 24–26 April 2000, *Hydrometeoizdat*. **1**, 189–194.

Mironov, V.P., Matusevich, J.L., Kudrjashov, V.P., Boulyga, S.F., and Becker, J.S. (2002) Determination of irradiation reactor uranium in soil samples in Belarus using ^{236}U as a tracer of irradiated uranium, *J. Env. Monit*, **4**, 997–1002.

Mironov, V.P., Zhuravkov, V.V., Ananich, P.I., Boulyga, S.F., and Matusevich, J.L. (2005) Determination of total amount and burn-up fraction of uranium fuel in radioactive fallouts after the accident at ChNPP with the help of ^{236}U tracer, *Belarusian Acad Sci Bull Phys Ser*, **4**, 121–123.

Moriyama, H., Fujiwara, K., and Yamana, H. (2002) Systematics of complexation constants of actinide ions with inorganic ligands, *J Nucl Sci Technol*, (Suppl. 3), 246–250.

Patent C1 BY. Method for determination of destruction of fuel particles in environmental objects, Mironov V.P., Matusevich J.L., Konoplya E.F. – No 4634; Claimed 22.06.1998; Published 30.09.2002.

Pribylev, S.V., Novitskaya, O.N., and Mironov, V.P. (2005) Chernobyl uranium contamination of Belarus territory, *PGU Bull. B Ser*, **3**, 67–71.

Shagalova, E.K. (1986) ^{238}U Content in soils of Belarus, *Soil Sci*, **2**, 140–145.

PART 4: BIOLOGICAL UPTAKE AND RISK ASSESSMENT

HEALTH EFFECTS OF DOUNREAY HOT PARTICLES: A BENCHMARK FOR THE EVALUATION OF DOSES AND RISKS

MONTY W. CHARLES
School of Physics & Astronomy, University of Birmingham, Edgbaston, Birmingham B15 2TT, UK

Abstract: This paper summarises recent work undertaken for SEPA and UKAEA to provide estimates of potential doses and risks from exposure to fragments of irradiated nuclear fuel that have been found near to the Dounreay nuclear site in Scotland. It concentrates on exposure to the most commonly found particles which originate from the materials test reactor (MTR). Assessments based on the characteristics of these particles are likely to be conservative when applied to other particle types. Doses and risks from skin contact, ingestion and inhalation are considered. No account is taken of the extremely low probability of encountering a particle, which is the subject of a separate analysis. The results obtained suggest that exposure of members of the public to MTR particles of activities of less than about 10^5 Bq ^{137}Cs (with associated ^{90}Sr/^{90}Y and lower activities of actinide isotopes) are unlikely to result in observable biological effects. Exposure to particles of activities of 10^6 Bq ^{137}Cs and greater could result in clearly discernible biological effects. This conclusion is based on the possibility of localised ulceration of skin or of the mucosal lining of the colon or extra-thoracic airways. Prolonged contact with 10^5 Bq particles, for whatever reason, could result in localised ulceration. Committed doses and cancer risks are of secondary importance in this case due to the low solubility and low respirability of these particular types of hot particle. The very detailed study of the environmental and public health impacts of the Dounreay fuel fragments has involved an enormous scientific effort which has been fully documented. It could be usefully used as a benchmark for the consideration of other types of hot particles arising in the environment.

Keywords: hot particles, fuel fragments, health effects, Dounreay.

D.H. Oughton and V. Kashparov (eds.), *Radioactive Particles in the Environment*,

1. Introduction

About 250 discrete fragments of irradiated nuclear fuel have been discovered over the past 20 years on the foreshore at the Dounreay nuclear site in Scotland operated by the United Kingdom Atomic Energy Authority (UKAEA). These have activities (^{137}Cs) up to ~10^8 Bq. Almost 100 lower activity fragments have been found over the last 10 years at a nearby public access beach at Sandside Bay. The origin of the fragments is offshore in seabed sediments – resulting from historical accidental discharges from the nuclear site. The Scottish Environment Protection Agency (SEPA) has commissioned work to evaluate appropriate monitoring techniques and to estimate the potential for exposure of members of the public, and the health implications of such exposures. This paper briefly summarises the main aspects of work carried out to calculate doses and risks to individuals who might come into contact with a particle (Harrison et al., 2005). The methodology for evaluating the very low probability of actually encountering a particle has been summarised by Smith and Bedwell et al. (2005). An extensive summary report which deals with all public health aspects of the Dounreay fuel fragments has recently been published by SEPA (DPAG, 2006).

The fuel fragments vary substantially in size but are most typically similar in size to grains of sand (COMARE/RWMAC, 1995; COMARE, 1999). The principal radionuclides contained within the particles are the fission products ^{137}Cs and ^{90}Sr/^{90}Y and small amounts of ^{238}Pu, ^{239}Pu and ^{241}Am. The ^{137}Cs activity by which the particles are generally characterised is within the range of 10^3 to 10^8 Bq. The most active particles found at Sandside Bay contain ~ 10^5 Bq ^{137}Cs. An inventory of particles can be found at: http://www.dounreay.com/particle-cleanup/particles-recovered.

There are two main types of particles: fragments of materials test reactor (MTR) and Dounreay fast reactor (DFR) fuel. The more abundant particles originated as MTR swarf – generated during milling to remove aluminium cases from fuel elements (COMARE/RWMAC, 1995; COMARE, 1999). In some cases this process removed some of the underlying fuel plate so that the swarf contained fragments of fuel embedded in aluminium. In general DFR particles have smaller proportions of ^{90}Sr/^{90}Y and the actinides, relative to ^{137}Cs, than do MTR particles. They are also generally less soluble. In addition, a small number of particles retrieved from the Dounreay foreshore and from the seabed offshore also contain ^{60}Co. The dose and risk assessments presented in this paper relate primarily to MTR particles. The ^{137}Cs content of MTR particles has been found to be related almost linearly to dose/risk. The application of this relationship to DFR or the few particles which also contain ^{60}Co provides over-estimates of dose/risk.

Because of the high local dose rates delivered by the more active particles and because they are generally of low solubility, the principal concerns are possible skin contact resulting in localised damage, and possible ingestion and damage to the alimentary tract, particularly the colon (large intestine). A previous assessment (Wilkins et al., 1998) considered both these possibilities but the dosimetric approach resulted in over-estimates of doses and risks (Darley et al., 2003), since it took no account of self-absorption of energy within larger particles and it used the standard ICRP (1979) model of the alimentary tract. In our analyses, self-absorption within particles is taken into account and alimentary tract doses are calculated using the new ICRP model (ICRP, 2006). In addition, the possibility of inhalation of particles is addressed and absorption to blood and doses to all organs and tissues are considered in the context of risks of cancer and hereditary effects.

2. Skin Contact

2.1. DOSES

Measurements of dose distributions from Dounreay fuel fragments, over various skin areas and at several depths, were carried out using the radiochromic dye film (RDF) technique (Aydarous et al., 2001). The 37 fuel fragments selected for measurement had been subject to prior characterisation using scanning

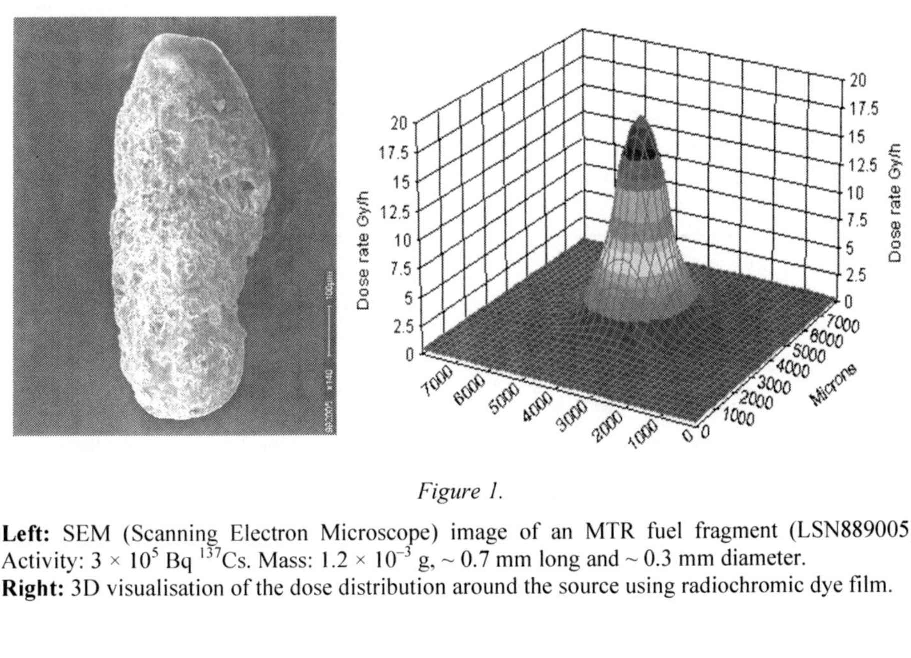

Figure 1.

Left: SEM (Scanning Electron Microscope) image of an MTR fuel fragment (LSN889005) Activity: 3×10^5 Bq ^{137}Cs. Mass: 1.2×10^{-3} g, ~ 0.7 mm long and ~ 0.3 mm diameter.
Right: 3D visualisation of the dose distribution around the source using radiochromic dye film.

electron microscopy and energy dispersive x-ray analysis (SEM/EDX) and ^{137}Cs activity determination using gamma spectrometry. This provided input data for Monte Carlo dose calculations and enabled comparison with the RDF measured doses. Calculations of skin dose were made using the MCNP (Briesmeister, 2000).

Monte Carlo radiation transport code, assuming spherical particles of uniform elemental composition of U/Al (15% U), density of 3.1 g cm^{-3}, specific activity 2 GBq ^{137}Cs g^{-1}, and a $^{90}Sr/^{90}Y$: ^{137}Cs activity ratio for MTR particles of 0.9. Figure 1 illustrates measurement data obtained for one of the highest activity MTR fragments found to date at Sandside Bay.

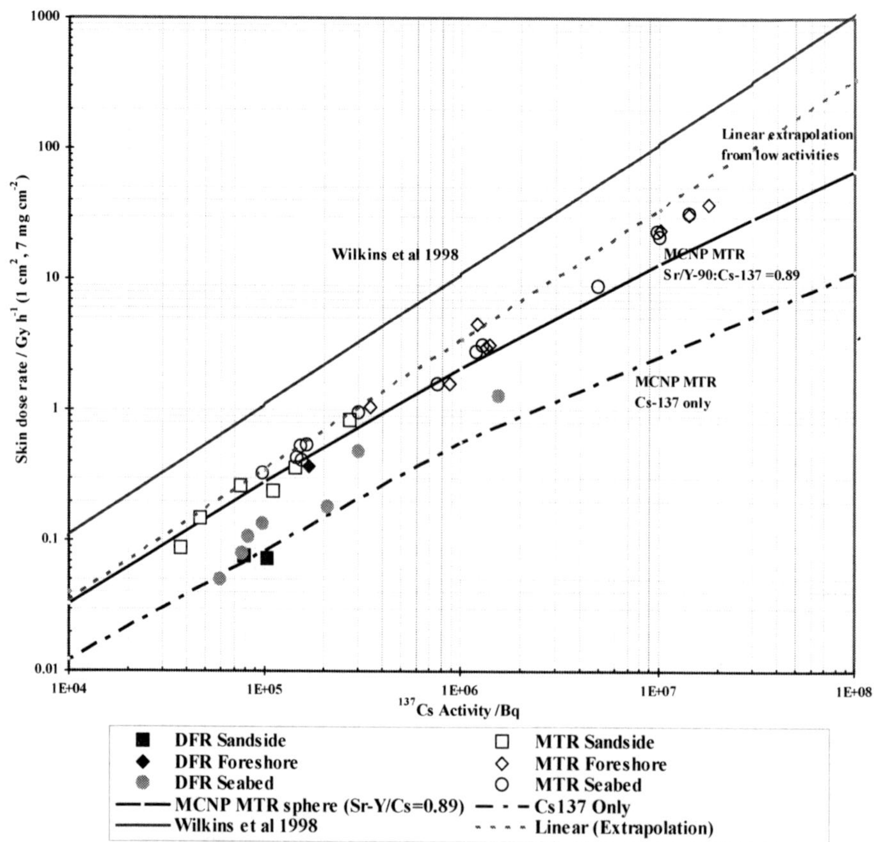

Figure 2. Skin dose rates for MTR and DFR Dounreay fuel fragments (Activity corrected to date of dose measurement). Skin dose is averaged over an area of 1 cm^2 at a depth of 70 microns – abbreviated to (1 cm^2, 70 μm) – as recommended by the ICRP (1991) and ICRU (1997).

Figure 2 shows MCNP calculations of doses from spherical fuel fragments. Calculations for non-spherical sources are somewhat higher. The measured and calculated dose rates are in good agreement for MTR particles. Measurements for DFR particles give lower dose rates than predicted by calculations for MTR particles of the same ^{137}Cs activity and most are close to calculations for particles with no ^{90}Sr/^{90}Y content (lower dashed line). Also shown in Fig. 2 are the dose rates assumed by Wilkins et al. (1998), taking no account of self-absorption of energy within larger particles and assuming a ^{90}Sr/^{90}Y: ^{137}Cs activity ratio of 2.1 (upper solid line). The linear extrapolation from small particles essentially represents a dose estimate which does not include self absorption.

Useful detailed spatial dose distributions can be measured using radiochromic dye film but the method can be time consuming (several days for low activity particles). A useful rapid evaluation of skin dose can however be obtained using portable dose rate meter survey instruments. The readings from these instruments significantly underestimate the skin dose (1 cm^2, 70 μm) because they use large volume detectors. The survey instruments based on ionisation chambers usually include a sliding beta shield in front of the chamber window which enables measurements to be made of gamma or (beta + gamma) radiations. Figure 3 shows the relationship between instrument reading (window open) and skin dose for three commonly employed survey instruments (Aydarous et al., 2008).

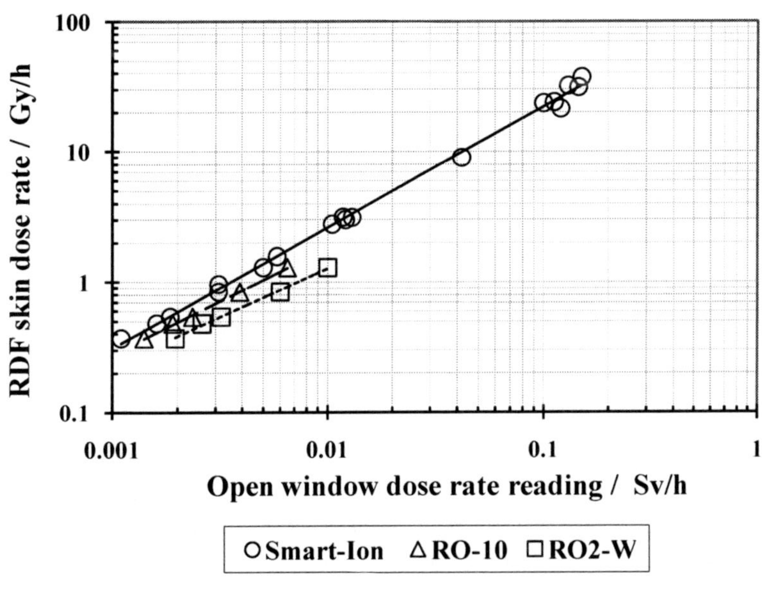

Figure 3. Open window dose rate reading using the SmartION, and Eberline RO2-W and RO10 dose rate meters vs. RDF skin dose rate (1 cm^2, 70 μm).

2.2. DETERMINISTIC EFFECTS (TISSUE DAMAGE)

There has been considerable research in the past 20 years in the UK and the USA to provide radiobiological data regarding the effect of skin exposure from hot particles. Data are available on pig skin exposures with various sizes of beta-emitting radionuclides. The incidence of early and late acute effects such as acute epidermal necrosis, ulceration and dermal thinning has been scored, as well as the time-course, severity and healing outcome.

The combination of UK and US data show remarkable consistency (Fig. 4). When doses are averaged over an area of 1 cm^2 at a depth of 70 µm the ED_{10} and ED_{50} values (the doses to produce an incidence in effect of 10% and 50% respectively) for acute epidermal necrosis or acute ulceration vary by only a factor of ±2 across a wide beta energy range.

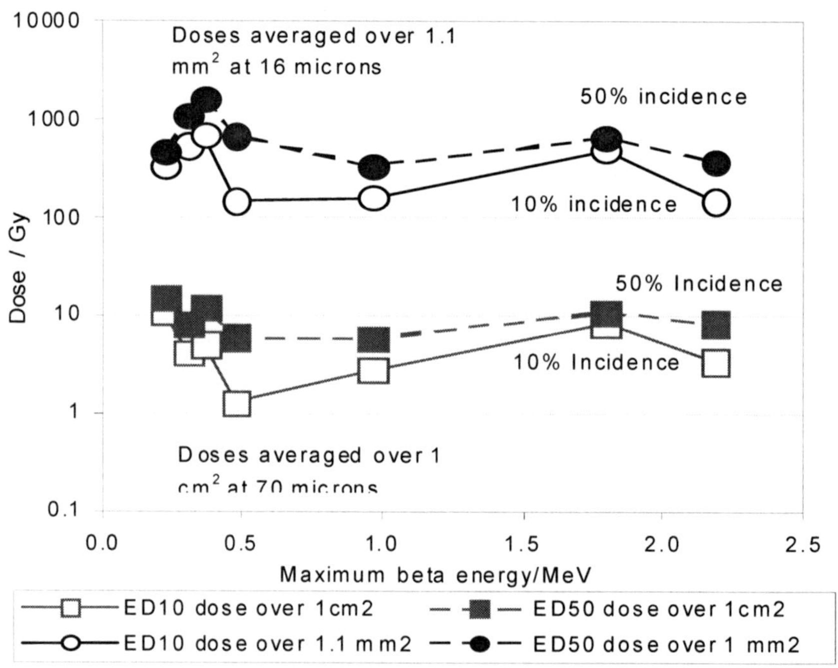

Figure 4. Variation in doses for 50% and 10% incidence (ED_{50} and ED_{10}) of acute ulceration of pig skin following hot particle exposures from specific radionuclides as a function of the maximum beta energy. The radionuclides, in order of increasing maximum beta energy are: ^{147}Pm, ^{60}Co, ^{46}Sc, ^{175}Yb, ^{170}Tm, ^{235}U (fissioned) and $^{90}Sr/^{90}Y$. Data are shown for skin doses evaluated as averages over areas of 1.1 mm^2 and a depth of 16 µm (as used in some publications of experimental data) and averaged over an area of 1 cm^2 and a depth of 70 µm (as recommended by the ICRP and ICRU). Data are from ICRP (1991), NCRP (1999) and Kaurin et al. (2001a,b).

The ED_{50} values are approximately 15, 8, 12, 6, 6, 11, and 8 Gy for ^{147}Pm (beta energy, Emax ~ 0.25 MeV), ^{60}Co (Emax ~ 0.32 MeV), ^{46}Sc (Emax ~ 0.35 MeV), ^{175}Yb (Emax ~ 0.47 MeV), ^{170}Tm (Emax ~0.97 MeV), fissioned $^{235}UC_2$ microspheres (Emax ~ 1.8 MeV), and ^{90}Sr/^{90}Y (Emax ~ 2.3 MeV), respectively. These data are for small sources (dimensions <1 mm) except for ^{147}Pm where the smallest source was 2 mm diameter. For sources larger than ~1 mm the ED values are increased (Charles, 1990, 1991). In the case of higher energy beta sources with diameters of ~3 mm (corresponding to the largest and most active of the Dounreay fuel fragments, 10^8 Bq ^{137}Cs) the increase in ED_{10} value is by a factor of about 2–3.

For the examples of 10^5 and 10^8 Bq ^{137}Cs MTR particles (diameters of about 300 μm and 3 mm), dose rates were estimated as about 0.3 Gy h^{-1} and about 70–140 Gy h^{-1}, respectively, averaged over 1 cm^2 at a depth of 70 μm. On the basis of a threshold for skin damage of about 2 Gy, 10^5 Bq ^{137}Cs particles would have to remain in stationary contact with skin for at least about 7 h to cause any discernible effect. Movements of even a few mm during contact would increase the threshold. However, 10^5 Bq particles trapped for longer periods of a day or two may cause small ulcerations that extend over areas of less than 1 cm^2 and heal within a few weeks. Such ulceration could occur following short periods of contact of 15–30 min or more to 10^8 Bq ^{137}Cs particles and exposures of several hours would result in more serious ulceration that would take longer to heal. Similar considerations apply to particles in contact with the skin of the external auditory canal of the ear and the corneal surface of the eye (Harrison et al., 2005) – but the probability of such exposures is exceedingly low.

2.3. CANCER INDUCTION

The ICRP has maintained its advice over many years that the use of mean organ or tissue dose is appropriate for the evaluation of carcinogenic risk for radiological protection purposes. However, there has been ongoing controversy regarding this assumption and there have been claims that the use of average organ dose could very considerably under-estimate the cancer risk. In particular, in the context of ^{239}Pu inhalation, claims have been made (Geesaman, 1968; Tamplin and Cochran, 1974; Tamplin and Cochran, 1974) that lung irradiation by 'hot particles' is several orders of magnitude more carcinogenic than estimates made by assuming a uniform lung exposure. The claim of increased carcinogenicity was referred to as the 'hot particle hypothesis'. After a decade of intensive experimental and theoretical investigation an ICRP (1980) review of the biological effects of inhaled radionuclides in 1980 refuted this claim in the context of inhaled 'hot particles'. The 'hot particle hypothesis' in the context of skin exposure has been refuted by a series of skin cancer induction experiments in mice and in cell studies of induced malignant transformation (Mill et al.,

1996; Charles et al., 1988, 2003). The ICRP has thus maintained its advice that the use of mean organ or tissue dose is appropriate for the evaluation of carcinogenic risk for radiological protection purposes.

3. Ingestion

An assessment has been made of possible acute radiation damage to the alimentary tract following the ingestion of fuel fragments, concentrating on the colon (large intestine) as the region receiving the greatest doses. Doses have been calculated using the new ICRP model of the human alimentary tract (ICRP, 2006). The model has explicit consideration of doses to target cells in the various regions and includes age-related parameter values for dimensions and transit times (Harrison et al., 2004).

Doses averaged over the rectosigmoid region of the colon were calculated for MTR particles in transit through the region using the MCNP code, taking account of self-absorption within particles, with assumptions regarding particle composition and specific activities as for skin dose calculations. A target cell depth of 290 μm was used (ICRP, 2006). Table 1 gives absorbed dose estimates for a particle moving randomly (random transit) through the lumen of the rectosigmoid, with no fixed position relative to the wall, and the higher values (max) doses for a particle moving always in contact with the wall. Doses from a 10^8 Bq ^{137}Cs particle moving randomly were calculated as 0.3–0.4 Gy in adults and 1 Gy in 1 year-old children. The corresponding maximum estimated doses were about 1–2 Gy in adults and 4 Gy in 1 year-old children. Estimated doses for particles with a ^{137}Cs activity of 10^5 Bq, typical of the highest activity particles found at Sandside Bay, were less than 10 mGy in adults and 20 mGy in 1 year old children. These doses compare with an estimated threshold dose for lethal damage to the colon, following protracted irradiation from ingested radionuclides, of 20 Gy and an LD_{50} of 35 Gy (Stather, et al., 1988; NRPB, 1996) (doses to target cells averaged over the whole colon).

Consideration of the possible effect on doses of changes in assumptions on particle composition, colon dimensions and transit times, and the depth of target cells, showed that variations in transit times are of greatest importance. Doses to the rectosigmoid from a fuel fragment in transit will be directly proportional to the transit time. The standard assumptions are transit times of 12 h in 1 year-old children and adult males and 16 h in adult females. Transit times and doses may be increased by factors of up to 2–3 in constipated individuals but in extreme cases may be increased by factors of up to 10 (Waller, 1975; Benninga et al., 1996; Evans et al., 1998).

For 10^8 Bq ^{137}Cs particles, average rectosigmoid colon absorbed doses might typically be increased by constipation to 1–5 Gy in adults and 2–10 Gy in 1 year-old children. There is the possibility, however unlikely, that doses could approach 10–20 Gy in adults and 10–35 Gy in 1 year-old children. On the basis of these dose estimates, it appears very unlikely that ingestion of even 10^8 Bq ^{137}Cs particles would result in death. Doses from 10^5 Bq ^{137}Cs particles and intermediate activity particles (10^6 and 10^7 Bq) can be regarded as below the threshold for lethality resulting from damage to the colon.

TABLE 1. Estimated average absorbed doses to the rectosigmoid colon from an ingested particle, mGy

Activity Bq ^{137}Cs[a]	Dia.(μm)[b]	Adult male		Adult female		One year old child	
		Random transit[c]	Max[d]	Random transit	Max	Random transit	Max
10^3	67	0.01	0.08	0.01	0.1	0.04	0.2
10^4	150	0.1	0.7	0.1	1	0.4	2
10^5	310	0.9	6	1	8	3	16
10^6	680	7	40	10	60	27	110
10^7	1,500	46	230	67	340	185	640
10^8	3,100	290	1,200	420	1,800	1,200	3,500

[a]With $^{90}Sr/^{90}Y:^{137}Cs$ ratio of 0.9.
[b]Assuming a specific activity of 2 GBq ^{137}Cs g^{-1}, fragment density 3.1 g cm^{-3}.
[c]Randon transit – expectation value of colon dose based on random movement through the lumen.
[d]Max – average colon dose for movement of particle always immediately adjacent to the mucosal lining.

An additional consideration for ingested particles is that the flow of material through the colon is highly variable. Movement in the rectosigmoid in particular does not occur as a constant flow but rather as mass movements resulting from periodic contractions between longer periods of quiescence. Local doses within the rectosigmoid may therefore be substantially greater than the average dose within the region. For example, 10^5 Bq and 10^8 Bq particles held stationary against the lumenal wall of the rectosigmoid for 6 h would deliver doses to 1 cm^2 of tissue of about 0.6 Gy and 240 Gy, respectively, at a depth of 300 μm. Local doses of 200–300 Gy are likely to cause ulceration that might not repair readily in the environment of the large intestine, while doses of less than 1 Gy would result in localised crypt sterilisation that should be replaceable by regeneration of new crypts (Potten, 1990).

4. Inhalation

The ICRP Human Respiratory Tract Model (ICRP, 1994) was used to calculate doses from inhaled particles, using the model to determine the probability of particle deposition in different regions. The analysis showed that inhaled particles of sufficient activity to cause acute damage to the lungs would be too large to reach the airways of the lungs and would deposit in the extra-thoracic airways. Doses from particles deposited in the anterior nasal passages were estimated as around 300 mGy for a 300 μm particle (10^5 Bq ^{137}Cs) and 300 Gy for a 3 mm particle (10^8 Bq ^{137}Cs). These estimates are based on standard ICRP model assumptions including a clearance half-time of one day and represent an average dose to the target tissue in the region. Potentially more important is the possibility of high local doses which may be substantially greater than the average dose within the region. For the purposes of assessment of possible effects of local irradiation, the epithelial lining of the extra-thoracic airways can be regarded as similar to skin. It should be noted that particles of 3 mm diameter are substantially larger than the range of particle sizes normally considered to be respirable; the probability of their inhalation will be very low and might only be expected to occur if the particles are airborne as a result for example of high winds.

5. Committed doses

The calculation of committed equivalent doses to all body tissues and committed effective doses requires information on absorption of radionuclides to blood following particle ingestion or inhalation. The work commissioned by SEPA included *in vitro* and *in vivo* measurements of the solubility of ingested particles and the availability of their radionuclide content for intestinal absorption. Dissolution was measured *in vitro* in simulated stomach and intestinal fluids and absorption was measured directly in rats given particles by gastric intubation. On the basis of these studies, typical values for absorption to blood from MTR particles were taken to be 1% for ^{137}Cs, 0.01% for ^{90}Sr and 0.001% for ^{238}Pu, ^{239}Pu, ^{241}Am and ^{90}Y. One MTR particle was found to exhibit uncharacteristic solubility in the *in vitro* study, with dissolution of more than half of its radionuclide content compared with typical values of less than 1% dissolution. For this particle, absorption to blood was assumed to be 60% for ^{137}Cs, 15% for ^{90}Sr and 0.03% for the actinides and ^{90}Y.

Committed equivalent doses to all body tissues and committed effective doses were calculated for ingestion of MTR particles of typical solubility and for the very soluble particle, using standard ICRP models but taking account of self-absorption of energy within particles throughout the alimentary tract. For typical particles, committed effective doses following ingestion of 10^5 and 10^8 Bq ^{137}Cs particles were estimated as 0.1 mSv and 80 mSv, respectively, for an adult male, and 0.5 mSv and 300 mSv, respectively, for a 1 year-old child (Table 2). In each case, these estimates of dose are dominated (>70%) by contributions from committed equivalent doses to the alimentary tract, particularly the colon. A threefold increase in the transit time of a 10^5 Bq ^{137}Cs particle through the colon of a 1 year-old child, due to constipation, would increase the committed effective dose by a factor of two to 1 mSv and a 10-fold increase would increase the dose by a factor of 6 to 3 mSv. The committed effective dose to an adult male following ingestion of a 10^5 Bq ^{137}Cs particle of high solubility (see above) was estimated as about 2 mSv. In this case, committed equivalent doses to the alimentary tract contributed about 15% of the committed effective dose while skeletal tissues (red bone marrow and bone surfaces) contributed about 60%. Doses to a 1 year-old child would be about two to four times greater than to an adult male.

TABLE 2. Committed equivalent doses to organs/tissues and committed effective dose (E) to adult males and 1 year-old children from the ingestion of fuel fragments, mSv[a]

Organ/tissue	10^5 Bq ^{137}Cs particle (300 μm)		10^8 Bq ^{137}Cs particle (3mm)	
	Adult	One year-old	Adult	One year-old
Stomach	0.22	0.94	120	460
Colon	0.61	2.4	350	1,300
RBM[b]	0.03	0.06	34	57
BS[c]	0.11	0.13	110	130
Liver	0.03	0.07	33	72
Gonads	0.08	0.31	77	310
E[d]	0.1	0.5	78	290

[a]Calculated assuming ratios of 0.9 for ^{90}Sr/^{90}Y:^{137}Cs, 0.003 for ^{238}Pu:^{137}Cs, 0.001 for ^{239}Pu:^{137}Cs and ^{241}Am:^{137}Cs. Absorption to blood was assumed to be 1% for ^{137}Cs, 0.01% for ^{90}Sr and 0.001% for ^{238}Pu, ^{239}Pu, ^{241}Am and ^{90}Y.
[b]RBM: red bone marrow.
[c]BS: bone surfaces.
[d]E: committed effective dose.

For inhaled particles deposited in the extra-thoracic regions of the respiratory tract (see above), it appears reasonable to assume that absorption to blood will be very low except for particles that are subsequently transferred to the alimentary tract, for which the doses shown in Table 2 would apply. Effective doses from particles on skin would be very low. For example, maximum values of effective dose corresponding to a local dose of 2 Gy to 1 cm^2 of skin were calculated as about 4 μSv in adults and 25 μSv in 1 year-old children.

6. Discussion and conclusions

Analyses of possible doses and risks from exposure to Dounreay fuel fragments indicate that the principal concern following skin contact, ingestion or inhalation is the possibility of localised ulceration of skin or of the mucosal lining of the colon or extra-thoracic airways. A 10^5 Bq ^{137}Cs MTR particle, typical of the more active particles found at Sandside Bay, would deliver skin doses of about 0.3 Gy h^{-1} (or 0.3 Sv h^{-1}; 1 cm^2, 70 μm). A 3–6 h exposure to such a particle is not expected to result in any visible skin lesions. Stationary contact for longer periods of 15 h or more might result in small lesions. More active particles have the potential to cause more serious ulceration.

Dose limits have been recommended by the ICRP[12] for localised skin exposure and committed effective dose for workers and members of the public. These dose limits apply to controlled sources and are not intended to apply to existing situations in which the only available protective action takes the form of intervention, such as is the case with these fuel fragments. Nevertheless, the limits provide values with which to compare possible doses from a fuel fragment. The annual dose limit for localised skin exposure of workers of 0.5 Sv (1 cm^2, 70 μm) can be regarded as conservative when applied to hot particle exposures (NRPB, 1997) since the threshold for effects is around 2 Sv. For the public, ICRP reduces the dose limit by a factor of ten to 50 mSv. However, this reduction has no scientific basis and it can be argued that no special limit for members of the public is necessary (NRPB, 1997). At a skin dose rate of 0.3 Sv h^{-1}, a 10^5 Bq ^{137}Cs MTR particle held stationary on the skin would deliver doses corresponding to the ICRP worker and public dose limits in <2 h and 10 min, respectively. The limit on committed effective dose to members of the public is 1 mSv. Doses from ingestion of 10^5 Bq ^{137}Cs MTR particle are likely to be <1 mSv in adults and children, but slower colonic transit or uncharacteristic particle solubility could result in doses of a few mSv.

It can be concluded that local skin dose from a 10^5 Bq ^{137}Cs MTR particle is unlikely to cause ulceration but the possibility of a small lesion cannot be ruled out for long residence times. More active particles are more likely to cause serious ulceration. Committed doses and cancer risks are of secondary importance.

The very detailed study of the environmental and public health impact of the Dounreay fuel fragments has involved an enormous scientific effort which has been fully documented and peer-reviewed (DPAG, 2006). This has considered the origin of the particles and their transport in the marine and terrestrial environments. The probability of a member of the public actually encountering a particle has been evaluated using habit surveys and detailed marine and beach monitoring. The probability and severity of the range of possible health effects has also been evaluated using the latest ICRP metabolic models and our latest understanding of the induction of deterministic and carcinogenic effects for hot particles. At the current time there are ongoing discussions, with full stakeholder involvement, regarding the most appropriate remedial action to restore the contaminated environment. The study could be usefully used as a benchmark for consideration of other types of hot particle arisings in the environment.

References

Aydarous, A.S., Charles, M.W., and Darley, P.J. (2008) Dose distribution measurements and calculations for Dounreay hot particles. *Radiation Protection Dosimetry*, **128** 146–158.

Aydarous, A.S., Darley, P.J., and Charles, M.W. (2001) A wide dynamic range, high spatial resolution readout system for radiochromic dye films. *Physics in Medicine and Biology*, **46** (5) 1379–1389.

Benninga, M.A., Buller, H.A., Tygat, G.N., Akkermans, L.M., Bossuyt, P.M., and Taminiau, J.A. (1996) Colonic transit time in constipated children: does pediatric slow-transit constipation exist? *Journal of Pediatric Gastroenterolgic Nutrition*, **23** 241–251.

Briesmeister, J.F. (2000) *MCNP – A General Monte Carlo N-Particle Transport Code, version 4C*. Los Alamos National Laboratory LA-13709-M, Los Alamos, New Mexico.

Charles, M.W., Mill, A.J., and Darley, P.J. (2003) Carcinogenic risk of hot particle exposures. *Journal of Radiological Protection*, **23** (1) 5–28.

Charles, M.W., Williams, J.P., and Coggle, J.E. (1988) Skin carcinogenesis following uniform and non-uniform beta Irradiation. *Health Physics*, **55** (2) 399–406.

Charles, M.W. (1990) General considerations of the choice of dose limits, averaging areas and weighting factors for the skin in the light of revised skin cancer risk figures and experimental data on non-stochastic effects. *International Journal of Radiation Biology*, **57** (4) 841–858.

Charles, M.W. (1991) The hot particle problem. *Radiation Protection Dosimetry*, **39** (1–3) 39–47.

COMARE. Committee on Medical Aspects of Radiation in the Environment. Sixth Report: A reconsideration of the possible health implications of the radioactive particles found in the general environment around the Dounreay Nuclear Establishment in the light of the work undertaken since 1995 to locate their source. Produced by the *National Radiological Protection Board*, (1999).

COMARE/RWMAC. Potential Health Effects and Possible Sources of Radioactive Particles Found in the Vicinity of the Dounreay Nuclear Establishment. A joint report by the Committee on Medical Aspects of Radiation in the Environment (COMARE), and the Radioactive Waste Management Advisory Committee (RAWMAC). *HMSO Publication*. (1995)

Darley, P.J., Charles, M.W., Fell, T.P., and Harrison, J.D. (2003) Doses and risks from the ingestion of Dounreay fuel fragments. *Radiation Protection Dosimetry*, **105** (1–4) 49–54.

DPAG. (2006) Dounreay Particles Advisory Group, Third Report. *Scottish Environmental Protection Agency* (SEPA) September 2006. ISBN: 1-901322-64-5.

Evans, J.M., Fleming, K.C., Talley, N.J., Schleck, C.D., Zinsmeister, A.R., and Melton, L.J. (1998) Relation of colonic transit to functional bowel disease in older people: A population-based study. *Journal of American Geriatric Society*, **46** 83–87.

Geesaman, D.P. (1968) An Analysis of the Carcinogenic Risk from an Insoluble Alpha Emitting Aerosol Deposited in Deep Respiratory Tissue. UCRL-50387 and Addendum.

Harrison, J.D., Fell, T.P., Phipps, A.W., and Smith, T.J. (2004) The new ICRP model of the human alimentary tract: Target cell assumptions and colon doses from ingested alpha and beta-emitting radionuclides. In *International Conference*: Health effects of incorporated radionuclides: Emphasis on radium, thorium, uranium and their daughter products. Neuherberg, Germany, November.

Harrison, J.D., Fell, T., Phipps, A.W., Smith, T., Ellender, M., Ham, G., Hodgson, A., Wilkins, B.T., Charles, M.W., Darley, P.J., and Aydarous, A.S. (2005) Health implications of Dounreay fuel fragments: Estimates of doses and risks. Radiation Protection Division (RPD), *Health Protection Agency* (HPA). RPD-RE-11.

ICRP. 1990 Recommendations of the International Commission on Radiological Protection. ICRP Publication 60. *Annals of the ICRP*, **21** (1–3) (1991).

ICRP. (1979) ICRP Publication 30: Limits of intakes of radionuclides by workers: Part I. *Annals of the ICRP*, **2** (3/4).

ICRP. (2006) International Commission on Radiological Protection, Publication 100. Human alimentary tract model for radiological protection. *Annals of the ICRP*, **36** (1–2).

ICRP. (1980) International Commission on Radiological Protection. ICRP Publication 31. Biological effects of inhaled radionuclides. *Annals of the ICRP*, **4** (1/2).

ICRP. (1991) The biological basis for skin dose limitation. ICRP publication 59. *Annals of the ICRP*, **22** (2).

ICRU. (1997) International Commission on Radiation Units and Measurements. ICRU Report No 56. Dosimetry of external beta rays for radiation protection. Bethesda, Maryland, USA.

Kaurin, D.G.L., Baum, J.W., Carsten, A.L., and Schaefer, C.W. (2001a) Scab incidence on pig skin resulting from hot particle exposures under varying conditions. *Health Physics*, **81** (1) 47–56.

Kaurin, D.G.L., Baum, J.W., Carsten, A.L., Schaefer, C.W., and Forbes, P.D. (2001b) Scab diameters on pig skin resulting from hot particle exposures under varying conditions. *Health Physics*, **81** (1) 35–46.

Mill, A.J., Darley, P.J., Allen, L.A., and Charles, M.W. (1996) Oncogenic transformation in C3H 10T 1/2 cells after irradiation by radioactive 'hot-particles'. *International Journal of Radiation Biology*, **70** (6) 775–788.

NCRP. (1999) Biological effects and exposure limits for 'hot particles'. NCRP Report No. 130. NCRP Report No. 130. National Council on Radiation Protection and Measurements, Bethesda, Washington DC.

NRPB. (1997) Assessment of skin doses. *Documents of the NRPB*, **8**(3).

NRPB. (1996) Risk from deterministic effects of ionising radiation. *Documents of the NRPB* **7**(3). *National Radiological Protection Board,* 1996.

Potten, C.S. (1990) A comprehensive study of the radiobiological response of the murine (BDF1) small intestine. *International Journal of Radiation Biology*, **58** 925–973.

Smith, K.R. and Bedwell, P. (2005) Public health implications of fragments of Irradiated fuel. Module 3: The likelihood of encountering a fuel fragment on Sandside beach. Radiation Protection Division (RPD), *Health Protection Agency* (HPA). RPD-EA-9.

Stather, J.W., Muirhead, C.R., Edwards, A.A., Harrison, J.D., Lloyd, D.C., and Wood, N.R. (1988) Health effects models developed from the 1988 UNSCEAR report. NRPB-R226. *National Radiological Protection Board Report*, December 1988.

Tamplin, A.R. and Cochran, T.B. (1974) The hot particle issue: A critique of WASH-1320 as it relates to the hot particle hypothesis. National Resources Defense Council, Washington DC. November.

Tamplin, A.R. and Cochran, T.B. (1974) A report on the inadequacy of existing radiation standards related to internal exposure of man to insoluble particles of plutonium and other alpha emitting hot particles. National Resources Defense Council, Washington, DC.

Waller, S.L. (1975) Differential measurement of small and large bowel transit times in constipation and diarrhoea: A new approach. *Gut*, **16** 372–378.

THE INFLUENCE OF HOT PARTICLE CONTAMINATION ON MODELS FOR RADIATION EXPOSURES VIA THE AQUATIC PATHWAY

JIM T. SMITH[1], ALEXEI V. KONOPLEV[2],
OLEG V. VOITSEKHOVITCH[3], GENNADY V. LAPTEV[3]
[1]School of Earth and Environmental Sciences, Burnaby Building, University of Portsmouth, Portsmouth, PO1 3QL, UK
[2]Centre for Environmental Chemistry, SPA Typhoon, Obninsk, 249020, Russia
[3]Ukrainian Institute of Hydrometeorology, Nauki Prospect, 252650 Kiev, Ukraine

Abstract: Most models for transfers of radionuclides through the food chain assume that the radioactivity is initially deposited in chemically available forms. However, it is known that releases of radionuclides in the form of hot particles may significantly influence their environmental transfers and uptake to the food chain. This study presents models for radionuclide activity concentrations in water and uptake to fish which incorporate hot particle contamination using observed rates of hot particle dissolution in terrestrial and aquatic systems following the Chernobyl accident . As expected from previous work, fallout of hot particles significantly influences radionuclide activity concentrations in drinking water and aquatic foodstuffs. The remaining significant uncertainties in dose estimates are discussed, particularly those concerning the role of direct ingestion of hot particles.

Keywords: hot particle, fuel particle, dissolution, radiocaesium, radiostrontium, Chernobyl, aquatic, cooling pond, fish.

1. Introduction

Most models for transfers of radionuclides through the food chain assume that the radioactivity is initially deposited in chemically available forms. However, it is known that the accidental release of radionuclides in the form of hot particles may significantly influence their environmental transfers and uptake to the food chain (e.g., Konoplev and Bobovnikova, 1991; Salbu et al., 1994;

D.H. Oughton and V. Kashparov (eds.), *Radioactive Particles in the Environment*,

Krouglov et al., 1997). Release of a proportion of fallout in chemically less available hot particles was observed in releases from the Windscale reactor in the UK during 1952–1957 (Salbu et al., 1994) as well as from the Chernobyl accident (e.g., Bobovnikova et al., 1991; Kashparov et al., 1999).

In this study, historical data on time changes in ^{90}Sr and ^{137}Cs in surface water and fish of the Chernobyl cooling pond will be modelled under two alternative assumptions (1) that hot particle contamination is not significant and; (2) that hot particle contamination is at a level at which it was observed after the Chernobyl accident. The model will be used to quantify the influence of hot particle contamination, at levels observed after Chernobyl, on radionuclide uptake via the aquatic pathway.

The initial explosion and subsequent fire at Chernobyl deposited fuel particles principally within an area of radius of 30 km around the reactor. Within this area, the majority of fallout was in the form of these hot particles (Bobovnikova et al., 1991).

Rates of degradation of fuel particles can be estimated (Konoplev et al., 1992; Krouglov et al., 1997; Kashparov et al., 1999) using a simple decay equation:

$$A(t) = A_0 \exp(-k_p t) \tag{1}$$

where $A(t)$ and A_0 are the radionuclide activities in particles at time t and at the time of deposition respectively. The half-life of fuel particle degradation,

$T^{p}_{1/2}$ is given by ($T^{p}_{1/2} = \ln 2/k_p$).

The rates of fuel particle decay estimated by Kashparov et al. (1999) imply that at present the majority of Chernobyl-derived fuel particles in terrestrial systems have dissolved. By 1997, less than 20% of ^{90}Sr remained in fuel particle form (Kashparov et al., 2004). In aquatic systems, however, fuel particle dissolution has been much slower, being of order $f_p \approx 0.02$ year^{-1} (Konoplev et al., unpubl. res.).

2. Modelling

The "AQUASCOPE" model (Smith et al., 2005a) was used to predict ^{137}Cs and ^{90}Sr activity concentrations in water and fish of the Chernobyl cooling pond. We will here briefly describe the model (see Fig. 1): further details of model development and parameterisation can be found in Smith et al. (2005a).

The rate of removal of radionuclides to the lake bed sediments and outflow occurs at rate K, where

$$K = \frac{1}{T_w} + \frac{1}{T_L} \tag{2}$$

T_w (y^{-1}) is the water residence time of the lake defined as the ratio of lake volume to rate of water inflow or outflow, estimated from the net rainfall to the lake and catchment:

$$T_w \approx \frac{V_L}{R.A_c} \tag{3}$$

where R (m $year^{-1}$) is the net rainfall, V_L (m^3) is the lake volume and A_c (m^2) is the catchment area.

T_L, the residence time for transfers of radionuclides to bed sediments (y^{-1}), is given by:

$$\frac{1}{T_L} = \frac{f_p v_p}{d} \tag{4}$$

where f_p is the particulate sorbed fraction, v_p (m $year^{-1}$) is the particle settling velocity and d (m) is the mean depth. The particulate sorbed fraction, f_p can be calculated from the radionuclide distribution coefficient, K_d and the suspended solids concentration (s, mg l^{-1}) of the lake water.

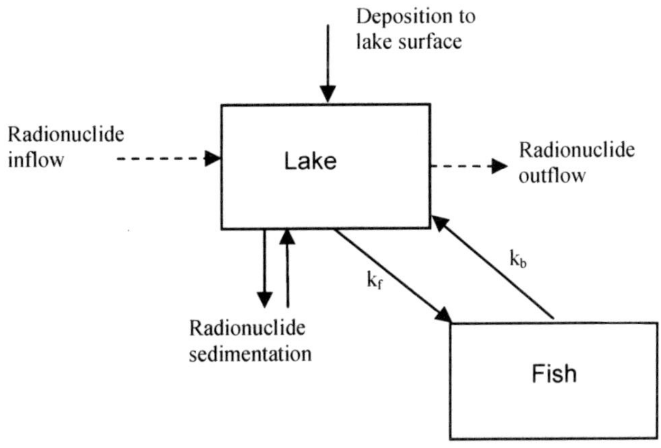

Figure 1. Simplified schematic of the AQUASCOPE lake model.

For "closed" lake systems (defined here as lakes with water residence time >1 year and mean depth <7 m), such as the Chernobyl cooling pond, it is assumed that the input of ^{137}Cs from the catchment can be neglected. It can be assumed (Bulgakov et al., 2002) that the remobilisation rate of radiocaesium from contaminated sediments declines inversely with time after fallout. The AQUASCOPE model (Smith et al., 2005a) simulates this decline over time using a series of two exponential equations:

$$\eta_1.e^{-(\lambda+k_2)t}+\eta_2.e^{-(\lambda+k_3)t} \tag{5}$$

Where λ is the physical decay constant and k_2 and k_3 (in year^{-1}) are empirically determined constants (Smith et al., 2005a). The constants η_1, η_2 (units: m^{-1}), are empirically determined from data on long term activity concentrations in closed lakes per unit of deposition to the lake surface, D_L (Smith et al. 2005a).

2.1. MODELS FOR FISH UPTAKE AND EXCRETION OF RADIONUCLIDES

It is known that the bioaccumulation of radioactivity in fish is determined by numerous ecological and environmental factors such as the trophic level of the fish species, the length of the food chain, water temperature and the water chemistry. Uptake may be via ingestion of contaminated food or direct transfers from the water via the gills. For most radionuclides, the food chain is the primary uptake pathway, so a food uptake model will be used to estimate uptake rates. For strontium isotopes, a model for direct uptake via the water (Chowdhury and Blust, 2001) will be used to estimate the intake rate. Though other studies (Kryshev, 2003) assume most uptake via the food pathway, in practice both assumed uptake routes lead to similar assimilation rates (Smith, 2006).

Note that uptake of radioactivity to biota does not significantly influence the amount of radioactivity in the water, so was ignored in the development of the physical transport models above.

The level of radioactive contamination of aquatic biota is commonly defined in terms of a concentration factor (*CF*) where

$$\text{CF} = \frac{\textit{Activity concentration per kg of fish (wet wt)}}{\textit{Activity concentration per cubic metre of water}} \ \text{m}^3\,\text{kg}^{-1} \tag{6}$$

Previous studies on the accumulation of radiocaesium in fish have focused on the prediction of *CF* (sometimes termed the bioaccumulation factor, *BAF*, or aggregated concentration factor, *ACF*). Some models (e.g., Rowan and Rasmussen, 1994; Smith et al., 2000) predict the water-fish *CF* for certain radionuclides using relationships with the water chemistry, for example using an inverse relation between the radiocaesium *CF* and the potassium concentration of the surrounding water.

For non-equilibrium conditions (e.g., in the short term after an accidental release), activity concentration of a radionuclide in fish, C_f (Bq kg^{-1}) may be modelled by a simple "two-box" model describing uptake from the water C_w (Bq m^{-3}) and release from the fish (Fig. 1):

$$\frac{dC_f}{dt} = k_f C_w - (k_b + \lambda) C_f \tag{7}$$

where k_f (m^3 kg^{-1} year^{-1}) is the rate constant describing transfers of ^{137}Cs to fish and k_b (y^{-1}) is the backward rate constant describing excretion of radioactivity from the fish. The ratio of these rate constants gives the equilibrium concentration factor, CF (m^3 kg^{-1}), of the radionuclide in fish relative to water:

$$\frac{k_f}{k_b + \lambda} = \frac{C_f}{C_w}(at\ equilibrium) = CF \tag{8}$$

2.2. SITE DESCRIPTION AND EMPIRICAL DATA

The Cooling Pond is classed as a closed lake. Although it has inflows from the NPP discharge canal, there is no surface outflow, with losses through groundwater flows into the Pripyat River being 1.2 × 10^5 m^3 year^{-1} (Kryshev, 1996). This is a small fraction of the total water volume (1.5 × 10^8 m^3) implying a long water residence time. The model default water residence time of 10 years has therefore been assumed.

From data in Kryshev (1996) the average fallout to the Cooling Pond surface was estimated to be 7.7 × 10^6 Bq m^{-2} of ^{137}Cs and 2.5 × 10^6 Bq m^{-2} of ^{90}Sr. A more recent assessment (Smith et al., 2005b), however, estimated the average fallout to be 1.92 × 10^7 Bq m^{-2} of ^{137}Cs and 3.0 × 10^6 Bq m^{-2} of ^{90}Sr: we will here use these more recent estimates. The majority of the release of ^{90}Sr, ^{141,144}Ce, Pu isotopes and ^{241}Am was in the form of fuel particles (Kashparov et al., 1999), and within 30 km of the plant most (approximately 50–75%) of the ^{137}Cs was in fuel particles (Krouglov et al., 1998; Kashparov et al., 1999). It is assumed that fuel particle associated radionuclides were rapidly transferred to sediments so (in the early period) all radionuclides in the sediment were assumed to be in the form of fuel particles. Therefore, from measurements of the relative amounts in bed sediments and water (Kryshev, 1996), 65% of ^{137}Cs and 89% of ^{90}Sr was estimated to be in the form of fuel particles. For the model prediction for the real scenario (2), therefore, fallout of chemically and biologically available radionuclides was estimated to be 6.7 × 10^6 Bq m^{-2} of ^{137}Cs and 3.3 × 10^5 Bq m^{-2} of ^{90}Sr.

3. Results

The AQUASCOPE model (Figs. 2 and 3) was run for the Cooling Pond scenario under two alternative assumptions (1) that hot particle contamination is not significant (i.e., all of the deposited radioactivity was in available forms) and; (2) that hot particle contaminiation is at the level observed after the Chernobyl accident (approx. 89% of ^{90}Sr and 65% of ^{137}Cs in fuel particle form).

(1) Assuming all ^{90}Sr in available forms **(2) Assuming 89% ^{90}Sr in fuel particles**

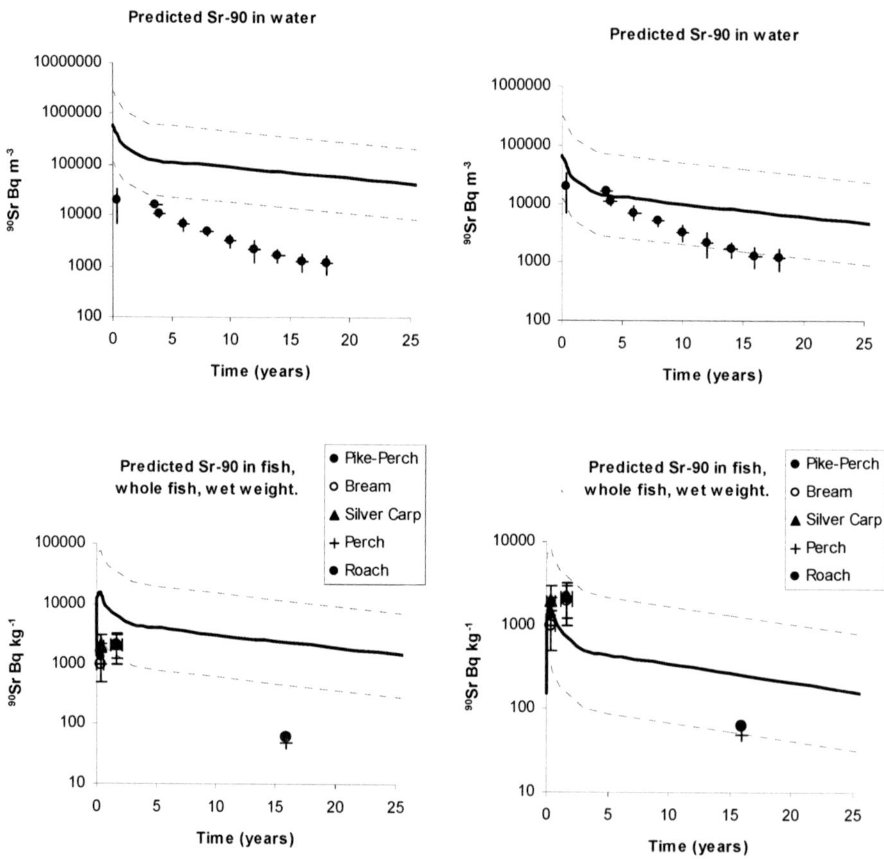

Figure 2. Comparing model predictions for ^{90}Sr in water and fish of the Chernobyl cooling pond assuming (1) all ^{90}Sr deposited in chemically available forms and (2) the actual case of 89% of ^{90}Sr deposited as fuel particles.

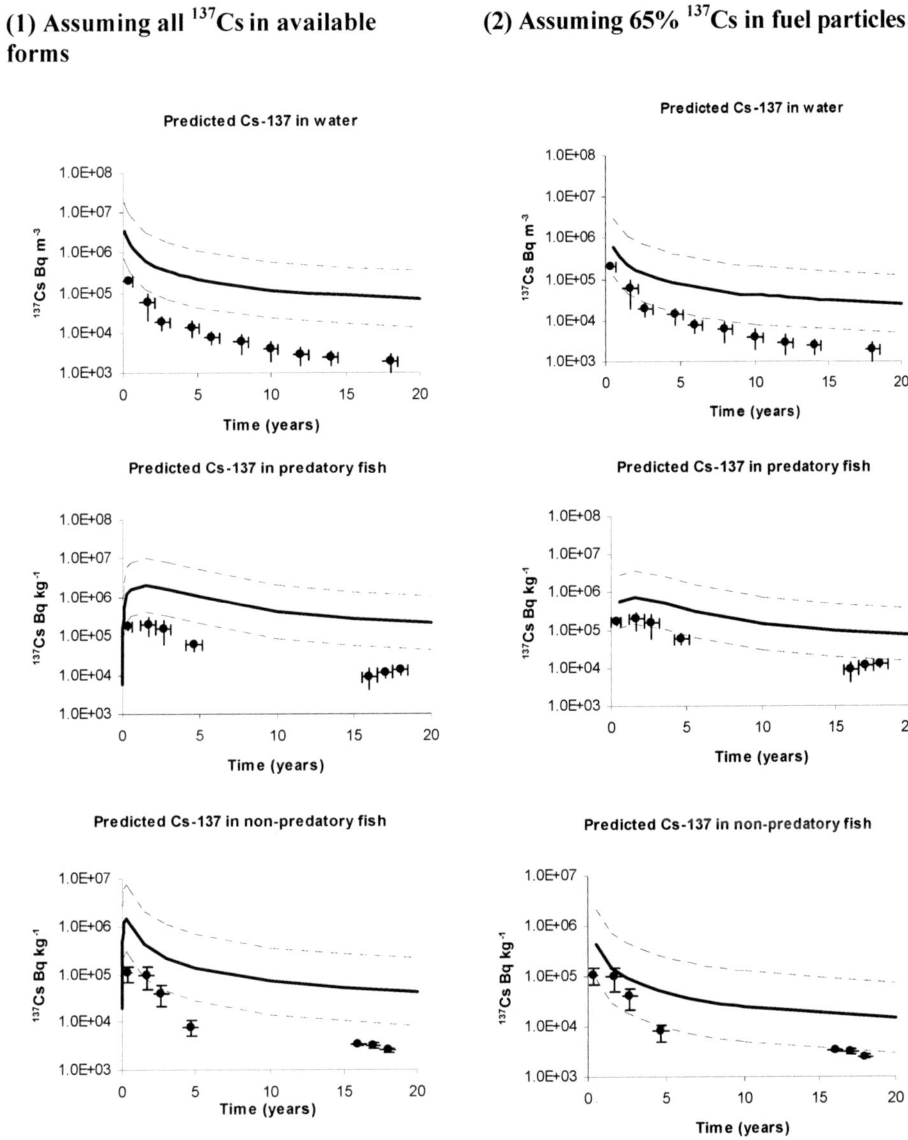

Figure 3. Comparing model predictions for ^{137}Cs in water and fish of the Chernobyl cooling pond assuming (1) all ^{137}Cs deposited in chemically available forms and (2) the actual case of 65% of ^{137}Cs deposited as fuel particles.

Note that the same total radionuclide deposition (1.92×10^7 Bq m^{-2} of ^{137}Cs and 3.0×10^6 Bq m^{-2} of ^{90}Sr) was used for both scenarios. The model was used to make "blind" predictions of ^{90}Sr and ^{137}Cs activity concentrations in water and fish for a period of up to 20 years after the accident for each case. The model results for each of the two scenarios are shown in Figs. 2 and 3.

4. Discussion

4.1. INFLUENCE OF HOT PARTICLES ON MODEL PREDICTIONS

As shown in Figs. 2 and 3, the AQUASCOPE model tends to over-estimate activity concentrations of ^{90}Sr and ^{137}Cs in water and fish of the Cooling Pond. However, for the real case of significant fallout in hot particle form, the model predictions are generally good and within the error bounds of the model. Note that the model is making "blind" predictions: i.e., the model is run in predictive mode as it would be in a real accident scenario and no model calibration to fit the data has been carried out.

For the case where fallout is assumed to be in wholly available form, the model significantly over-predicts activity concentrations in water and fish (see Figs. 2 and 3), and thus ingestion doses via the aquatic pathway.

The influence of fuel particle contamination on dose is dependent on the fraction of radioactivity initially deposited in fuel particle form and the rate of hot particle degradation. If a significant proportion of radioactivity is deposited as hot particles (and hence is unavailable for uptake via the aquatic pathway) then the time-integrated dose is reduced by that fraction of radioactivity which decays whilst still in unavailable forms. For the very slow rates of fuel particle degradation observed in aquatic systems ($f_p \approx 0.02$ year^{-1}; Konoplev et al., unpubl. res.), the fraction not contributing to dose during the first 20 years after fallout is approximately equal to that fraction deposited as fuel particles. In this case, therefore, deposition of radionuclides in fuel particle form is estimated to reduce ingestion doses via the aquatic pathway (drinking water and fish consumption) by approximately 90% for ^{90}Sr and by 65% for ^{137}Cs.

4.2. UNCERTAINTIES

Uncertainties in radionuclide deposition, adsorption to bed sediments and water-fish concentration ratio are major sources of uncertainty in predicting activity concentrations in water and uptake fish. Note also that predictions are made on an annual average basis – the models were calibrated using annual average data.

A further, important, source of uncertainty in models is the ingestion of hot particles either by fish or directly by humans. There is insufficient data at present to assess the uncertainty which this introduces into models. The fact that, for the real scenario, measured activity concentrations in fish bodies are lower than model-predicted activity concentrations implies that absorption of fuel particle incorporated ^{90}Sr and ^{137}Cs across the gut is not significant. However, it should be noted that the models described above do not account for inadvertent ingestion by humans of hot particles contained in the fish gut, or by direct ingestion of sediments by humans.

5. Conclusions

- As expected from previous work into the influence of hot particles on biological uptake (Salbu et al., 1994; Krouglov et al., 1997), fallout of hot particles significantly influences uptake of radionuclides to foodstuffs. The influence of this on model predictions has been quantified for the Chernobyl Cooling Pond scenario.
- For the real case of significant hot particle contamination after Chernobyl, model predictions of ^{90}Sr and ^{137}Cs in water and fish were generally good and within model uncertainty bounds. Where no hot particle contamination was assumed, the model significantly over-estimated activity concentrations in water and fish.
- Significant uncertainties in dose estimates remain, particularly concerning the role of direct ingestion of hot particles by fish and humans.

References

Bobovnikova, Ts.I., Virchenko, Ye.P., Konoplev, A.V., Siverina, A.A., and Shkuratova, I.G. (1991) Chemical forms of occurrence of long-lived radionuclides and their alteration in soils near the Chernobyl Nuclear Power Station. *Soviet Soil Sci* **23**, 52–57.

Bulgakov, A.A., Konoplev, A.V., Smith, J.T., Hilton, J., Comans, R.N.J., Laptev, G.V., and Christyuk, B.F. (2002) Modelling the long-term dynamics of radiocaesium in a closed lake system. *J Environ Radioact* **61**, 41–53.

Bulgakov, A., Konoplev, A., Smith, J.T., Laptev, G., and Voitsekhovich, O. (2009) Fuel particles in the Chernobyl cooling pond: current state and prediction for remediation options. *J Environ Radioact* in press.

Chowdhury, M.J. and Blust, R. (2001) A mechanistic model for the uptake of waterborne strontium in the common carp. *Environ Sci Technol* **35**, 669–675.

Kashparov, V.A., Oughton, D.H., Zvarich, S.I., Protsak, V.P., and Levchuk, S.E. (1999). Kinetics of fuel particle weathering and ^{90}Sr mobility in the Chernobyl 30 km exclusion zone. *Health Phys* **76**, 251–299.

Kashparov, V.A., Ahamdach, N., Zvarich, S.I., Yoschenko, V.I., Maloshtan, I.M., and Dewiere, L. (2004) Kinetics of dissolution of Chernobyl fuel particles in soil in natural conditions. *J Environ Radioact* **72**, 335–353.

Konoplev, A.V. and Bobovnikova, T.I. (1991) Comparative analysis of chemical forms of long-lived radionuclides and their migration and transformation in the environment following the Kyshtym and Chernobyl accidents. In: Proceedings of a seminar on comparative assessment of the environmental impact of radionuclides released during three major nuclear accidents: Kyshtym, Windscale and Chernobyl. EUR 13574, European Commission, Luxembourg, pp. 371–396.

Konoplev, A.V., Bulgakov, A.A., Popov, V.E., and Bobovnikova, Ts.I. (1992) Behaviour of long-lived radionuclides in a soil-water system. *Analyst* **117**, 1041–1047.

Krouglov, S.V., Filipas, A.S., Alexakhin, R.M., and Arkhipov, N.P. (1997) Long-term study on the transfer of ^{137}Cs and ^{90}Sr from Chernobyl-contaminated soils to grain crops. *J Environ Radioact* **34**, 267–286.

Krouglov, S.V., Kurinov, A.D., and Alexakhin, R.M. (1998) Chemical fractionation of ^{90}Sr, ^{106}Ru, ^{137}Cs and ^{144}Ce in Chernobyl-contaminated soils: an evolution in the course of time. *J Environ Radioact* **38**, 59–76.

Kryshev, A.I. (2003) Model reconstruction of ^{90}Sr concentrations in fish from 16 Ural lakes contaminated by the Kyshtym accident of 1957. *J Environ Radioact* **64**, 67–84.

Kryshev, I.I. (1996) Model testing using Chernobyl data: II. Assessment of the consequences of the radioactive contamination of the Chernobyl nuclear power plant cooling pond. *Health Phys* **70**, 13–17.

Rowan, D.J. and Rasmussen, J.B. (1994) Bioaccumulation of radiocesium by fish: The influence of physicochemical factors and trophic structure. *Can J Fish Aquat Sci* **51**, 2388–2410.

Salbu, B., Krekling, T., Oughton, D.H., Østby, G., Kashparov, V.A., Brand, T.L., and Day, J.P. (1994) Hot particles in accidental releases from Chernobyl and Windscale nuclear installations. *Analyst* **119**, 125–130.

Smith J.T., Kudelsky A.V., Ryabov I.N., and Hadderingh R.H. (2000) Radiocaesium concentration factors of Chernobyl-contaminated fish: a study of the influence of potassium, and "blind" testing of a previously developed model. *J Environ Radioact* **48**, 359–369.

Smith, J.T., A.A., Bulgakov, R.N.J., Comans, A.V., Konoplev, A.V., Kudelsky, M.J., Madruga, I.N., Ryabov, O.V., Voitsekhovitch, G., and Zibold, (2005a) The "AQUASCOPE" simplified model for predicting ^{89,90}Sr, ^{131}I and ^{134,137}Cs in surface waters after a large-scale radioactive fallout. *Health Physics* **89**, 628–644.

Smith, J.T., Adam, C., Belova, N., Bulgakov, A.A., Cailes, C., Laptev, G.V., Konoplev, A.V., Kudelsky, A.V., Murphy, J., Nasvit, O., and Voitsekhovitch, O.V. (2005b) *Radioecological Study of the Chernobyl Cooling Pond and Options for Remediation. Final Report to EU INTAS Programme*. Centre for Ecology and Hydrology, Dorset, 48 pp.

Smith, J.T. (2006) Modelling the dispersion of radionuclides following short duration releases to rivers: Part 2. Uptake by fish. *Sci Total Environ* **368**, 502–518.

Smith, J.T. (2009) The influence of hot particle contamination on ^{90}Sr and ^{137}Cs transfers to milk and on time-integrated ingestion doses. *J Environ Radioact* **100**, 322–328.

HOT PARTICLES BEHAVIOR IN COWS AFTER PERORAL INTAKE

VASYL YOSCHENKO*, VALERY KASHPAROV,
MYKOLA LAZAREV
*Ukrainian Institute of Agricultural Radiology,
Mashinobudivnykiv str.7, Chabany, Kyiv-Svjatoshin distr.,
Kyiv reg., 08162, Ukraine
phone: (380 44) 526 2444, e-mail: vasyl@uiar.kiev.ua

Abstract: The behavior of Chernobyl hot particles in the gastro-intestinal tract of cows after a single peroral intake has been studied. The parameters of radionuclide absorption into the organism and transfer into milk were determined. It was shown that the bioavailability of the radionuclides of cesium and strontium in the hot particle form was two orders of magnitude lower than in the water-soluble form.

Keywords: Chernobyl hot particles, peroral intake, absorption coefficient, bioavailability.

1. Introduction

The behaviour of radionuclide in animals has been studied in numerous studies. As a rule, these researches have focused on the determination of the radionuclide absorption in the gastro-intestinal tract (GIT) and their further redistribution in the organs and removal from the organism after the oral intake in the soluble forms (Sirotkin, 1973; Sirotkin et al., 1978; Korneev and Sirotkin, 1987). In some experiments, radioactive particles (products of the nuclear weapon tests) were introduced into the animals' GIT accompanied with the high doses of the external irradiation; and the possibility of the acute lesions of the GIT mucous membrane and radiation disease was reported (Budarkov et al., 1985).

As a result of the Chernobyl accident, a huge amount of fuel hot particles with a uranium oxide matrix and various sizes (from submicron to hundred micrometers) was released into the atmosphere and deposited in the Environment (Loshchilov et al., 1991; Kuriny et al., 1993). During the first post-accidental period the particles could enter the animal body along with the grass from the contaminated pastures, and with the soil particles in the followed period. However, the available data on hot particle behaviour in the animal body still remains limited and controversial. This report partly covers that gap and presents the results of our study of the Chernobyl hot particles after peroral intake into cows.

2. Methods

In order to study the fuel particles behavior in the cow organism after the single peroral intake we injected the prepared fuel particles (analogous to the Chernobyl FP) into the two mature lactating cows (Holstein, 4 year old, 400 kg and 425 kg respectively). Two types of the particles were applied: with and without the condensed component of the radionuclides. Each cow received the particles of only one type.

The main aims of the study were to follow the dynamics of the fuel particles in the GIT including their possible retention at the walls and to determine the radionuclides absorption coefficients and transfer into the animal organs.

2.1. PREPARATION OF THE PARTICLES

The fuel particles have been prepared in accordance with the methods developed by UIAR (Kashparov et al., 1994), which included the 4-h annealing at 600°C of the Chernobyl nuclear fuel in air media. For this purpose, a piece (0.1 g) of the UO_2 fuel element with the bulk density of 10.5 ± 0.9 g cm^{-3}, burn-up of 6.5 ± 0.9 MW day kg^{-1} and ^{137}Cs specific activity of 660 MBq g^{-1} was chosen. As a result of the annealing, the initial fuel sample has destroyed into the fine (mainly in the range of 5–10 μm) particles.

Obtained particles were divided into two portions. The first portion, FPI, was applied directly. These particles contained the condensed component of the radiocesium at their surfaces. The residual particles, FPII portion, were treated with 1 mM HCl and distilled water in order to wash out the condensed radionuclides, and than dried with alcohol. This treatment resulted in washing out 4.8% and 5.3% of initial activities of ^{134}Cs and ^{137}Cs in the particles, respectively.

2.2. FUEL PARTICLES PARAMETERS

γ-Spectrometric measurements were applied to determine the radionuclides activities in FPI and FPII portions. Activities of transuranium elements and ^{90}Sr were derived from the fixed ratios between the radionuclides specific activities in the Chernobyl nuclear fuel. Obtained results are presented in Table 1.

Dispersal composition of the fuel particles was described by the lognormal distribution with the median diameter of 9.5 μm and STD of the diameter logarithm of 0.81. Thus, the ^{137}Cs activity in the spherical 9.5 μm particle could be estimated as 3 Bq; activities of other radionuclides could be calculated using the data presented in Table 1 in assumption of the uniform spatial activity distribution within the particle matrix (except the condensed phase).

TABLE 1. Radionuclides activities, MBq, in the fuel particles at the moment of intake

Radionuclide	FPI	FPII
^{90}Sr	28.4	30.1
^{106}Ru	1.28	1.36
^{134}Cs	0.93	0.94 (0.988)[a]
^{137}Cs	32.85	32.8 (34.63)[a]
^{144}Ce	2.03	2.16
^{147}Pm	18.3	19.4
^{125}Sb	0.45	0.57
^{154}Eu	0.58	0.61
^{155}Eu	0.92	0.97
^{239}Pu	0.21	0.22
^{240}Pu	0.11	0.12

[a]Values in brackets represent the activity before the washing.

2.3. INTAKE AND POST-INTAKE PROCEDURES

Whole FPI and FPII portions were placed each into the separate gelatin capsule and then the capsules were inserted into the cows using the special tool for the boluses. Intake was carried out during the cows feeding in the specially designed stalls. Thus the possibility of the activity losing was minimized.

The experiment duration from intake to slaughter was 9 days. In this period the cows were kept in the stalls and fed with hay and mixed fodder in accordance with the norms.

Twice a day during the observation period the samples of milk, blood and faeces were taken for the spectrometric measurements. Also, the hematological analyses were carried out. In parallel, the exposure dose rates were periodically measured in the control points at the animal bodies using the field dosimeter equipped with the led collimator. The results of these measurements were used for correction of the experiment duration. The duration should be long enough for removal of the fuel particles from GIT. In the same time, the absorbed radionuclides should remain in the cows' organs, thus the experiment duration could not be too long. Therefore the decision about slaughter was taken when the dosimetric measurements had shown absence of the FP in GIT of cows.

After the slaughter, the representative samples of blood, organs, tissues and the GIT content were collected.

3. Experimental results and discussion

3.1. DYNAMICS OF THE FUEL PARTICLES REMOVAL FROM THE COWS

According to the measurement results, the radionuclides activities in the GIT and organs of the cows after the slaughter were very small as compared to the initial activities FPI and FPII (Table 2).

TABLE 2. Total activity in the organ A, kBq, and specific activity C, (% of injected) $kg^{-1} \cdot 10^{-3}$, at the moment of slaughter. Measurement uncertainty for $^{137}Cs \leq 10\%$, for $^{144}Ce \leq 25\%$

Organ/tissue	Intake:					
	FPI			FPII		
	A ^{137}Cs	C ^{137}Cs	A ^{144}Ce	A ^{137}Cs	C ^{137}Cs	A ^{144}Ce
Lung	2.9	2.5		2.4	2.1	
Heart	1.6	2.8		1.5	2.7	
Spleen	1.3	3.5		1.1	2.8	
Kidney	1.4	4.1		1.1	3.4	
Liver	4.9	2.6		3.5	1.9	
Udder	2.8	1.7		2.4	1.5	
Muscle	450	5.5		296	3.6	
Upper intestine	3.8			3.7		
Lower intestine	12.8		0.7	17.5		0.7
Paunch	23.1		1.7	26.9		
Psalterium	4.3		0.3	7.6		0.4
Honeycomb bag	2.2			6.8		
Abomasums	0.71			1.3		0.08
Total	512		2.7	372		1.2

Thus, the total activities of the most reliable measured radionuclides ^{137}Cs and ^{134}Cs in the GIT of the first cow after the slaughter reached 0.14–0.16% of their activities in FPI, and for the second cow these values were 0.19–0.22% of FPII. Remaining fractions of ^{144}Ce in the GIT were 0.14% of FPI and 0.05% of FPII, respectively.

Dynamics of the radionuclides specific activities in faeces during the observation period is presented in Fig. 1.

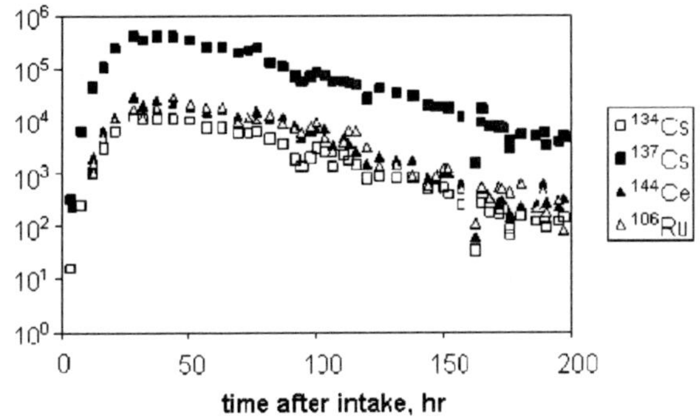

a)

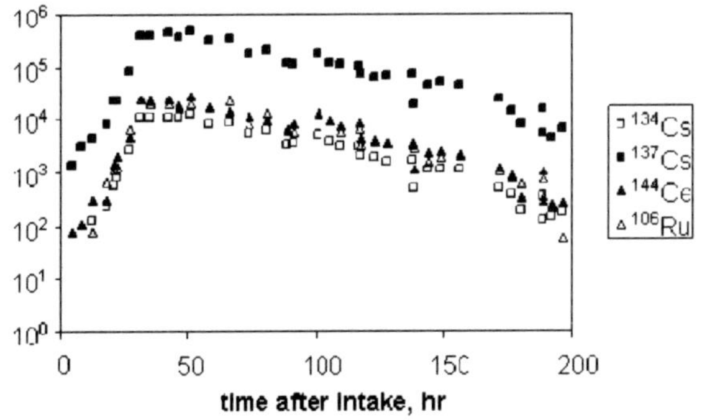

b)

Figure 1. The radionuclides specific activities, Bq kg^{-1}, in faeces of the first (a, FPI) and second (b, FPII) cows during the observation period.

The radionuclides specific activities in faeces reached the maximum values 30–40 h and 40–50 h after the radionuclides injection in the first and second cow, respectively. In the followed days the specific activities exponentially decreased with the half-decrease period $T_{1/2}$ of about 35 h (rate $\lambda \approx 0.02$ h^{-1}). The exponential decrease might indicate that the significant fraction of the fuel particles had transported to the lower departments of the GIT in approx 40 h after their intake.

Obtained data showed that the main fractions of the injected radionuclide activities were removed from the cows' bodies with faeces. Removal with urine and milk was very small. However, since the radionuclide transfer into milk is an important radioecological issue, it must be considered separately.

3.2. RADIONUCLIDES TRANSFER FROM THE FUEL PARTICLES INTO ORGANISM

A clear peak of the radiocesium specific activity in milk is observed in the period 44–68 h after the fuel particles intake (Fig. 2). In the followed period the specific activities exponentially decreased ($T_{1/2}$ = 58–69 h, λ = 0.01–0.12 h^{-1}). Total removal of the radionuclide with milk during 9 days could be estimated as 0.17% of FPI and 0.11% of FPII.

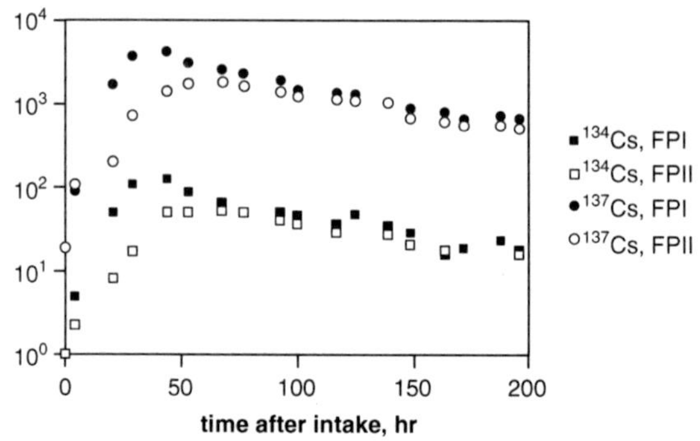

Figure 2. The radiocesium specific activities, Bq l^{-1}, in the cows' milk after the fuel particles intake.

Therefore, taking into account the radionuclides activities in the cows' organs at the moment of slaughter and the activities removed with milk, the total absorption of $^{134,137}Cs$ in GIT could be estimated as 1.3–1.4% of FPI and 0.8–0.9% of FPII. Since the main parts of these amounts were located in the muscle tissues, removal with urine could give only a small addition to these values.

Dynamics of ^{90}Sr specific activities in milk is presented in Fig. 3. Although the variations of the measured values, one can identify the specific activity peaks at 70–80 h for the first cow and 90–120 h for the second cow, which proves the fact of the radionuclide leaching from the fuel particle matrix and transfer into the cows. The background ^{90}Sr specific activities in milk of the cows were about 20 Bq l^{-1}, in maximum they raised to approx 50 Bq l^{-1} and 35 Bq l^{-1} for the first and second cows, respectively, and then decreased to the background values by the end of observation. The total transfer of ^{90}Sr into milk from the fuel particles was some hundreds Bq, or 0.00 n% of the radionuclide activities in FPI and FPII.

Since the ^{90}Sr activities in the cows at the moment of slaughter were comparable small (approx 600 Bq in the first cow and 800 Bq in the second cow), and because of the absence of data on the activities before the experiment, the estimates of the radionuclide transfer from the fuel particles into the cows can not be accurate. Probably the mentioned values can not be unambiguously related to the radionuclide transfer from the fuel particles. If the other probable ways of the radionuclide intake, activities deposited in bone tissues and removed with urine are ignored, its transfer from the fuel particles will be estimated as 0.004–0.005% of the intake activity. Finally, taking into account the literature data concerning ^{90}Sr muscle/bone activity ratio (Korneev and Sirotkin, 1987), the total transfer likely is one order of magnitude higher, i.e., approx 0.05% of the intake activity.

The radiocesium specific activities in blood varied at the background (pre-intake) levels. Other radionuclides in blood were not observed.

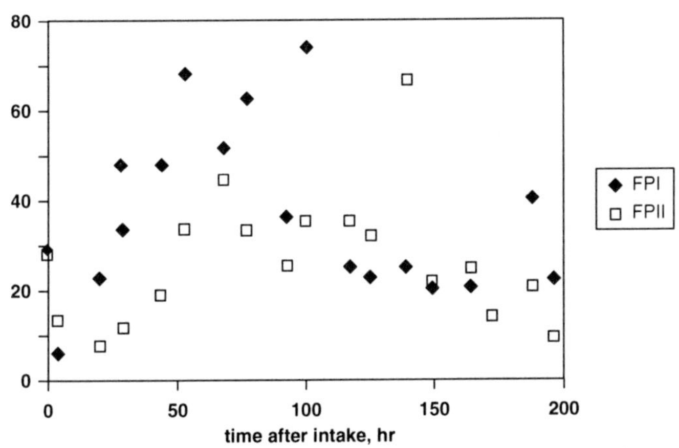

Figure 3. The ^{90}Sr specific activities, Bq l^{-1}, in the cows' milk after the fuel particles intake.

3.3. CLINICOPATHOLOGIC STUDIES

Intake of FPI and FPII into the cows have not resulted in the significant changes of their physiological state. The cows did not lose the appetite and their GIT worked properly. However, the second cow's pose at 4–6 days after the intake was bent, which could indicate some kind of inflammation or irritation in the GIT.

According to the hematological analyses, during the experiment there was an increase of the leucocytes concentration in blood at the expense of neutrophils, while indexes of erythrocytes and platelets remained almost stable.

Autopsy showed absence of the visual changes of the GIT mucous surfaces.

3.4. DISCUSSION

First of all, the fact of the radionuclides transfer from FP into the cow organism must be noted. Both FPI and FPII dissolved in the cow GIT, and therefore we can state the dissolution of the particle matrix in the GIT.

The absorption coefficient of radiocesium (1.3–1.4% for FPI and 0.8–0.9% for FPII) was two order of magnitude lower than the literature data for its intake in the soluble forms (50–75%, Sirotkin et al., 1978, Korneev and Sirotkin, 1987). Incorporation in the FP matrix significantly decreases the radionuclide bioavailability, while the presence of the condensed fraction at the particle surface increases it.

Lower bioavailability of radiocesium in the FP form also determines its lower transfer into milk. Sirotkin (1973) reportyв removal with milk during 9 days of 18% of ^{137}Cs intake activity in the soluble form, while in the present study this value reached only 0.17% and 0.11% for FPI and FPII, respectively. The peak values in milk in our study were observed later (44–68 h after the beginning of the experiment) than at the soluble form intake (30 h, the cited source). It is interesting to note that presence of the condensed form (FPI) caused the earlier attaining the maximum specific activity. The removal rate of radiocesium with milk in our study was also lower than the literature value: 0.01–0.012 h^{-1} vs. 0.018 h^{-1}.

Thus, the radiocesium transfer from FP into the cow body is lower and slower as compared to the transfer at the soluble form intake. It can be a possible reason of absence of the radionuclide in the blood samples: due to the slow absorption from the particles in the GIT, the radionuclide was effectively filtrated in the organs and tissues and could not be detected in the samples of the venous blood.

Literature data (Sirotkin et al., 1978; Korneev and Sirotkin, 1987) recommend the ^{90}Sr soluble form absorption coefficient in the GIT as 6.4–15.5%. Thus, we can state that the radionuclide absorption in the fuel particle form is two order of magnitude lower, 0.05%. Peak concentration of the radionuclide in milk at the FP intake is reached 3 days later than at the soluble form intake (Sirotkin, 1973), and the removal with milk is three orders of magnitude lower. Probably, some fraction of the radionuclide activity in FP can be presented in the condensed form; at least the transfer into the cows of both radiocesium and radiostrontium from FPI was higher than from FPII.

According to the experimental results, a possibility of the long-term retardation of the fuel particles at the GIT membranes can not be fully excluded (Salbu, 2007); however, one can state that the vast majority of the particles were removed from the GIT during the observation period (Fig. 1). It is important to note that in general the fuel particles transportation in the GIT compartments corresponds to the dynamics of the forage transportation (the average period of the forage transportation through the GIT is 2–3 days according to Fedorov et al. (1973)).

4. Conclusions

Taking into account that bioavailability of both radiocesium and radiostrontium in our study was two orders of magnitude lower than at their intake in the soluble forms, and assuming the radionuclides uniform spatial distribution in the fuel particles, the FP matrix dissolution in the cows GIT during the observation period did not exceed 1%.

Thus the radionuclide behavior in the cows following peroral intake significantly depends on the radionuclide intake form. Intake in the FP matrix decreases the radionuclides bioavailability. The problem of the long-term retardation of FP in the GIT remains un-solved, as well as the consequences of such possible retardation.

References

Budarkov, V.A., Arkhipov, N.I., and Kirshin, V.A. (1985) Concerning the classification of the acute impacts of sheep due to the radiation factors of a local trace of the terrestrial nuclear explosion, in: *Impact of the Ionizing Irradiation to the Agricultural Animal Organisms*, VNIISHR, pp. 6–8 (In Russian).

Fedorov, E.A., Prister, B.S., Burov, N.I., Romanov, G.N., Buldakov, L.A., Panchenko, I.Ya., and Osanov, D.P. (1973) Biological impact of the fresh fission products to the milking cattle and their transfer into the agricultural production, in: *Radiobiology and Radioecology of Agricultural Animals*, B.N. Annenkov, I.K. Dibibes, and R.M. Aleksakhin, eds., Book Company Atomizdat, Moscow, pp. 70–139 (In Russian).

Kashparov, V.A., Ivanov, Yu.A., Zvarich, S.I., Protsak, V.P., Khomutinin, Yu.V., Polyakov, V.D., Gudkov, A.N., Kurepin, A.D., and Pazukhin, E.M. (1994) Modeling the hot particles formation during the ChNPP accident, *Radiohimiya (Radiochemistry)*, 1: 87–94 (In Russian).

Korneev, N.A. and Sirotkin, A.N. (1987) *Principles of Radioecology of Agricultural Animals*, Book Company Energoatomizdat, Moscow, 208 p. (In Russian).

Kuriny, V.D., Ivanov, Yu.A., Kashparov, V.A., Loshchilov, N.A., Protsak, V.P., Yudin, E.B., Zhyrba, M.A., and Parshakov, A.E. (1993) Particle-associated Chernobyl fall-out in the local and intermediate zones, *Annals of Nuclear Energy*, 20(6): 415–420.

Loshchilov, N.A., Kashparov, V.A., Yudin, Ye.B., Protsak, V.P., Zhurba, M.A., and Parshakov, A.E. (1991) Experimental assessment of radioactive fallout from the Chernobyl accident, *Sicurezza e Protezione*, 25–26: 46–49.

Salbu, B. (2007) Discussion at the NATO ARW Hot Particles.

Sirotkin, A.N. (1973) Intake of the fission products into the agricultural animal organisms and the radionuclides transfer into the agricultural production, in: *Radiobiology and Radioecology of Agricultural Animals*, B.N. Annenkov, I.K. Dibibes and R.M. Aleksakhin, eds., Book Company Atomizdat, Moscow, pp. 140–171 (In Russian).

Sirotkin, A.N., Burov, N.I., and Fedorov, E.A. (1978) Intake and exchange of the radioisotopes in the agricultural animals, in: *Radioecology of the vertebral animals*, A.I. Ilyenko, ed., Book Company Nauka, Moscow, pp. 103–123 (In Russian).

SOURCES AND TYPES OF UNCERTAINTIES ASSOCIATED WITH RADIOACTIVE PARTICLES

DEBORAH H. OUGHTON, BRIT SALBU
Department of Plant and Environmental Sciences,
Norwegian University of Life Sciences,
P.O. Box 5003, 1432 Aas, Norway

Abstract: The environmental release of radionuclides associated with hot particles is more frequent than commonly recognised, and should also be expected in future nuclear events. Hence the presence of radioactive particles can be expected to influence the environmental impact and risk assessments associated with contaminated areas. This paper addresses the different types of uncertainty arising from hot particles. The first part considers the different categories and dimensions of uncertainty, and the second part illustrates the way that hot particles can impact on the source term, transfer, and biological effects of radionuclides. While it is certainly not the case that increased knowledge will always result in reduced uncertainty (for example the presence of hot particles can lead to greater variability in parameters), it should be clear that assessments will benefit greatly from a better understanding of the way hot particles can impact on the behaviour of radionuclides in the environment.

Keywords: hot particles, environmental impact assessment, uncertainty, variability.

1. Introduction

A series of sources have contributed to the release of radioactive particles to the environment in sizes that range from sub-micron to fragments (see for example the review in this volume (Salbu, 2009)). Indeed, almost all the major sources of radionuclides to the environment include particle associated radionuclides. These include particles formed following destruction of macro amounts of radioactive materials, such as nuclear weapons tests (Crocker et al., 1966; Cooper et al., 1994; Simon et al., 1995), use of depleted uranium ammunitions (Salbu et al., 2003, 2005) and conventional detonation of nuclear weapons (Lind et al., 2007), as well as nuclear reactor accidents including (e.g., Chernobyl) (Devell et al., 1986; Kuriny et al., 1993) and fires in nuclear reactors (e.g.,

D.H. Oughton and V. Kashparov (eds.), *Radioactive Particles in the Environment*,

Windscale) (Chamberlain, 1987). Alternatively, the release of colloids and particles is associated with authorised discharges from fuel reprocessing (Salbu et al., 1993, 2003) and waste disposal (IAEA CRP). Despite this, world-wide, measurements of environmental radioactivity and the associated risk assessments are usually based on the average bulk mass or surface concentration, under the assumption that radionuclides in sample matrices are homogeneously distributed as simple ionic species. If the releases of radioactive particles have occurred more frequently than previously anticipated, and should also be expected in future nuclear events, how will the presence of radioactive particles influence the impact and risk assessments associated with contaminated areas? With this in mind this paper will review the main sources and types of uncertainties associated with hot particles. Following a definition of what hot particles are, the second part of the paper will give an overview of the different types of uncertainties, building largely on previous reviews of the categorisation of uncertainties in environmental impact assessment (Oughton, 2006; Oughton et al., 2008). Finally, the paper concludes with a discussion of the possible impact of these uncertainties on the behaviour of particles in the environment.

2. Definition

According to the IAEA Coordinated Research programme, radioactive particles in the environment are defined as *localised aggregates of radioactive atoms that give rise to an inhomogeneous distribution of radionuclides significantly different from that of the matrix background* (IAEA CRP). In water, particles are defined as entities having diameters larger than 0.45 μm, i.e., entities that will settle due to gravity. Nanoparticles (engineered or natural) or radioactive colloids are entities larger than 1 nm (with nominal molecular mass larger than about 10^3 Da) that do not settle in water due to Brownian movements. Particles larger than 1 mm in size are referred to as fragments. In air, radioactive particles ranging from submicrons in aerosols to fragments are classified according to the aerodynamic diameters, where particles less than 10 μm are considered respiratory. Following the Chernobyl accident, radioactive particles and fragments were identified close to the source, while micrometer sized particles and submicrons settled in Scandinavia and Norway more than 2,000 km from the damaged reactor (Salbu, 2000). However, the transition between size categories is gradual, and the cut-offs are to a certain extent defined according to available physical separation methods. Furthermore, the identification of heterogeneities with radioactivity significantly different from that of the background will depend on the sensitivity of the measurement technique. Today, the presence of alpha or beta emitting heterogeneities with activities higher than about 0.1 Bq ranging from particles to fragments can be identified by autoradiography (Fig. 1) (Salbu,

2000). This implies that low level particles, especially submicron sized particles, can be easily overlooked.

3. Categorisation of Uncertainties

Scientific uncertainties can be categorised in various ways. One of the most conventional and widespread distinctions is that between uncertainty and variability (Suter, 1993; USEPA, 1998).

Knowledge uncertainty (Type I uncertainty) – arising from lack of scientific knowledge about specific factors, parameters or models (that can partly be reduced through further study). This includes parameter, model and scenario uncertainties. It can be expressed by the uncertain belief about the likelihood of the variable (random variable) having different values represented by probability distribution.

Variability (Type II uncertainty) – arising from natural variability due to true heterogeneity that is not usually reduced through further study. Variability is characterised by frequency distribution (discrete random variable) or through a probability density function. This includes actual differences that occur between different environments or individuals.

Figure 1. Autoradiography (P imaging) of samples collected at Mayak PA reflects the presence of radioactive particles.

While this distinction is a useful theoretical construct, it can often be difficult to make in practice and it is often the case that some parameters demonstrate extrinsic uncertainty – due to limitations of measurements and models – and intrinsic variability. The difference between uncertainty and variability are thus not always straightforward and it should be appreciated that variability in input parameters can be a legitimate component in the uncertainty in outputs. However this is only one of the dimensions of uncertainties.

3.1. DIMENSIONS OF UNCERTAINTY

Historically, the focus of uncertainty analysis has often been on the quantifiable aspects. In many cases, this type of assessment represents the implicit assumption that uncertainty can be quantified and expressed as an error, that the quantity has a central value within a definable range, and that the systems being described are limited. This type of uncertainty can be represented using statistics by assigning simple standard deviations or probability distributions to the parameters and the input data. The parameter/data uncertainties can then be propagated through the models by performing probabilistic simulations. More recent work has focused on broader (generally less quantifiable) aspects of uncertainty – particularly related to using uncertain information in decision-making and in communicating uncertainties. The additional sub-categories of uncertainty have also been established to allow uncertainties to be identified in a more systematic fashion (Walker et al., 2003; van der Sluijs, 2003).

Measurement uncertainty refers to the uncertainty in field or laboratory data on which models are based. This includes lack of precision, inaccuracy, sampling and analysis errors. Radioactive particles are unevenly distributed and often can carry a significant fraction of the bulk sample activity (Bunzl, 1997; Lind et al., 2007), hence the presence of hot particles in samples can give rise to a range of different types of measurement uncertainty, including errors in:

1. Representative sampling due to non-homogenity in deposition
2. Counting due to geometry errors from the non-homogenity of activity in samples
3. Underestimation of radionuclides due to incomplete digestion of samples

For example, studies at UMB showed that the sampling variability, based on Cs-134 in 5 soil samples (1–5 cm) collected within 1 m^2 situated in a particle affected area inside the Chernobyl 30 km zone was a factor of 10^3–10^4 (unpublished results). Since inert radioactive particles can be difficult to dissolve, and analytical results may only reflect partial leaching of radionuclides and can lead to large sources of errors if it assumed that the radionuclide is present in ionic form. The case of ^{90}Sr measurement is a classic example, which has been shown to have relevance for releases from Chernobyl and Windscale in the 1960s. In both cases, use of the standard HCl extraction procedure that has been developed for ionic sources of Sr, such as from weapons fallout, underestimated Sr-90 by up to 40%, due to the fact that HCl is a non-oxidising acid and will not dissolve uranium oxides (Oughton et al., 1993).

Data or Numerical Uncertainty arises from uncertainties in the values of physical quantities used in calculations, most obviously in the data for input to models, but also in the parameters used within the models themselves, for example for calculation of pollutant uptake factors, biological half-lives, etc. and in the input data (concentrations in soil, water, sediments and the organisms). This category also includes intrinsic characteristics of the environment and organisms living within it. Measurement, data and numerical uncertainties have been described as imprecision or inexactness. In Environmental Risk Assessment the most common form of errors due to the presence of particles are in the transfer and bio-concentration factors (BCF) used in models, where particles can lead to large differences in these parameters as compared to ionic species. This can includes variations due to the different chemical form (i.e., ^{90}Sr which is much more mobile as Sr^{2+}), or increased retention of colloids and particles in filter feeders – leading to a greater BCF.

Model (mechanistic or computational) and Scenario uncertainties arise from the (simplified) mathematical representation of the conceptual models and the imprecision in numerical solutions implicit in mathematical models. It includes model structural errors. This type of uncertainty is usually assessed by performing inter-comparisons between alternative models and between model predictions and empirical observations. One example is the increased mobility of radionuclides associated with colloidal species in water systems – where monitoring has shown greater transport of Pu than would have been expected from ionic species (Lind et al., 2006). Scenario uncertainty refers to uncertainty in the states of the system under analysis, including not only the situation at the moment of the assessment, but also the situation in the past and in the future. It includes uncertainty in the environmental properties and how these change, in the sources of contamination, etc. The increased mobility of ^{90}Sr with particle weathering in Chernobyl contaminated areas is one example (Kashparov et al., 1999, 2000). Both model and scenario uncertainties can be described as unreliability.

Social, Political and Ethical Uncertainties. Evaluation of uncertainties in a societal context raises a number of philosophical, ethical, social and economic issues, many associated with the perception of harm – What is an acceptable risk for humans? For environmental risks, what is it we are trying to protect and why? How much "damage" to a species can be tolerated in an ecosystem? Do the benefits of a new technology outweigh the risks? And who decides that? (Calow, 1998; Oughton, 2003). These may manifest themselves as uncertainties surrounding the interpretation of legislation (including the applicability of the precautionary principle), acceptability of the methodology and the results of the assessment to stakeholders, economic costs and benefits, burden of proof for demonstrating harm, and the perception of the importance of uncertainty itself. For example, the public may lose trust in authorities and scientists if they predict that doses will be reduced and things get better with time, only to be proved wrong due to particle weathering.

4. Impact of hot particles on environmental impact assessment

To assess the environmental impact and risk associated with contaminated areas, information is needed on the source or release term, deposition, mobility (K_d) and ecosystem transfer (TC) as well as biological uptake, accumulation (BCF) and effect on exposed organisms. Thus, information on particle characteristics, weathering and remobilization kinetics influencing mobility and biological uptake would be essential to address uncertainties associated with impact and risk assessments.

4.1. SOURCE TERM AND INVENTORY ESTIMATES

The source or release terms and/or estimates of the inventory of the contaminated sites represent essential input to dose, impact and risk models. These estimates are usually based on total activity concentrations (Bq) of radionuclides released or deposited in an affected ecosystem, without taking the radionuclide speciation into account. As discussed above, radioactive particles present in the environment are unevenly distributed and often contain a significant fraction of the bulk sample activity. For particle contaminated areas, estimated inventories, impact and risk assessments can therefore be underestimated (Bunzl, K. 1997; Kashparov et al., 1999; Salbu et al., 1998, 2001; Salbu B. 2000). According to IAEA (IAEA CRP): "*The failure to recognize the heterogeneity of radionuclide distribution in the environment leads to analytical inconsistencies, irreproducible results, and erratic conclusions*". Consequently, the estimated inventories for particle contaminated areas can be underestimated.

International research, especially at UMB, has shown that the particle composition will depend on the emitting source (e.g., matrix, burn-up), while the release scenario (e.g., temperature and pressure and the presence of air) dictates particle characteristics such as particle size distributions, crystallographic structures and oxidation states, i.e. key variables of importance for particle weathering, remobilisation of associated radionuclides and metals, ecosystem transfer and biological effects (Kashparov et al., 1999; Salbu, 2000; Salbu et al., 2001). To characterize such particle properties, advanced techniques such as scanning electron microscopy in BEI mode combined with synchrotron radiation X-ray micro-techniques have therefore been developed (Salbu et al., 1994, 1998; Salbu, 2000; Salbu et al., 2005) (Fig. 2). Releases at high temperature and pressures will influence the structure, while the presence of oxygen/air will influence on the oxidation state of radionuclide matrix such as U and Pu (Kashparov et al., 1999; Salbu et al., 2001). Thus, there is a need to summarize existing knowledge on particle characteristics and link particle characteristics to specific source and release conditions.

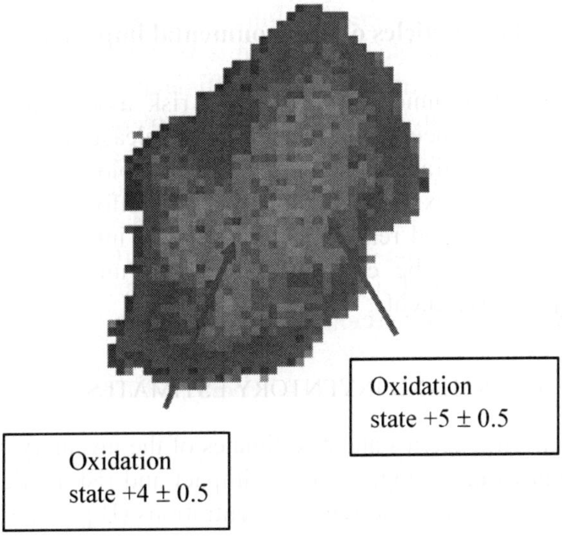

Figure 2. Uranium fuel particle released during the Chernobyl explosion, having a UO_2 core with oxidized U_3O_8 surface layers (synchrotron radiation μ-XANES) (Salbu et al., 2001).

4.2. TRANSFER, MOBILITY AND BIOLOGICAL UPTAKE

Radioactive particles can carry substantial amounts of refractory fission products, activation products and transuranics as well as stable metals, and represent point sources. Following deposition of radioactive particles, soils and sediment can act as sinks, and compared to mobile radionuclide species, the ecosystem transfer will be delayed. In transport models, soil–water or sediment–water interactions are usually described by distribution coefficients, Kd, assumed to be constants at equilibrium. With particles present, this assumption fails. Due to weathering of particles, associated radionuclides are mobilised and the ecosystem transfer increases over time. Weathering rates will depend on the particle composition (e.g., UO_2 fuel), structural changes occurring during the event (e.g., transformation from UO_2 to U_3O_8), and on local chemical conditions and transformation processes occurring after deposition (e.g., pH, redox conditions, bioerosion) (Kashparov et al., 1999, 2000). These processes will be particularly important in controlling the speciation, solubilisation and migration of mobile radionuclides, thus radionuclides such as ^{90}Sr would be expected to show the greatest change in mobility with particle weathering.

It follows that the distribution of radionuclides between solid and solution is a time dependent process and the thermodynamic constant should be replaced by a time-function. Information on retention of particles, weathering rates and binding mechanisms in soils and sediments is essential for estimating potential remobilisation and valuable information is attained from leaching experiments (Lind et al., 2009). Similarly, transfer coefficients, TC (m^2/kg), utilized for soil-to-plant transfer modelling and aggregated transfer coefficients, Tag (m^2/kg), utilized for soil-plant-animal transfer modelling are "constants" which are valid under equilibrium conditions. Similarly to Kd, particles will confound the equilibrium concept and constants should be replaced by time-dependent functions. Particles can be retained on plant surfaces, in filtering organisms, retained in GI tract of animals, and inhaled. In particle affected areas, the apparent bioconcentration factor (BCF) can therefore be significantly higher than for ionic species.

4.3. BIOLOGICAL EFFECTS

Biological effects following the exposure of any organism and tissue to radioactive particles will depend on a number of factors, all of which can show significant deviation from ionic sources. First, the exposure route and biological uptake and metabolism of particles can be different to ions: for example with high retention in filtering organisms. As hot particles include high-specific activity isotopes there is a risk from inhalation, dermal absorption, wound exposure and ingestion of the particles. For filter-feeders (e.g., molluscs), aggregates may be retained by the organism and eventually be ingested by man. Other organisms may show different exposure routes, for example to gills or roots. The size of particles will determine the type of uptake; particles less that 10 μm can be retained by inhalation; while nanoparticles can cross skin and other biological membranes (Oberdorster, 2004; Oughton et al., 2008). Thus, it the particle oxidation state can impact on the subsequent metabolism and half-life; and changes in oxidation state of particles can contribute to biological effects from ROS (see below).

Secondly, the dose and dose rate will depend on the internal distribution and biological half-life; and problems associated with heterogenic exposure represent an important source of uncertainty in internal dosimetry (CERRIE, 2004). To date most models are based on the assumption of ionic species, and the role of speciation, exposure time, and accumulation/excretion is not properly accounted for. It is likely that particles will lead to inhomogeneous doses and dose rates. Thirdly, the biological response can be confounded by the particles acting in effect as three different types of stressors; radioactivity, heavy metals and particles. This can lead to a multiple stressor situation. These issues are essential for risk characterization, where risks are estimated by integrating the results of hazard identification, effects assessment and exposure assessment.

5. Conclusion

The inherent differences in the transport and bioavailability of particle bound radionuclides compared with those existing as simple ions, have largely been ignored in radioecology/radiation dosimetry/risk estimates. As a result, there is a high degree of scientific uncertainty about the levels of risk to human health and the long-term ecological consequences of radioactive particles present in the environment (IAEA CRP, Salbu, 2000). Research is clearly needed to address these concerns, and provide fundamental data and information on the different physical forms, size and frequency distribution, transport and behaviour of radioactive aggregates on which sound environmental assessments can be made.

Of course, it is not always the case that increased knowledge will result in reduced uncertainty. For example, increased sampling can show that the presence of hot particles results in a greater *variability* in concentration or transfer than models had previously predicted. But it should be clear that risk assessment will benefit greatly from a better understanding of the way hot particles can impact on the transfer, mobility and effect of radionuclides in the environment. Several years of research demonstrate that the particle composition will depend on the specific source, i.e., reflecting the matrix and the refractory radionuclide composition (e.g., burn-up). The release scenarios (e.g., temperature, pressures, redox conditions) will influence particle characteristics such as particle size distribution, crystallographic structures and oxidation states, being essential for the ecosystem behaviour. Furthermore, information on particle characteristics and processes influencing particle weathering and remobilisation of associated radionuclides is essential to assess long-term impact from radioactive particle contamination. Key chemical, physical and transport properties of radioactive aggregates could be used to help facilitate environmental management of contaminated sites. For example, knowledge related to particle size, solubility and transport of particles may provide a scientific basis for developing new cleanup standards and associated technologies. Fundamental knowledge on chemical and physical properties should help to identify those radionuclides that are most likely to be impacted by the presence in particles.

References

Bunzl, K. 1997. Probability of detecting hot particles in environmental samples by sample splitting, *Analyst* **122**: 653–656.

Calow, P. 1998. Ecological risk assessment: Risk of what? How do we decide? *Ecotox Environ Safety* **40**: 15–18.

CERRIE. 2004. Report of the Committee Examining Radiation Risks of Internal Emitters (CERRIE). London.

Chamberlain, AC. 1987. Environmental impact of particles emitted from windscale piles, 1954. *Sci Total Environ* **63**: 139–160.

Cooper MB, Burns PA, Tracy BL, Wilks MJ, Williams GA. 1994. Characterization of plutonium contamination at the former nuclear-weapons testing range at Maralinga in South-Australia. *J Radioanal Nucl Chem* **177**: 161–184.

Crocker GR, Oconnor JD, Freiling EC. 1966. Physical and radiochemical properties of fallout particles. *Health Phys* **12**: 1099.

Devell L, Tovedal H, Bergstrom U, Appelgren A, Chyssler J, Andersson L. 1986. Initial observations of fallout from the reactor accident at Chernobyl. *Nature* **321**: 192–193.

IAEA CRP. Radiochemical, Chemical and Physical Characterisation of radioactive particles in the environment. First RCM report. IAEA, Vienna.

Kashparov VA, Oughton DH, Protsak VP, Zvarisch SI, Levchuk SE. 1999. Kinetics of fuel particle weathering and 90Sr mobility in the Chernobyl 30 km exclusion zone. *Health Phys* **76**: 251–259.

Kashparov VA, Protsak VP, Ahamdach N, Stammose D, Peres JM, Yoshchenko VI, Zvarich SI. 2000. Dissolution kinetics of particles of irradiated Chernobyl nuclear fuel: influence of pH and oxidation state on the release of radionuclides in the contaminated soil of Chernobyl. *J Nucl Mater* **279**: 225–233.

Kuriny VD, Ivanov YA, Kashparov VA, et al. 1993. Particle-associated Chernobyl fallout in the local and intermediate zones. *Ann Nucl Energy* **20**: 415–420.

Lind OC, Salbu B, Janssens K, Proost K, Garcia-Leon M, Garcia-Tenorio R. 2007. Characterization of U/Pu particles originating from the nuclear weapon accidents at Palomares, Spain, 1966 and Thule, Greenland, 1968. *Sci Total Environ* **376**: 294–305.

Lind OC, Salbu B, Skipperud L, Janssens K, Jaroszewicz J, De Nolf W. 2009. Solid state speciation and potential bioavailability of depleted uranium particles from Kosovo and Kuwait. *J Environ Radioact* **100**: 301–307.

Lind OC, Salbu B, Janssens K, Proost K, Garcia-Leon M, Garcia-Tenorio R. 2007. Characterization of U/Pu particles originating from the nuclear weapon accidents at Palomares, Spain, 1966 and Thule, Greenland, 1968. *Sci Total Environ* **376**(1–3): 294–305.

Lind OL, Oughton DH, Salbu B, Skipperud L, Sickel MA, Brown JE, Fifield LK, Tims SG. 2006. Transport of low ^{240}Pu/^{239}Pu atom ratio plutonium species in the Ob and Yenisey Rivers to the Kara Sea. *Earth Planet Sci Lett* **251**: 33–43.

Morgan MG, Henrion M. 1990. *Uncertainty. A Guide to Dealing with Uncertainty in Quantitative Risk and Panalysis*. Cambridge University Press.

Oberdorster E. 2004. Manufactured nanomaterials (Fullerenes, C-60) induce oxidative stress in the brain of juvenile largemouth bass. *Environ Health Perspect* **112**: 1058–1062.

Oughton DH, Hertel-Aas T, Pellicer E, Mendoza E, Joner EJ. 2008. Neutron activation of engineered nanoparticles as a tool for tracing their environmental fate and uptake in organisms. *Environ Toxicol Chem* **27**: 1883–1887.

Oughton DH, Salbu B, Brand TL, Day JP, Aarkrog A. 1993. Under-determination of Sr-90 in soils containing particles of irradiated uranium oxide fuel. *Analyst* **118**: 1101–1105.

Oughton DH, Agüero A, Avila R, Brown JE, Copplestone D, Gilek M. 2008. Addressing uncertainties in the ERICA integrated approach. *J Environ Radioact* **99**: 1384–1392.

Oughton DH. 2003. Protection of the environment against ionising radiation: ethical issues. *J Environ Radioact* **66**: 3–18.

Oughton DH. 2006. Uncertainties from multiple stressors, in: *Multiple Stressors: A Challenge for the Future*, NATO Science for Peace and Security Series C: Environmental Security, pp. 455–465.

Salbu B, Bjornstad HE, Svaren I, Prosser SL, Bulman RA, Harvey BR, Lovett MB. 1993. Size Distribution of radionuclides in nuclear-fuel reprocessing liquids after mixing with seawater. *Sci Total Environ* **130**: 51–63.

Salbu B, Janssens K, Lind OC, Proost K, Danesi PR. 2003. Oxidation states of uranium in DU particles from Kosovo. *J Environ Radioact* **64**: 167–173.

Salbu B, Janssens K, Lind OC, Proost K, Gijsels L, Danesi PR. 2005. Oxidation states of uranium in depleted uranium particles from Kuwait. *J Environ Radioact* **78**: 125–135.

Salbu B, Krekling T, Lind OC, Oughton DH, et al. 2001. High energy X-ray microscopy for characterisation of fuel particles. *Nucl Instr and Meth A* **467**: 1249–1252.

Salbu B, Krekling T, Oughton DH, Ostby G, Kashparov VA, Brand TL, Day JP. 1994. Hot particles in accidental releases from Chernobyl and windscale nuclear installations. *Analyst* **119**: 125–130.

Salbu B, Krekling T, Oughton DH. 1998. Characterisation of radioactive particles in the environment. *Analyst* **123**: 843–849.

Salbu B, Skipperud L, Germain P, Guegueniat P, Strand P, Lind OC, Christensen G. 2003. Radionuclide speciation in effluent from La Hague reprocessing plant in France. *Health Phys* **85**: 311–322.

Salbu B. 2000. Speciation of radionuclides in the environment, in *Encyclopedia of Analytical Chemistry*, R.A. Meyers ed. John Wiley & Sons Ltd, Chichester, pp. 12993–13016.

Salbu, B. 2009. Sources of radioactive particles in the environment. This volume.

Simon SL, Jenner T, Graham JC, Borchert A. 1995. A comparison of macroscopic and microscopic measurements of plutonium in contaminated soil from the Republic-Of-The-Marshall-Islands. *J Radioanal Nucl Chem Art* **194**: 197–205.

Suter GW. 1993. *Ecological Risk Assessment*. Lewis Publishers.

USEPA (United States Environmental Protection Agency). 1998. Guidelines for Ecological Risk Assessment. EPA/630/R-95/002F, USEPA, Risk Assessment Forum.

van der Sluijs, J. 2003. RIVM. *MNP Guidance for Uncertainty Assessment and Communication Detailed Guidance* (RIVM/MNP Guidance for Uncertainty Assessment and Communication Series, Volume 3), RIVN, Bithoven, The Netherlands.

Walker WE, et al. 2003. Defining uncertainty: A conceptual basis for uncertainty management in model-based decision support. *Integr Assess* **4**: 5–17.

Lightning Source UK Ltd.
Milton Keynes UK
14 March 2010

151358UK00007B/49/P

9 789048 129478